# BIO 030
# Introduction to Biology

Duchess Community College

Cover Image credit: "Field Color Blue Sky." Provided by: Negative Space. Located at https://negativespace.co/field-color-blue-sky/.

Published by State University of New York Press, Albany

© 2021

Printed in the United States of America

The majority of this book was pulled from four primary sources, all of which are Open Access and can be downloaded for free in their original form at the links provided. The sources include: Open Stax Biology (CC BY 4.0) https://openstax.org/details/books/biology, Concepts of Biology (CC BY 4.0) http://cnx.org/content/col11487/latest/; Chemistry (CC BY 4.0) https://opentextbc.ca/chemistry/, and Introductory Chemistry (CC BY-NC-SA 3.0) By Saylor Academy 2012 https://opentextbc.ca/chemistry/. These materials were adapted and rearranged, with some additions made by the author.

Except where otherwise noted, the text of this book is licensed under a Creative Commons Attribution 4.0 International License (CC BY 4.0) https://creativecommons.org/licenses/by/4.0/.

# Contents

| | |
|---|---|
| Chapter 1: Measurement Basics | 1 |
|     Measurements | 3 |
|     Scientific notation | 9 |
|     Measurement Uncertainty, Accuracy, and Precision | 12 |
|     Basic Dimensional analysis | 19 |
|     Exercises | 25 |
|     Check Your Knowledge: Self-Test | 33 |
| Chapter 2: Introduction to Biology | 35 |
|     Themes and Concepts of Biology | 37 |
|     The Process of Science | 53 |
|     Exercises | 62 |
|     Check Your Knowledge: Self-Test | 63 |
| Chapter 3: Chemistry of Life | 65 |
|     Phases and Classification of Matter | 67 |
|     Physical and Chemical Properties | 79 |
|     Atoms, Isotopes, Ions, and Molecules: The Building Blocks | 84 |
|     Lewis Dot Formula | 103 |
|     Exercises | 107 |
|     Check Your Knowledge: Self-Test | 111 |
| Chapter 4-Stoichiometry and The Mole | 115 |
|     Chemical Equations | 117 |
|     Formula Mass and the Mole Concept | 120 |

| | |
|---|---|
| Stoichiometry | 130 |
| Determining Empirical and Molecular Formulas | 134 |
| Mole-Mass and Mole-Mole conversions | 141 |
| Exercises | 145 |
| Check Your Knowledge: Self-Test | 155 |
| **Chapter 5-Water** | **157** |
| Water | 159 |
| Molarity | 167 |
| Other Units for Solution Concentrations | 173 |
| Electrolytes | 178 |
| pH, Acids, Bases and Buffers | 182 |
| Exercises | 186 |
| Check Your Knowledge: Self-Test | 190 |
| **Chapter 6: Organic Molecules** | **193** |
| Introduction | 195 |
| Chemical Formulas | 196 |
| Carbon | 202 |
| Synthesis of Biological Macromolecules | 211 |
| Exercises | 214 |
| Check Your Knowledge: Self-Test | 217 |
| **Chapter 7: Biological Macromolecules** | **219** |
| Carbohydrates | 221 |
| Lipids | 231 |
| Proteins | 239 |

| | |
|---|---|
| Enzymes | 250 |
| Nucleic Acids | 259 |
| Exercises | 265 |
| Check Your Knowledge: Self-Test | 268 |
| **Chapter 8: Cell Structure and Function** | **271** |
| Introduction | 273 |
| Studying Cells | 274 |
| Prokaryotic Cells | 279 |
| Eukaryotic Cells | 282 |
| The Endomembrane System and Proteins | 294 |
| The Cytoskeleton | 299 |
| Exercises | 304 |
| Check Your Knowledge: Self-Test | 305 |
| **Chapter 9: Cell Transport** | **309** |
| Introduction | 311 |
| Components and Structure of the Membrane | 312 |
| Passive Transport | 320 |
| Active Transport | 331 |
| Connections between Cells and Cellular Activities | 336 |
| Exercises | 341 |
| Check Your Knowledge: Self-Test | 343 |

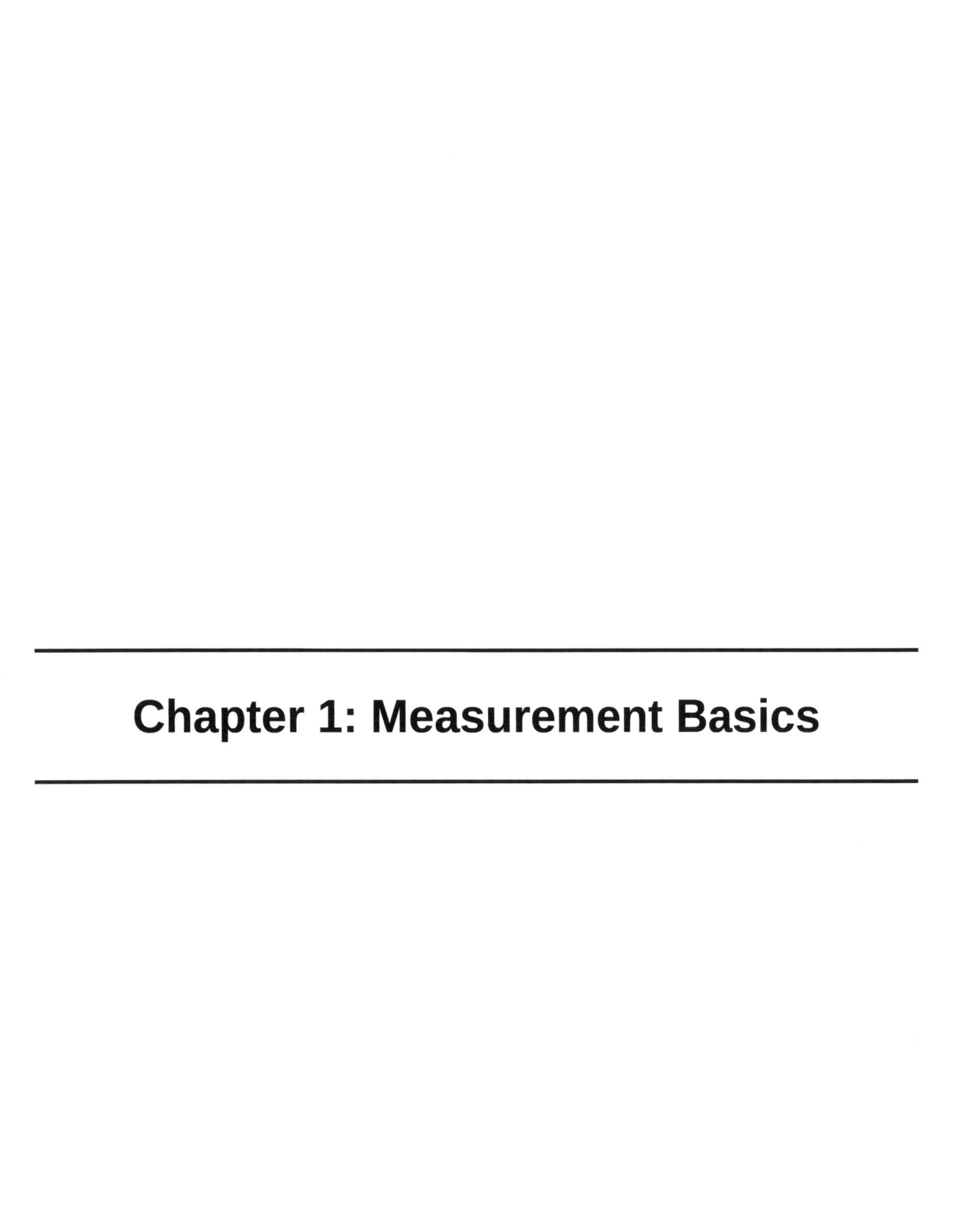

# Chapter 1: Measurement Basics

# Measurements

### Learning Objectives

By the end of this section, you will be able to:
- Explain the process of measurement
- Identify the three basic parts of a quantity
- Describe the properties and units of length, mass, volume, density, temperature, and time
- Perform basic unit calculations and conversions in the metric and other unit systems

Measurements provide the macroscopic information that is the basis of most of the hypotheses, theories, and laws that describe the behavior of matter and energy in both the macroscopic and microscopic domains of chemistry. Every measurement provides three kinds of information: the size or magnitude of the measurement (a number); a standard of comparison for the measurement (a unit); and an indication of the uncertainty of the measurement. While the number and unit are explicitly represented when a quantity is written, the uncertainty is an aspect of the measurement result that is more implicitly represented and will be discussed later.

The number in the measurement can be represented in different ways, including decimal form and scientific notation. (Scientific notation is also known as exponential notation; a review of this topic can be found in Essential Mathematics.) For example, the maximum takeoff weight of a Boeing 777-200ER airliner is 298,000 kilograms, which can also be written as $2.98 \times 10^5$ kg. The mass of the average mosquito is about 0.0000025 kilograms, which can be written as $2.5 \times 10^{-6}$ kg.

**Units**, such as liters, pounds, and centimeters, are standards of comparison for measurements. When we buy a 2-liter bottle of a soft drink, we expect that the volume of the drink was measured, so it is two times larger than the volume that everyone agrees to be 1 liter. The meat used to prepare a 0.25-pound hamburger is measured so it weighs one-fourth as much as 1 pound. Without units, a number can be meaningless, confusing, or possibly life threatening. Suppose a doctor prescribes phenobarbital to control a patient's seizures and states a dosage of "100" without specifying units. Not only will this be confusing to the medical professional giving the dose, but the consequences can be dire: 100 mg given three times per day can be effective as an anticonvulsant, but a single dose of 100 g is more than 10 times the lethal amount.

We usually report the results of scientific measurements in SI units, an updated version of the metric system, using the units listed in Table 1. Other units can be derived from these base units. The standards for these units are fixed by international agreement, and they are called the **International System of Units** or **SI Units** (from the French, *Le Système International d'Unités*). SI units have been used by the United States National Institute of Standards and Technology (NIST) since 1964.

# 4 Measurement Basics

*Table 1. Base Units of the SI System*

| Property Measured | Name of Unit | Symbol of Unit |
|---|---|---|
| length | meter | m |
| mass | kilogram | kg |
| time | second | s |
| temperature | kelvin | K |
| electric current | ampere | A |
| amount of substance | mole | mol |
| luminous intensity | candela | cd |

Sometimes we use units that are fractions or multiples of a base unit. Ice cream is sold in quarts (a familiar, non-SI base unit), pints (0.5 quart), or gallons (4 quarts). We also use fractions or multiples of units in the SI system, but these fractions or multiples are always powers of 10. Fractional or multiple SI units are named using a prefix and the name of the base unit. For example, a length of 1000 meters is also called a kilometer because the prefix *kilo* means "one thousand," which in scientific notation is $10^3$ (1 kilometer = 1000 m = $10^3$ m). The prefixes used and the powers to which 10 are raised are listed in Table 2.

*Table 2. Common Unit Prefixes*

| Prefix | Symbol | Factor | Example |
|---|---|---|---|
| femto | f | $10^{-15}$ | 1 femtosecond (fs) = $1 \times 10^{-15}$ m (0.000000000001 m) |
| pico | p | $10^{-12}$ | 1 picometer (pm) = $1 \times 10^{-12}$ m (0.000000000001 m) |
| nano | n | $10^{-9}$ | 4 nanograms (ng) = $4 \times 10^{-9}$ g (0.000000004 g) |
| micro | μ | $10^{-6}$ | 1 microliter (μL) = $1 \times 10^{-6}$ L (0.000001 L) |
| milli | m | $10^{-3}$ | 2 millimoles (mmol) = $2 \times 10^{-3}$ mol (0.002 mol) |
| centi | c | $10^{-2}$ | 7 centimeters (cm) = $7 \times 10^{-2}$ m (0.07 m) |
| deci | d | $10^{-1}$ | 1 deciliter (dL) = $1 \times 10^{-1}$ L (0.1 L) |
| kilo | k | $10^3$ | 1 kilometer (km) = $1 \times 10^3$ m (1000 m) |
| mega | M | $10^6$ | 3 megahertz (MHz) = $3 \times 10^6$ Hz (3,000,000 Hz) |
| giga | G | $10^9$ | 8 gigayears (Gyr) = $8 \times 10^9$ yr (8,000,000,000 Gyr) |
| tera | T | $10^{12}$ | 5 terawatts (TW) = $5 \times 10^{12}$ W (5,000,000,000,000 W) |

> Need a refresher or more practice with scientific notation? Visit Math Skills Review: Scientific Notation to go over the basics of scientific notation.

## SI BASE UNITS

The initial units of the metric system, which eventually evolved into the SI system, were established in France during the French Revolution. The original standards for the meter and the kilogram were adopted there in 1799 and eventually by other countries. This section introduces four of the SI base units commonly used in chemistry. Other SI units will be introduced in subsequent chapters.

### Length

The standard unit of **length** in both the SI and original metric systems is the **meter (m)**. A meter was originally specified as 1/10,000,000 of the distance from the North Pole to the equator. It is now defined as the distance light in a vacuum travels in 1/299,792,458 of a second. A meter is about 3 inches longer than a yard (Figure 1); one meter is about 39.37 inches or 1.094 yards. Longer distances are often reported in kilometers (1 km = 1000 m = $10^3$ m), whereas shorter distances can be reported in centimeters (1 cm = 0.01 m = $10^{-2}$ m) or millimeters (1 mm = 0.001 m = $10^{-3}$ m).

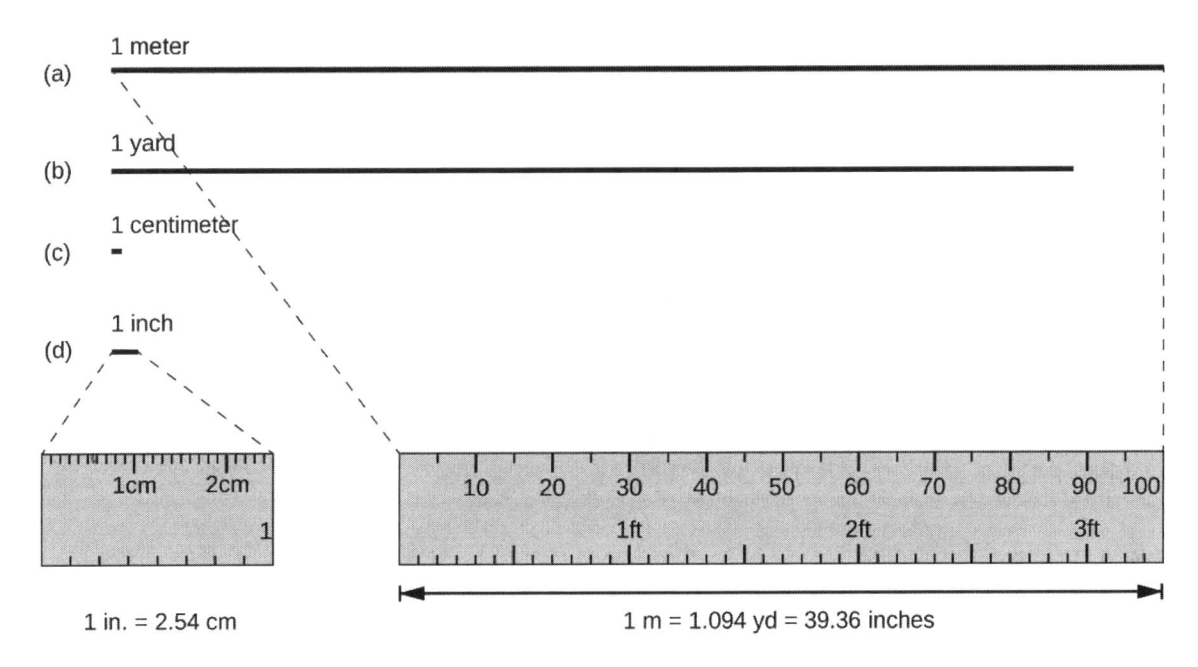

*Figure 1. The relative lengths of 1 m, 1 yd, 1 cm, and 1 in. are shown (not actual size), as well as comparisons of 2.54 cm and 1 in., and of 1 m and 1.094 yd.*

## Mass

The standard unit of mass in the SI system is the **kilogram (kg)**. A kilogram was originally defined as the mass of a liter of water (a cube of water with an edge length of exactly 0.1 meter). It is now defined by a certain cylinder of platinum-iridium alloy, which is kept in France (Figure 2). Any object with the same mass as this cylinder is said to have a mass of 1 kilogram. One kilogram is about 2.2 pounds. The gram (g) is exactly equal to 1/1000 of the mass of the kilogram ($10^{-3}$ kg).

## Temperature

Temperature is an intensive property. The SI unit of temperature is the **kelvin (K)**. The IUPAC convention is to use kelvin (all lowercase) for the word, K (uppercase) for the unit symbol, and neither the word "degree" nor the degree symbol (°). The degree **Celsius (°C)** is also allowed in the SI system, with both the word "degree" and the degree symbol used for Celsius measurements. Celsius degrees are the same magnitude as those of kelvin, but the two scales place their zeros in different places. Water freezes at 273.15 K (0 °C) and boils at 373.15 K (100 °C) by definition, and normal human body temperature is approximately 310 K

*Figure 2. This replica prototype kilogram is housed at the National Institute of Standards and Technology (NIST) in Maryland. (credit: National Institutes of Standards and Technology)*

6   Measurement Basics

(37 °C). The conversion between these two units and the Fahrenheit scale will be discussed later in this chapter.

### Time

The SI base unit of time is the second (s). Small and large time intervals can be expressed with the appropriate prefixes; for example, 3 microseconds = 0.000003 s = 3 × 10$^{-6}$ and 5 megaseconds = 5,000,000 s = 5 × 10$^6$ s. Alternatively, hours, days, and years can be used.

## DERIVED SI UNITS

We can derive many units from the seven SI base units. For example, we can use the base unit of length to define a unit of volume, and the base units of mass and length to define a unit of density.

### Volume

Volume is the measure of the amount of space occupied by an object. The standard SI unit of volume is defined by the base unit of length (Figure 3). The standard volume is a **cubic meter (m$^3$)**, a cube with an edge length of exactly one meter. To dispense a cubic meter of water, we could build a cubic box with edge lengths of exactly one meter. This box would hold a cubic meter of water or any other substance.

A more commonly used unit of volume is derived from the decimeter (0.1 m, or 10 cm). A cube with edge lengths of exactly one decimeter contains a volume of one cubic decimeter (dm$^3$). A **liter (L)** is the more common name for the cubic decimeter. One liter is about 1.06 quarts.

A **cubic centimeter (cm$^3$)** is the volume of a cube with an edge length of exactly one centimeter. The abbreviation **cc** (for cubic centimeter) is often used by health professionals. A cubic centimeter is also called a **milliliter (mL)** and is 1/1000 of a liter.

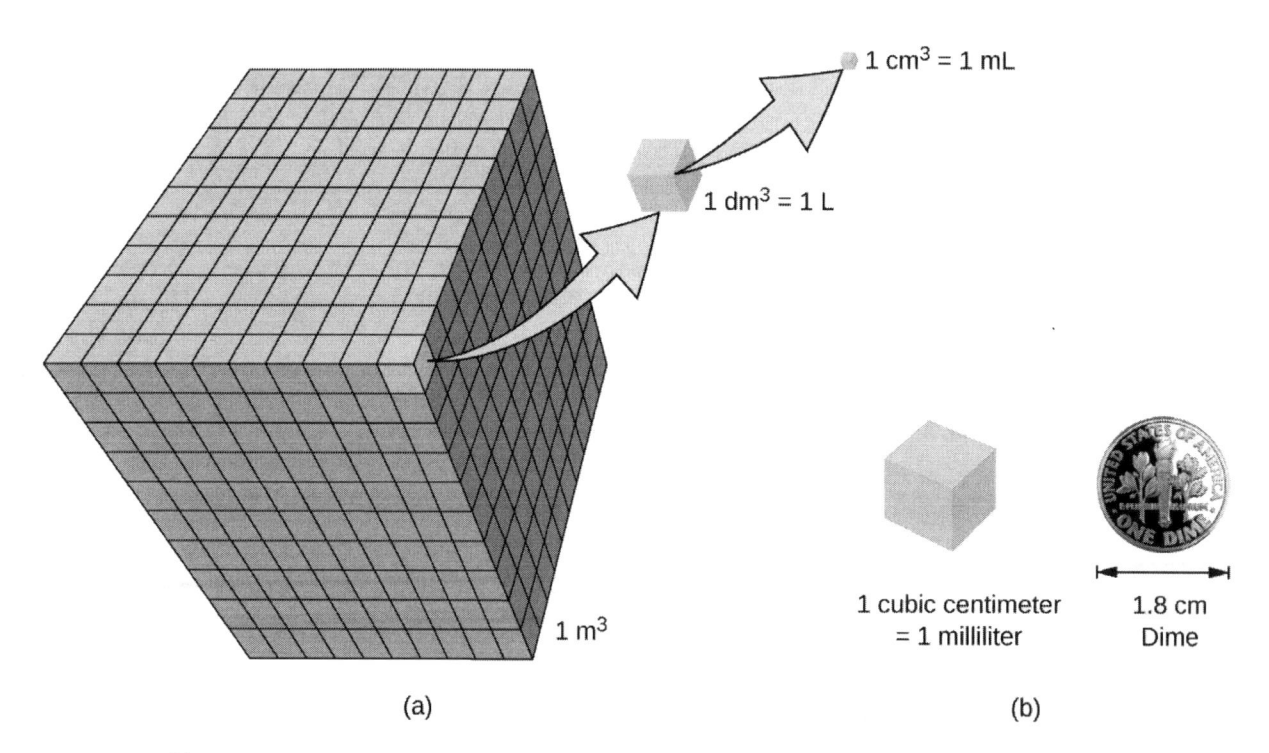

*Figure 3.* (a) The relative volumes are shown for cubes of 1 m$^3$, 1 dm$^3$ (1 L), and 1 cm$^3$ (1 mL) (not to scale). (b) The diameter of a dime is compared relative to the edge length of a 1-cm$^3$ (1-mL) cube.

## Density

We use the mass and volume of a substance to determine its density. Thus, the units of density are defined by the base units of mass and length.

The **density** of a substance is the ratio of the mass of a sample of the substance to its volume. The SI unit for density is the kilogram per cubic meter ($kg/m^3$). For many situations, however, this as an inconvenient unit, and we often use grams per cubic centimeter ($g/cm^3$) for the densities of solids and liquids, and grams per liter (g/L) for gases. Although there are exceptions, most liquids and solids have densities that range from about 0.7 $g/cm^3$ (the density of gasoline) to 19 $g/cm^3$ (the density of gold). The density of air is about 1.2 g/L. Table 3 shows the densities of some common substances.

Table 3. Densities of Common Substances

| Solids | Liquids | Gases (at 25 °C and 1 atm) |
|---|---|---|
| ice (at 0 °C) 0.92 $g/cm^3$ | water 1.0 $g/cm^3$ | dry air 1.20 g/L |
| oak (wood) 0.60–0.90 $g/cm^3$ | ethanol 0.79 $g/cm^3$ | oxygen 1.31 g/L |
| iron 7.9 $g/cm^3$ | acetone 0.79 $g/cm^3$ | nitrogen 1.14 g/L |
| copper 9.0 $g/cm^3$ | glycerin 1.26 $g/cm^3$ | carbon dioxide 1.80 g/L |
| lead 11.3 $g/cm^3$ | olive oil 0.92 $g/cm^3$ | helium 0.16 g/L |
| silver 10.5 $g/cm^3$ | gasoline 0.70–0.77 $g/cm^3$ | neon 0.83 g/L |
| gold 19.3 $g/cm^3$ | mercury 13.6 $g/cm^3$ | radon 9.1 g/L |

While there are many ways to determine the density of an object, perhaps the most straightforward method involves separately finding the mass and volume of the object, and then dividing the mass of the sample by its volume. In the following example, the mass is found directly by weighing, but the volume is found indirectly through length measurements.

$$\text{density} = \text{mass/volume}$$

### Example 1: Calculation of Density

Gold—in bricks, bars, and coins—has been a form of currency for centuries. In order to swindle people into paying for a brick of gold without actually investing in a brick of gold, people have considered filling the centers of hollow gold bricks with lead to fool buyers into thinking that the entire brick is gold. It does not work: Lead is a dense substance, but its density is not as great as that of gold, 19.3 $g/cm^3$. What is the density of lead if a cube of lead has an edge length of 2.00 cm and a mass of 90.7 g?

---

To learn more about the relationship between mass, volume, and density, use this PhET Density Simulator to explore the density of different materials, like wood, ice, brick, and aluminum.

## Example 2: Using Displacement of Water to Determine Density

This PhET simulation illustrates another way to determine density, using displacement of water. Determine the density of the red and yellow blocks.

## Key Concepts and Summary

Measurements provide quantitative information that is critical in studying and practicing chemistry. Each measurement has an amount, a unit for comparison, and an uncertainty. Measurements can be represented in either decimal or scientific notation. Scientists primarily use the SI (International System) or metric systems. We use base SI units such as meters, seconds, and kilograms, as well as derived units, such as liters (for volume) and g/cm³ (for density). In many cases, we find it convenient to use unit prefixes that yield fractional and multiple units, such as microseconds ($10^{-6}$ seconds) and megahertz ($10^6$ hertz), respectively.

**Key Equations**

- $\text{density} = \dfrac{\text{mass}}{\text{volume}}$

## Glossary

**Celsius (°C):** unit of temperature; water freezes at 0 °C and boils at 100 °C on this scale

**cubic centimeter (cm³ or cc):** volume of a cube with an edge length of exactly 1 cm

**cubic meter (m³):** >SI unit of volume

**density:** ratio of mass to volume for a substance or object

**kelvin (K):** SI unit of temperature; 273.15 K = 0° C

**kilogram (kg):** standard SI unit of mass; 1 kg = approximately 2.2 pounds

**length:** measure of one dimension of an object

**liter (L):** (also, cubic decimeter) unit of volume; 1 L = 1,000 cm³

**meter (m):** standard metric and SI unit of length; 1 m = approximately 1.094 yards

**milliliter (mL):** 1/1,000 of a liter; equal to 1 cm³

**second (s):** SI unit of time

**SI units (International System of Units):** standards fixed by international agreement in the International System of Units (*Le Système International d'Unités*)

**unit:** standard of comparison for measurements

**volume:** amount of space occupied by an object

# Scientific Notation

### Learning Objectives

1. Learn to express numbers properly.

Quantities have two parts: the number and the unit. The number tells "how many." It is important to be able to express numbers properly so that the quantities can be communicated properly.

Standard notation is the straightforward expression of a number. Numbers such as 17, 101.5, and 0.00446 are expressed in standard notation. For relatively small numbers, standard notation is fine. However, for very large numbers, such as 306,000,000, or for very small numbers, such as 0.000000419, standard notation can be cumbersome because of the number of zeros needed to place nonzero numbers in the proper position.

Scientific notation is an expression of a number using powers of 10. Powers of 10 are used to express numbers that have many zeros:

| | |
|---|---|
| $10^0$ | = 1 |
| $10^1$ | = 10 |
| $10^2$ | = 100 = 10 × 10 |
| $10^3$ | = 1,000 = 10 × 10 × 10 |
| $10^4$ | = 10,000 = 10 × 10 × 10 × 10 |

and so forth. The raised number to the right of the 10 indicating the number of factors of 10 in the original number is the exponent. (Scientific notation is sometimes called *exponential notation*.) The exponent's value is equal to the number of zeros in the number expressed in standard notation.

Small numbers can also be expressed in scientific notation but with negative exponents:

| | |
|---|---|
| $10^{-1}$ | = 0.1 = 1/10 |
| $10^{-2}$ | = 0.01 = 1/100 |
| $10^{-3}$ | = 0.001 = 1/1,000 |
| $10^{-4}$ | = 0.0001 = 1/10,000 |

and so forth. Again, the value of the exponent is equal to the number of zeros in the denominator of the associated fraction. A negative exponent implies a decimal number less than one.

A number is expressed in scientific notation by writing the first nonzero digit, then a decimal point, and then the rest of the digits. The part of a number in scientific notation that is multiplied by a power of 10 is called the coefficient. Then determine the power of 10 needed to make that number into the original number and multiply the written number by the proper power of 10. For example, to write 79,345 in scientific notation,

79,345 = 7.9345 × 10,000 = 7.9345 × $10^4$

Thus, the number in scientific notation is 7.9345 × $10^4$. For small numbers, the same process is used, but the exponent for the power of 10 is negative:

0.000411 = 4.11 × 1/10,000 = 4.11 × $10^{-4}$

Typically, the extra zero digits at the end or the beginning of a number are not included. (See Figure 2.1 "Using Scientific Notation".)

### Example 1

Express these numbers in scientific notation.

1. 306,000
2. 0.00884
3. 2,760,000
4. 0.000000559

Solution

1. The number 306,000 is 3.06 times 100,000, or 3.06 times $10^5$. In scientific notation, the number is 3.06 × $10^5$.
2. The number 0.00884 is 8.84 times 1/1,000, which is 8.84 times $10^{-3}$. In scientific notation, the number is 8.84 × $10^{-3}$.
3. The number 2,760,000 is 2.76 times 1,000,000, which is the same as 2.76 times $10^6$. In scientific notation, the number is written as 2.76 × $10^6$. Note that we omit the zeros at the end of the original number.
4. The number 0.000000559 is 5.59 times 1/10,000,000, which is 5.59 times $10^{-7}$. In scientific notation, the number is written as 5.59 × $10^{-7}$.

Another way to determine the power of 10 in scientific notation is to count the number of places you need to move the decimal point to get a numerical value between 1 and 10. The number of places equals the power of 10. This number is positive if you move the decimal point to the right and negative if you move the decimal point to the left:

$$56{,}900 = 5.69 \times 10^4 \qquad 0.000028 = 2.8 \times 10^{-5}$$

Many quantities in chemistry are expressed in scientific notation. When performing calculations, you may have to enter a number in scientific notation into a calculator. Be sure you know how to correctly enter a number in scientific notation into your calculator. Different models of calculators require different actions for properly entering scientific notation. If in doubt, consult your instructor immediately. (See Figure 2.2 "Scientific Notation on a Calculator".)

> ### Key Takeaways
>
> - Standard notation expresses a number normally.
> - Scientific notation expresses a number as a coefficient times a power of 10.
> - The power of 10 is positive for numbers greater than 1 and negative for numbers between 0 and 1.

# Measurement Uncertainty, Accuracy, and Precision

### Learning Objectives

By the end of this section, you will be able to:
- Define accuracy and precision
- Distinguish exact and uncertain numbers
- Correctly represent uncertainty in quantities using significant figures
- Apply proper rounding rules to computed quantities

Counting is the only type of measurement that is free from uncertainty, provided the number of objects being counted does not change while the counting process is underway. The result of such a counting measurement is an example of an exact number. If we count eggs in a carton, we know *exactly* how many eggs the carton contains. The numbers of defined quantities are also exact. By definition, 1 foot is exactly 12 inches, 1 inch is exactly 2.54 centimeters, and 1 gram is exactly 0.001 kilogram. Quantities derived from measurements other than counting, however, are uncertain to varying extents due to practical limitations of the measurement process used.

## SIGNIFICANT FIGURES IN MEASUREMENT

The numbers of measured quantities, unlike defined or directly counted quantities, are not exact. To measure the volume of liquid in a graduated cylinder, you should make a reading at the bottom of the meniscus, the lowest point on the curved surface of the liquid.

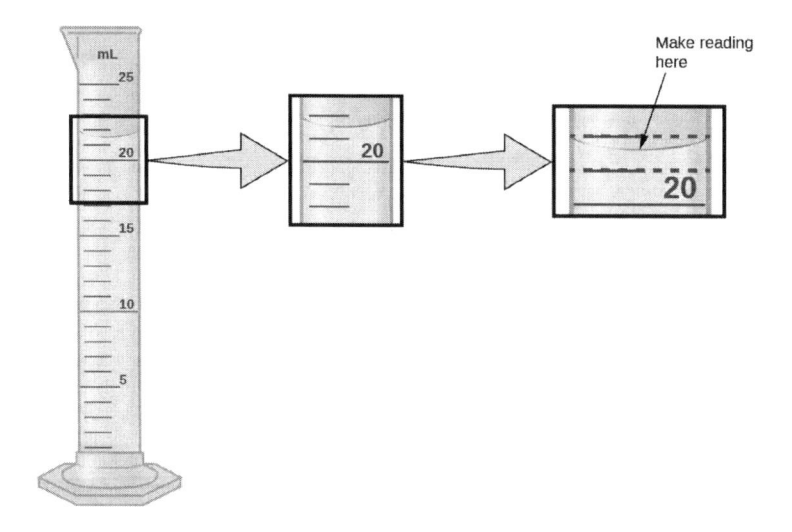

*Figure 1. To measure the volume of liquid in this graduated cylinder, you must mentally subdivide the distance between the 21 and 22 mL marks into tenths of a milliliter, and then make a reading (estimate) at the bottom of the meniscus.*

Refer to the illustration in Figure 1. The bottom of the meniscus in this case clearly lies between the 21 and 22 markings, meaning the liquid volume is *certainly* greater than 21 mL but less than 22 mL. The meniscus appears to be a bit closer to the 22-mL mark than to the 21-mL mark, and so a reasonable estimate of the liquid's volume would be 21.6 mL. In the number 21.6, then, the digits 2 and 1 are certain, but the 6 is an estimate. Some people might estimate the meniscus position to be equally distant from each of the markings and estimate the tenth-place digit as 5, while others may think it to be even closer to the 22-mL mark and estimate this digit to be 7. Note that it would be pointless to attempt to estimate a digit for the hundredths place, given that the tenths-place digit is uncertain. In general, numerical scales such as the one on this graduated cylinder will permit measurements to one-tenth of the smallest scale division. The scale in this case has 1-mL divisions, and so volumes may be measured to the nearest 0.1 mL.

This concept holds true for all measurements, even if you do not actively make an estimate. If you place a quarter on a standard electronic balance, you may obtain a reading of 6.72 g. The digits 6 and 7 are certain, and the 2 indicates that the mass of the quarter is likely between 6.71 and 6.73 grams. The quarter weighs *about* 6.72 grams, with a nominal uncertainty in the measurement of ± 0.01 gram. If we weigh the quarter on a more sensitive balance, we may find that its mass is 6.723 g. This means its mass lies between 6.722 and 6.724 grams, an uncertainty of 0.001 gram. Every measurement has some **uncertainty**, which depends on the device used (and the user's ability). All of the digits in a measurement, including the uncertain last digit, are called **significant figures** or **significant digits**. Note that zero may be a measured value; for example, if you stand on a scale that shows weight to the nearest pound and it shows "120," then the 1 (hundreds), 2 (tens) and 0 (ones) are all significant (measured) values.

Whenever you make a measurement properly, all the digits in the result are significant. But what if you were analyzing a reported value and trying to determine what is significant and what is not? Well, for starters, all nonzero digits are significant, and it is only zeros that require some thought. We will use the terms "leading," "trailing," and "captive" for the zeros and will consider how to deal with them.

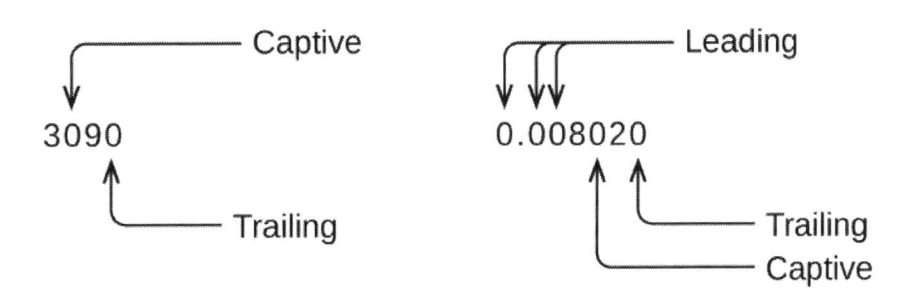

Starting with the first nonzero digit on the left, count this digit and all remaining digits to the right. This is the number of significant figures in the measurement unless the last digit is a trailing zero lying to the left of the decimal point.

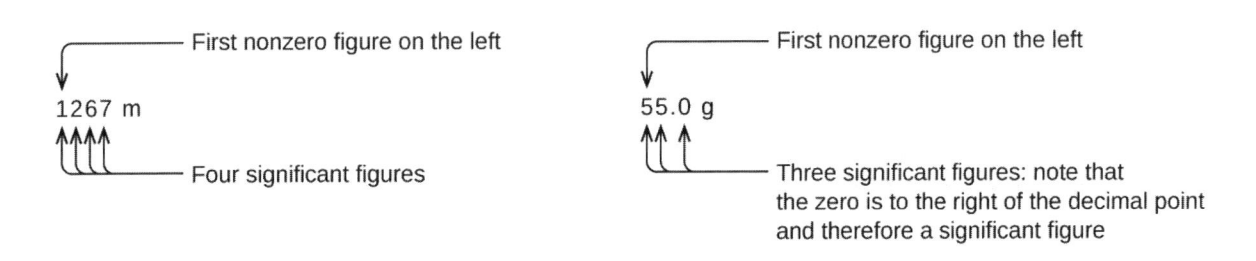

Captive zeros result from measurement and are therefore always significant. Leading zeros, however, are never significant—they merely tell us where the decimal point is located.

70.607 mL — First nonzero figure on the left; Five significant figures: all figures are measured including the two zeros

0.00832407 mL — First nonzero figure on the left; Six significant figures

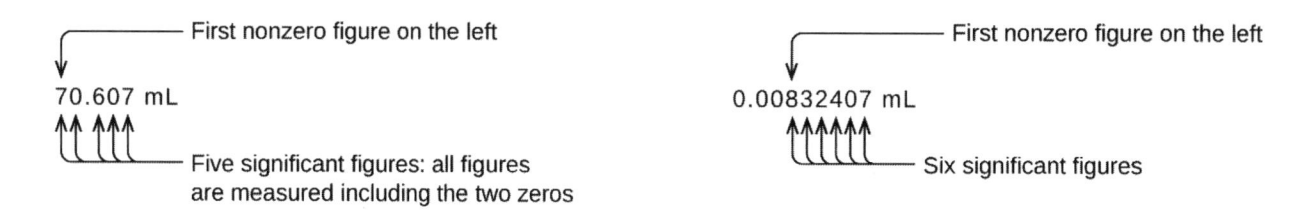

The leading zeros in this example are not significant. We could use exponential notation (as described in Appendix B) and express the number as $8.32407 \times 10^{-3}$; then the number 8.32407 contains all of the significant figures, and $10^{-3}$ locates the decimal point.

The number of significant figures is uncertain in a number that ends with a zero to the left of the decimal point location. The zeros in the measurement 1,300 grams could be significant or they could simply indicate where the decimal point is located. The ambiguity can be resolved with the use of exponential notation: $1.3 \times 10^3$ (two significant figures), $1.30 \times 10^3$ (three significant figures, if the tens place was measured), or $1.300 \times 10^3$ (four significant figures, if the ones place was also measured). In cases where only the decimal-formatted number is available, it is prudent to assume that all trailing zeros are not significant.

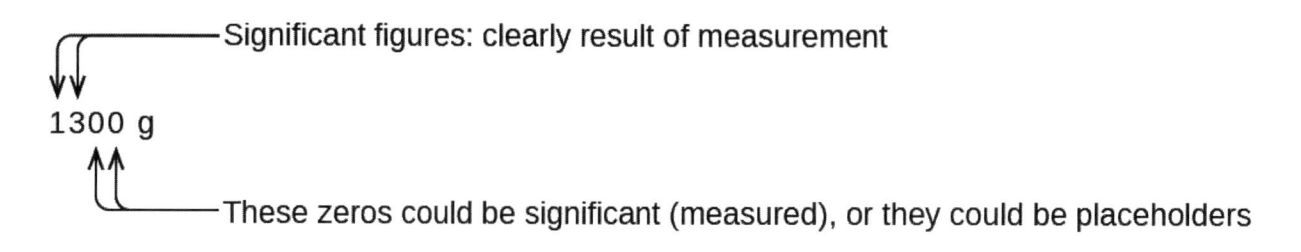

1300 g — Significant figures: clearly result of measurement; These zeros could be significant (measured), or they could be placeholders

When determining significant figures, be sure to pay attention to reported values and think about the measurement and significant figures in terms of what is reasonable or likely when evaluating whether the value makes sense. For example, the official January 2014 census reported the resident population of the US as 317,297,725. Do you think the US population was correctly determined to the reported nine significant figures, that is, to the exact number of people? People are constantly being born, dying, or moving into or out of the country, and assumptions are made to account for the large number of people who are not actually counted. Because of these uncertainties, it might be more reasonable to expect that we know the population to within perhaps a million or so, in which case the population should be reported as $3.17 \times 10^8$ people.

## SIGNIFICANT FIGURES IN CALCULATIONS

A second important principle of uncertainty is that results calculated from a measurement are at least as uncertain as the measurement itself. We must take the uncertainty in our measurements into account to avoid misrepresenting the uncertainty in calculated results. One way to do this is to report the result of a calculation with the correct number of significant figures, which is determined by the following three rules for rounding numbers:

1. When we add or subtract numbers, we should round the result to the same number of decimal places as the number with the least number of decimal places (the least precise value in terms of addition and subtraction).

2. When we multiply or divide numbers, we should round the result to the same number of digits as the number with the least number of significant figures (the least precise value in terms of multiplication and division).

3. If the digit to be dropped (the one immediately to the right of the digit to be retained) is less than 5, we "round down" and leave the retained digit unchanged; if it is more than 5, we "round up" and increase the retained digit by 1; if the dropped digit *is* 5, we round up or down, whichever yields an even value for the retained digit. (The last part of this rule may strike you as a bit odd, but it's based on reliable statistics and is aimed at avoiding any bias when dropping the digit "5," since it is equally close to both possible values of the retained digit.)

The following examples illustrate the application of this rule in rounding a few different numbers to three significant figures:

- 0.028675 rounds "up" to 0.0287 (the dropped digit, 7, is greater than 5)
- 18.3384 rounds "down" to 18.3 (the dropped digit, 3, is lesser than 5)
- 6.8752 rounds "up" to 6.88 (the dropped digit is 5, and the retained digit is even)
- 92.85 rounds "down" to 92.8 (the dropped digit is 5, and the retained digit is even)

Let's work through these rules with a few examples.

---

### Example 1: Rounding Numbers

Round the following to the indicated number of significant figures:

1. 31.57 (to two significant figures)
2. 8.1649 (to three significant figures)
3. 0.051065 (to four significant figures)
4. 0.90275 (to four significant figures)

---

### Example 2: Addition and Subtraction with Significant Figures

Rule: When we add or subtract numbers, we should round the result to the same number of decimal places as the number with the least number of decimal places (i.e., the least precise value in terms of addition and subtraction).

1. Add 1.0023 g and 4.383 g.
2. Subtract 421.23 g from 486 g.

16  Measurement Basics

---

### Example 3: Multiplication and Division with Significant Figures

Rule: When we multiply or divide numbers, we should round the result to the same number of digits as the number with the least number of significant figures (the least precise value in terms of multiplication and division).

1. Multiply 0.6238 cm by 6.6 cm.
2. Divide 421.23 g by 486 mL.

---

In the midst of all these technicalities, it is important to keep in mind the reason why we use significant figures and rounding rules—to correctly represent the certainty of the values we report and to ensure that a calculated result is not represented as being more certain than the least certain value used in the calculation.

---

### Example 4: Calculation with Significant Figures

One common bathtub is 13.44 dm long, 5.920 dm wide, and 2.54 dm deep. Assume that the tub is rectangular and calculate its approximate volume in liters.

---

### Example 5: Experimental Determination of Density Using Water Displacement

A piece of rebar is weighed and then submerged in a graduated cylinder partially filled with water, with results as shown.

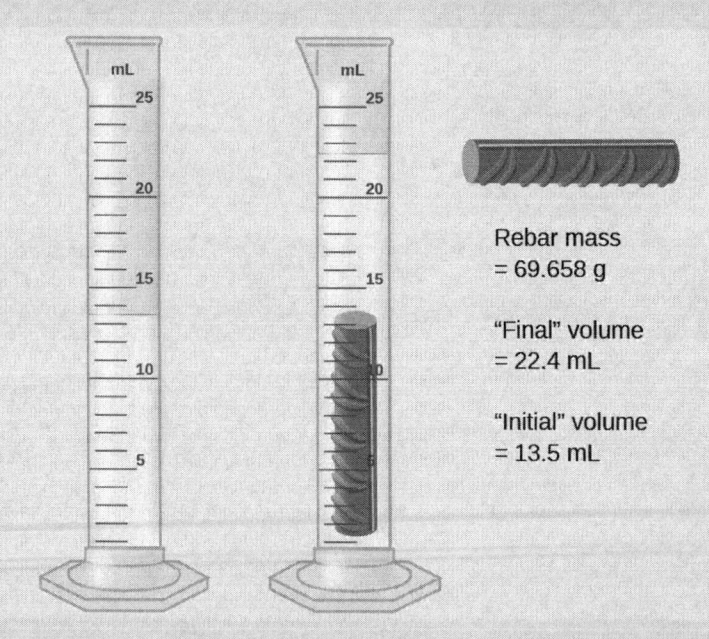

Rebar mass = 69.658 g

"Final" volume = 22.4 mL

"Initial" volume = 13.5 mL

1. Use these values to determine the density of this piece of rebar.
2. Rebar is mostly iron. Does your result in number 1 support this statement? How?

## ACCURACY AND PRECISION

Scientists typically make repeated measurements of a quantity to ensure the quality of their findings and to know both the **precision** and the **accuracy** of their results. Measurements are said to be precise if they yield very similar results when repeated in the same manner. A measurement is considered accurate if it yields a result that is very close to the true or accepted value. Precise values agree with each other; accurate values agree with a true value. These characterizations can be extended to other contexts, such as the results of an archery competition (Figure 2).

Accurate and precise
(a)

Precise, not accurate
(b)

Not accurate, not precise
(c)

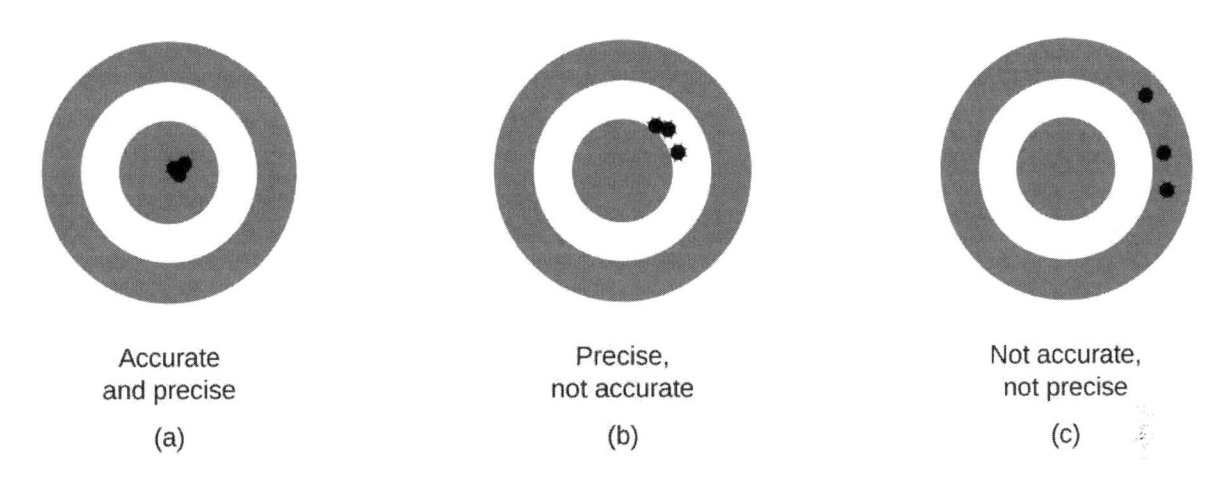

*Figure 2. (a) These arrows are close to both the bull's eye and one another, so they are both accurate and precise. (b) These arrows are close to one another but not on target, so they are precise but not accurate. (c) These arrows are neither on target nor close to one another, so they are neither accurate nor precise.*

Suppose a quality control chemist at a pharmaceutical company is tasked with checking the accuracy and precision of three different machines that are meant to dispense 10 ounces (296 mL) of cough syrup into storage bottles. She proceeds to use each machine to fill five bottles and then carefully determines the actual volume dispensed, obtaining the results tabulated in Table 1.

*Table 1. Volume (mL) of Cough Medicine Delivered by 10-oz (296 mL) Dispensers*

| Dispenser #1 | Dispenser #2 | Dispenser #3 |
|---|---|---|
| 283.3 | 298.3 | 296.1 |
| 284.1 | 294.2 | 295.9 |
| 283.9 | 296.0 | 296.1 |
| 284.0 | 297.8 | 296.0 |
| 284.1 | 293.9 | 296.1 |

Considering these results, she will report that dispenser #1 is precise (values all close to one another, within a few tenths of a milliliter) but not accurate (none of the values are close to the target value of 296 mL, each being more than 10 mL too low). Results for dispenser #2 represent improved accuracy (each volume is less than 3 mL away from 296 mL) but worse precision (volumes vary by more than 4 mL). Finally, she can report that dispenser #3 is working well, dispensing cough syrup both accurately (all volumes within 0.1 mL of the target volume) and precisely (volumes differing from each other by no more than 0.2 mL).

## Key Concepts and Summary

Quantities can be exact or measured. Measured quantities have an associated uncertainty that is represented by the number of significant figures in the measurement. The uncertainty of a calculated value depends on the uncertainties in the values used in the calculation and is reflected in how the value is rounded. Measured values can be accurate (close to the true value) and/or precise (showing little variation when measured repeatedly).

## Glossary

**accuracy:** how closely a measurement aligns with a correct value

**exact number:** number derived by counting or by definition

**precision:** how closely a measurement matches the same measurement when repeated

**rounding:** procedure used to ensure that calculated results properly reflect the uncertainty in the measurements used in the calculation

**significant figures:** (also, significant digits) all of the measured digits in a determination, including the uncertain last digit

**uncertainty:** estimate of amount by which measurement differs from true value

# Basic Dimensional analysis

### Learning Objectives

By the end of this section, you will be able to:

- Explain the dimensional analysis (factor label) approach to mathematical calculations involving quantities
- Use dimensional analysis to carry out unit conversions for a given property and computations involving two or more properties

It is often the case that a quantity of interest may not be easy (or even possible) to measure directly but instead must be calculated from other directly measured properties and appropriate mathematical relationships. For example, consider measuring the average speed of an athlete running sprints. This is typically accomplished by measuring the *time* required for the athlete to run from the starting line to the finish line, and the *distance* between these two lines, and then computing *speed* from the equation that relates these three properties:

$$\text{speed} = \frac{\text{distance}}{\text{time}}$$

An Olympic-quality sprinter can run 100 m in approximately 10 s, corresponding to an average speed of $\frac{100 \text{ m}}{10 \text{ s}} = 10$ m/s.

Note that this simple arithmetic involves dividing the numbers of each measured quantity to yield the number of the computed quantity (100/10 = 10) *and likewise* dividing the units of each measured quantity to yield the unit of the computed quantity (m/s = m/s). Now, consider using this same relation to predict the time required for a person running at this speed to travel a distance of 25 m. The same relation between the three properties is used, but in this case, the two quantities provided are a speed (10 m/s) and a distance (25 m). To yield the sought property, time, the equation must be rearranged appropriately:

$$\text{time} = \frac{\text{distance}}{\text{speed}}$$

The time can then be computed as $\frac{25 \text{ m}}{10 \text{ m/s}} = 2.5$ s.

Again, arithmetic on the numbers (25/10 = 2.5) was accompanied by the same arithmetic on the units (m/m/s = s) to yield the number and unit of the result, 2.5 s. Note that, just as for numbers, when a unit is divided by an identical unit (in this case, m/m), the result is "1"—or, as commonly phrased, the units "cancel."

20  Measurement Basics

These calculations are examples of a versatile mathematical approach known as **dimensional analysis** (or the **factor-label method**). Dimensional analysis is based on this premise: *the units of quantities must be subjected to the same mathematical operations as their associated numbers.* This method can be applied to computations ranging from simple unit conversions to more complex, multi-step calculations involving several different quantities.

## CONVERSION FACTORS AND DIMENSIONAL ANALYSIS

A ratio of two equivalent quantities expressed with different measurement units can be used as a **unit conversion factor**. For example, the lengths of 2.54 cm and 1 in. are equivalent (by definition), and so a unit conversion factor may be derived from the ratio,

$$\frac{2.54 \text{ cm}}{1 \text{ in.}} (2.54 \text{ cm} = 1 \text{ in.}) \text{ or } 2.54 \frac{\text{cm}}{\text{in.}}$$

Several other commonly used conversion factors are given in Table 1.

*Table 1. Common Conversion Factors*

| Length | Volume | Mass |
|---|---|---|
| 1 m = 1.0936 yd | 1 L = 1.0567 qt | 1 kg = 2.2046 lb |
| 1 in. = 2.54 cm (exact) | 1 qt = 0.94635 L | 1 lb = 453.59 g |
| 1 km = 0.62137 mi | 1 ft$^3$ = 28.317 L | 1 (avoirdupois) oz = 28.349 g |
| 1 mi = 1609.3 m | 1 tbsp = 14.787 mL | 1 (troy) oz = 31.103 g |

When we multiply a quantity (such as distance given in inches) by an appropriate unit conversion factor, we convert the quantity to an equivalent value with different units (such as distance in centimeters). For example, a basketball player's vertical jump of 34 inches can be converted to centimeters by:

$$34 \text{ in.} \times \frac{2.54 \text{ cm}}{1 \text{ in.}} = 86 \text{ cm}$$

Since this simple arithmetic involves *quantities*, the premise of dimensional analysis requires that we multiply both *numbers and units*. The numbers of these two quantities are multiplied to yield the number of the product quantity, 86, whereas the units are multiplied to yield $\frac{\text{in.} \times \text{cm}}{\text{in.}}$. Just as for numbers, a ratio of identical units is also numerically equal to one, $\frac{\text{in.}}{\text{in.}} = 1$, and the unit product thus simplifies to cm. (When identical units divide to yield a factor of 1, they are said to "cancel.") Using dimensional analysis, we can determine that a unit conversion factor has been set up correctly by checking to confirm that the original unit will cancel, and the result will contain the sought (converted) unit.

### Example 1: Using a Unit Conversion Factor

The mass of a competition frisbee is 125 g. Convert its mass to ounces using the unit conversion factor derived from the relationship 1 oz = 28.349 g (Table 1).

Beyond simple unit conversions, the factor-label method can be used to solve more complex problems involving computations. Regardless of the details, the basic approach is the same—all the *factors* involved in the calculation must be appropriately oriented to insure that their *labels* (units) will appropriately cancel and/or combine to yield the desired unit in the result. This is why it is referred to as the factor-label method. As your study of chemistry continues, you will encounter many opportunities to apply this approach.

> ### Example 2: Computing Quantities from Measurement Results and Known Mathematical Relations
>
> What is the density of common antifreeze in units of g/mL? A 4.00-qt sample of the antifreeze weighs 9.26 lb.

> ### Example 3: Computing Quantities from Measurement Results and Known Mathematical Relations
>
> While being driven from Philadelphia to Atlanta, a distance of about 1250 km, a 2014 Lamborghini Aventador Roadster uses 213 L gasoline.
>
> 1. What (average) fuel economy, in miles per gallon, did the Roadster get during this trip?
> 2. If gasoline costs $3.80 per gallon, what was the fuel cost for this trip?

## CONVERSION OF TEMPERATURE UNITS

We use the word temperature to refer to the hotness or coldness of a substance. One way we measure a change in temperature is to use the fact that most substances expand when their temperature increases and contract when their temperature decreases. The mercury or alcohol in a common glass thermometer changes its volume as the temperature changes. Because the volume of the liquid changes more than the volume of the glass, we can see the liquid expand when it gets warmer and contract when it gets cooler.

To mark a scale on a thermometer, we need a set of reference values: Two of the most commonly used are the freezing and boiling temperatures of water at a specified atmospheric pressure. On the Celsius scale, 0 °C is defined as the freezing temperature of water and 100 °C as the boiling temperature of water. The space between the two temperatures is divided into 100 equal intervals, which we call degrees. On the **Fahrenheit** scale, the freezing point of water is defined as 32 °F and the boiling temperature as 212 °F. The space between these two points on a Fahrenheit thermometer is divided into 180 equal parts (degrees).

Defining the Celsius and Fahrenheit temperature scales as described in the previous paragraph results in a slightly more complex relationship between temperature values on these two scales than for different units of measure for other properties. Most measurement units for a given property are directly proportional to one another (y = mx). Using familiar length units as one example:

$$\text{length in feet} = \left(\frac{1 \text{ ft}}{12 \text{ in.}}\right) \times \text{length in inches}$$

## 22  Measurement Basics

where y = length in feet, x = length in inches, and the proportionality constant, m, is the conversion factor. The Celsius and Fahrenheit temperature scales, however, do not share a common zero point, and so the relationship between these two scales is a linear one rather than a proportional one (y = mx + b). Consequently, converting a temperature from one of these scales into the other requires more than simple multiplication by a conversion factor, m, it also must take into account differences in the scales' zero points (b).

The linear equation relating Celsius and Fahrenheit temperatures is easily derived from the two temperatures used to define each scale. Representing the Celsius temperature as *x* and the Fahrenheit temperature as *y*, the slope, *m*, is computed to be:

$$m = \frac{\Delta y}{\Delta x} = \frac{212° \text{ F} - 32° \text{ F}}{100° \text{ C} - 0° \text{ C}} = \frac{180° \text{ F}}{100° \text{ C}} = \frac{9° \text{ F}}{5° \text{ C}}$$

The y-intercept of the equation, *b*, is then calculated using either of the equivalent temperature pairs, (100 °C, 212 °F) or (0 °C, 32 °F), as:

$$b = y - mx = 32° \text{ F} - \frac{9° \text{ F}}{5° \text{ C}} \times 0° \text{ C} = 32° \text{ F}$$

The equation relating the temperature scales is then:

$$T_{°F} = \left(\frac{9° \text{ F}}{5° \text{ C}} \times T_{°C}\right) + 32° \text{ C}$$

An abbreviated form of this equation that omits the measurement units is:

$$T_{°F} = \frac{9}{5} \times T_{°C} + 32$$

Rearrangement of this equation yields the form useful for converting from Fahrenheit to Celsius:

$$T_{°C} = \frac{5}{9}(T_{°F} - 32)$$

As mentioned earlier in this chapter, the SI unit of temperature is the kelvin (K). Unlike the Celsius and Fahrenheit scales, the kelvin scale is an absolute temperature scale in which 0 (zero) K corresponds to the lowest temperature that can theoretically be achieved. The early 19th-century discovery of the relationship between a gas's volume and temperature suggested that the volume of a gas would be zero at −273.15 °C. In 1848, British physicist William Thompson, who later adopted the title of Lord Kelvin, proposed an absolute temperature scale based on this concept (further treatment of this topic is provided in this text's chapter on gases).

The freezing temperature of water on this scale is 273.15 K and its boiling temperature 373.15 K. Notice the numerical difference in these two reference temperatures is 100, the same as for the Celsius scale, and so the linear relation between these two temperature scales will exhibit a slope of $1\frac{\text{K}}{°\text{C}}$. Following the same approach, the equations for converting between the kelvin and Celsius temperature scales are derived to be:

$$T_K = T_{°C} + 273.15$$

$$T_{°C} = T_K - 273.15$$

The 273.15 in these equations has been determined experimentally, so it is not exact. Figure 1 shows the relationship among the three temperature scales. Recall that we do not use the degree sign with temperatures on the kelvin scale.

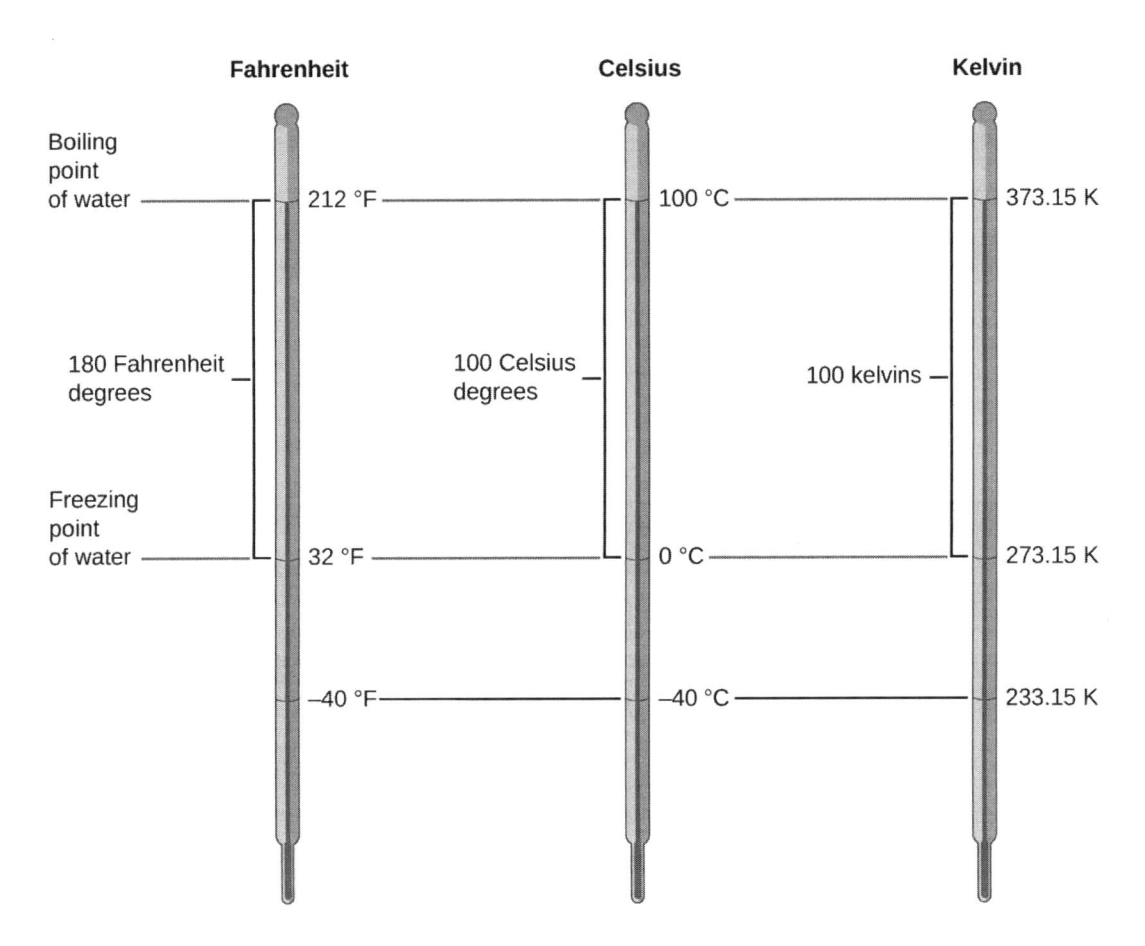

*Figure 1. The Fahrenheit, Celsius, and kelvin temperature scales are compared.*

Although the kelvin (absolute) temperature scale is the official SI temperature scale, Celsius is commonly used in many scientific contexts and is the scale of choice for nonscience contexts in almost all areas of the world. Very few countries (the U.S. and its territories, the Bahamas, Belize, Cayman Islands, and Palau) still use Fahrenheit for weather, medicine, and cooking.

---

### Example 4: Conversion from Celsius

Normal body temperature has been commonly accepted as 37.0 °C (although it varies depending on time of day and method of measurement, as well as among individuals). What is this temperature on the kelvin scale and on the Fahrenheit scale?

**Check Your Learning**

Convert 80.92 °C to K and °F.

## Example 5: Conversion from Fahrenheit

Baking a ready-made pizza calls for an oven temperature of 450 °F. If you are in Europe, and your oven thermometer uses the Celsius scale, what is the setting? What is the kelvin temperature?

## Key Concepts and Summary

Measurements are made using a variety of units. It is often useful or necessary to convert a measured quantity from one unit into another. These conversions are accomplished using unit conversion factors, which are derived by simple applications of a mathematical approach called the factor-label method or dimensional analysis. This strategy is also employed to calculate sought quantities using measured quantities and appropriate mathematical relations.

**Key Equations**

- $T_{°C} = \frac{5}{9} \times T_{°F} - 32$
- $T_{°F} = \frac{9}{5} \times T_{°C} + 32$
- $T_K = °C + 273.15$
- $T_{°C} = K - 273.15$

## Glossary

**dimensional analysis:** (also, factor-label method) versatile mathematical approach that can be applied to computations ranging from simple unit conversions to more complex, multi-step calculations involving several different quantities

**Fahrenheit:** unit of temperature; water freezes at 32 °F and boils at 212 °F on this scale

**unit conversion factor:** ratio of equivalent quantities expressed with different units; used to convert from one unit to a different unit

# Exercises

## PART 1

1. Is one liter about an ounce, a pint, a quart, or a gallon?
2. Is a meter about an inch, a foot, a yard, or a mile?
3. Indicate the SI base units or derived units that are appropriate for the following measurements:

    a. the length of a marathon race (26 miles 385 yards)

    b. the mass of an automobile

    c. the volume of a swimming pool

    d. the speed of an airplane

    e. the density of gold

    f. the area of a football field

    g. the maximum temperature at the South Pole on April 1, 1913

4. Indicate the SI base units or derived units that are appropriate for the following measurements:

    a. the mass of the moon

    b. the distance from Dallas to Oklahoma City

    c. the speed of sound

    d. the density of air

    e. the temperature at which alcohol boils

    f. the area of the state of Delaware

    g. the volume of a flu shot or a measles vaccination

5. Give the name and symbol of the prefixes used with SI units to indicate multiplication by the following exact quantities.

    a. $10^3$

    b. $10^{-2}$

    c. 0.1

    d. $10^{-3}$

    e. 1,000,000

    f. 0.000001

26   Measurement Basics

6. Give the name of the prefix and the quantity indicated by the following symbols that are used with SI base units.

   a. c

   b. d

   c. G

   d. k

   e. m

   f. n

   g. p

   h. T

7. A large piece of jewelry has a mass of 132.6 g. A graduated cylinder initially contains 48.6 mL water. When the jewelry is submerged in the graduated cylinder, the total volume increases to 61.2 mL.

   a. Determine the density of this piece of jewelry.

   b. Assuming that the jewelry is made from only one substance, what substance is it likely to be? Explain.

## PART 2

1. Visit this PhET density simulation and select the Same Volume Blocks.

   a. What are the mass, volume, and density of the yellow block?

   b. What are the mass, volume and density of the red block?

   c. List the block colors in order from smallest to largest mass.

   d. List the block colors in order from lowest to highest density.

   e. How are mass and density related for blocks of the same volume?

2. Visit this PhET density simulation and select Custom Blocks and then My Block.

   a. Enter mass and volume values for the block such that the mass in kg is *less than* the volume in L. What does the block do? Why? Is this always the case when mass < volume?

   b. Enter mass and volume values for the block such that the mass in kg is *more than* the volume in L. What does the block do? Why? Is this always the case when mass > volume?

   c. How would (a) and (b) be different if the liquid in the tank were ethanol instead of water?

   d. How would (a) and (b) be different if the liquid in the tank were mercury instead of water?

3. Visit this PhET density simulation and select Mystery Blocks.

   a. Pick one of the Mystery Blocks and determine its mass, volume, density, and its likely identity.

   b. Pick a different Mystery Block and determine its mass, volume, density, and its likely identity.

   c. Order the Mystery Blocks from least dense to most dense. Explain.

## PART 3

1. Express each of the following numbers in scientific notation with correct significant figures:

   a. 711.0

   b. 0.239

   c. 90743

   d. 134.2

   e. 0.05499

   f. 10000.0

   g. 0.000000738592

2. Express each of the following numbers in exponential notation with correct significant figures:

   a. 704

   b. 0.03344

   c. 547.9

   d. 22086

   e. 1000.00

   f. 0.0000000651

   g. 0.007157

3. Indicate whether each of the following can be determined exactly or must be measured with some degree of uncertainty:

   a. the number of eggs in a basket

   b. the mass of a dozen eggs

   c. the number of gallons of gasoline necessary to fill an automobile gas tank

   d. the number of cm in 2 m

   e. the mass of a textbook

   f. the time required to drive from San Francisco to Kansas City at an average speed of 53 mi/h

4. Indicate whether each of the following can be determined exactly or must be measured with some degree of uncertainty:

   a. the number of seconds in an hour

   b. the number of pages in this book

   c. the number of grams in your weight

   d. the number of grams in 3 kilograms

   e. the volume of water you drink in one day

   f. the distance from San Francisco to Kansas City

5. How many significant figures are contained in each of the following measurements?

   a. 38.7 g

   b. $2 \times 10^{18}$ m

## 28  Measurement Basics

   c. 3,486,002 kg

   d. 9.74150 × 10⁻⁴ J

   e. 0.0613 cm³

   f. 17.0 kg

   g. 0.01400 g/mL

6. How many significant figures are contained in each of the following measurements?

   a. 53 cm

   b. 2.05 × 10⁸ m

   c. 86,002 J

   d. 9.740 × 10⁴ m/s

   e. 10.0613 m³

   f. 0.17 g/mL

   g. 0.88400 s

7. The following quantities were reported on the labels of commercial products. Determine the number of significant figures in each.

   a. 0.0055 g active ingredients

   b. 12 tablets

   c. 3% hydrogen peroxide

   d. 5.5 ounces

   e. 473 mL

   f. 1.75% bismuth

   g. 0.001% phosphoric acid

   h. 99.80% inert ingredients

8. Round off each of the following numbers to two significant figures:

   a. 0.436

   b. 9.000

   c. 27.2

   d. 135

   e. 1.497 × 10⁻³

   f. 0.445

9. Round off each of the following numbers to two significant figures:

   a. 517

   b. 86.3

   c. 6.382 × 10³

   d. 5.0008

e. 22.497

f. 0.885

10. Perform the following calculations and report each answer with the correct number of significant figures.

    a. 628 × 342

    b. (5.63 × 10$^2$) × (7.4 × 10$^3$)

    c. $\dfrac{28.0}{13.483}$

    d. 8119 × 0.000023

    e. 14.98 + 27,340 + 84.7593

    f. 42.7 + 0.259

11. Perform the following calculations and report each answer with the correct number of significant figures.

    a. 62.8 × 34

    b. 0.147 + 0.0066 + 0.012

    c. 38 × 95 × 1.792

    d. 15 − 0.15 − 0.6155

    e. $8.78 \times \left(\dfrac{0.0500}{0.478}\right)$

    f. 140 + 7.68 + 0.014

    g. 28.7 − 0.0483

    h. $\dfrac{(88.5 - 87.57)}{45.13}$

12. Consider the results of the archery contest shown in this figure.

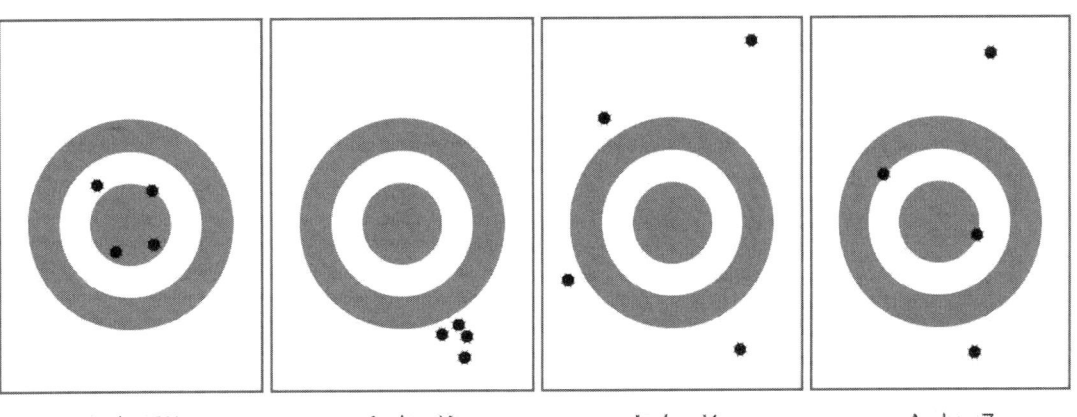

    Archer W    Archer X    Archer Y    Archer Z

    a. Which archer is most precise?

    b. Which archer is most accurate?

    c. Who is both least precise and least accurate?

30   Measurement Basics

13. Classify the following sets of measurements as accurate, precise, both, or neither.

    a. Checking for consistency in the weight of chocolate chip cookies: 17.27g, 13.05g, 19.46g, 16.92g

    b. Testing the volume of a batch of 25-mL pipettes: 27.02 mL, 26.99 mL, 26.97 mL, 27.01 mL

    c. Determining the purity of gold: 99.9999%, 99.9998%, 99.9998%, 99.9999%

## PART 4

1. Write conversion factors (as ratios) for the number of:

    a. yards in 1 meter

    b. liters in 1 liquid quart

    c. pounds in 1 kilogram

2. Write conversion factors (as ratios) for the number of:

    a. kilometers in 1 mile

    b. liters in 1 cubic foot

    c. grams in 1 ounce

3. Soccer is played with a round ball having a circumference between 27 and 28 in. and a weight between 14 and 16 oz. What are these specifications in units of centimeters and grams?

4. A woman's basketball has a circumference between 28.5 and 29.0 inches and a maximum weight of 20 ounces (two significant figures). What are these specifications in units of centimeters and grams?

5. Use scientific (exponential) notation to express the following quantities in terms of the SI base units in Table 1:

    a. 0.13 g

    b. 232 Gg

    c. 5.23 pm

    d. 86.3 mg

    e. 37.6 cm

    f. 54 µm

    g. 1 Ts

    h. 27 ps

    i. 0.15 mK

6. Complete the following conversions between SI units.

    a. 612 g = _____ mg

    b. 8.160 m = _____ cm

    c. 3779 µg = _____ g

    d. 781 mL = _____ L

    e. 4.18 kg = _____ g

f. 27.8 m = _____ km

g. 0.13 mL = _____ L

h. 1738 km = _____ m

i. 1.9 Gg = _____ g

7. Make the conversion indicated in each of the following:

   a. the men's world record long jump, 29 ft 4¼ in., to meters

   b. the greatest depth of the ocean, about 6.5 mi, to kilometers

   c. the area of the state of Oregon, 96,981 mi$^2$, to square kilometers

   d. the volume of 1 gill (exactly 4 oz) to milliliters

   e. the estimated volume of the oceans, 330,000,000 mi$^3$, to cubic kilometers.

   f. the mass of a 3525-lb car to kilograms

   g. the mass of a 2.3-oz egg to grams

8. Make the conversion indicated in each of the following:

   a. the length of a soccer field, 120 m (three significant figures), to feet

   b. the height of Mt. Kilimanjaro, at 19,565 ft the highest mountain in Africa, to kilometers

   c. the area of an 8.5 t 11-inch sheet of paper in cm$^2$

   d. the displacement volume of an automobile engine, 161 in.$^3$, to liters

   e. the estimated mass of the atmosphere, 5.6 t 10$^{15}$ tons, to kilograms

   f. the mass of a bushel of rye, 32.0 lb, to kilograms

   g. the mass of a 5.00-grain aspirin tablet to milligrams (1 grain = 0.00229 oz)

9. As an instructor is preparing for an experiment, he requires 225 g phosphoric acid. The only container readily available is a 150-mL Erlenmeyer flask. Is it large enough to contain the acid, whose density is 1.83 g/mL?

10. To prepare for a laboratory period, a student lab assistant needs 125 g of a compound. A bottle containing 1/4 lb is available. Did the student have enough of the compound?

11. 27 A chemistry student is 159 cm tall and weighs 45.8 kg. What is her height in inches and weight in pounds?

12. In a recent Grand Prix, the winner completed the race with an average speed of 229.8 km/h. What was his speed in miles per hour, meters per second, and feet per second?

13. Solve these problems about lumber dimensions.

    a. To describe to a European how houses are constructed in the US, the dimensions of "two-by-four" lumber must be converted into metric units. The thickness × width × length dimensions are 1.50 in. × 3.50 in. × 8.00 ft in the US. What are the dimensions in cm × cm × m?

    b. This lumber can be used as vertical studs, which are typically placed 16.0 in. apart. What is that distance in centimeters?

14. Calculate the density of aluminum if 27.6 cm$^3$ has a mass of 74.6 g.

15. Osmium is one of the densest elements known. What is its density if 2.72 g has a volume of 0.121 cm$^3$?

32  Measurement Basics

## PART 5

1. Convert the boiling temperature of gold, 2966 °C, into degrees Fahrenheit and kelvin.
2. Convert the temperature of scalding water, 54 °C, into degrees Fahrenheit and kelvin.
3. Convert the temperature of the coldest area in a freezer, −10 °F, to degrees Celsius and kelvin.
4. Convert the temperature of dry ice, −77 °C, into degrees Fahrenheit and kelvin.
5. Convert the boiling temperature of liquid ammonia, −28.1 °F, into degrees Celsius and kelvin.
6. The label on a pressurized can of spray disinfectant warns against heating the can above 130 °F. What are the corresponding temperatures on the Celsius and kelvin temperature scales?
7. The weather in Europe was unusually warm during the summer of 1995. The TV news reported temperatures as high as 45 °C. What was the temperature on the Fahrenheit scale?

# Check Your Knowledge: Self-Test

1. To three decimal places, what is the volume of a cube (cm³) with an edge length of 0.843 cm?
2. If the cube in part 1 is copper and has a mass of 5.34 g, what is the density of copper to two decimal places?
3. Remove all of the blocks from the water and add the green block to the tank of water, placing it approximately in the middle of the tank. Determine the density of the green block.
4. An irregularly shaped piece of a shiny yellowish material is weighed and then submerged in a graduated cylinder, with results as shown.

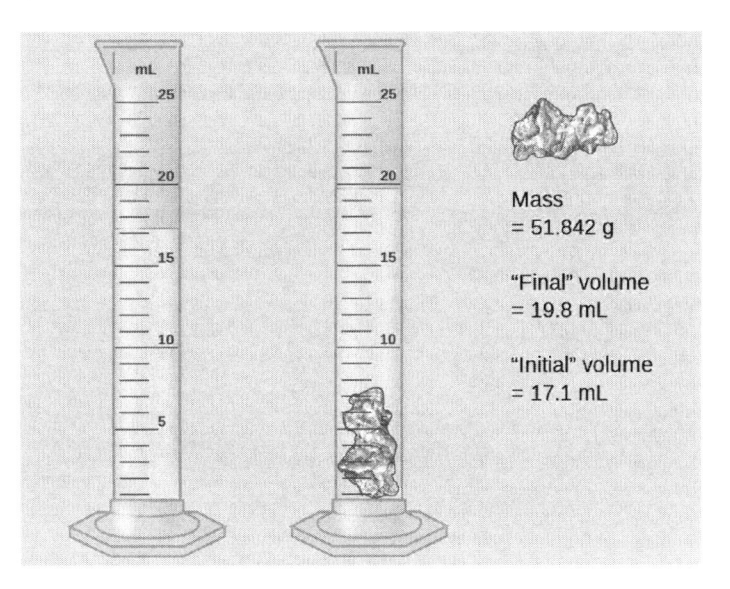

   Mass = 51.842 g

   "Final" volume = 19.8 mL

   "Initial" volume = 17.1 mL

   a. Use these values to determine the density of this material.

   b. Do you have any reasonable guesses as to the identity of this material? Explain your reasoning.

5. Add 2.334 mL and 0.31 mL.
6. Subtract 55.8752 m from 56.533 m.
7. Multiply 2.334 cm and 0.320 cm.
8. Divide 55.8752 m by 56.53 s.
9. Convert a volume of 9.345 qt to liters.
10. What is the volume in liters of 1.000 oz, given that 1 L = 1.0567 qt and 1 qt = 32 oz (exactly)?
11. A Toyota Prius Hybrid uses 59.7 L gasoline to drive from San Francisco to Seattle, a distance of 1300 km (two significant digits).

    1. What (average) fuel economy, in miles per gallon, did the Prius get during this trip?

    2. If gasoline costs $3.90 per gallon, what was the fuel cost for this trip?

12. Convert 80.92 °C to K and °F.
13. Convert 50 °F to °C and K.

# Chapter 2: Introduction to Biology

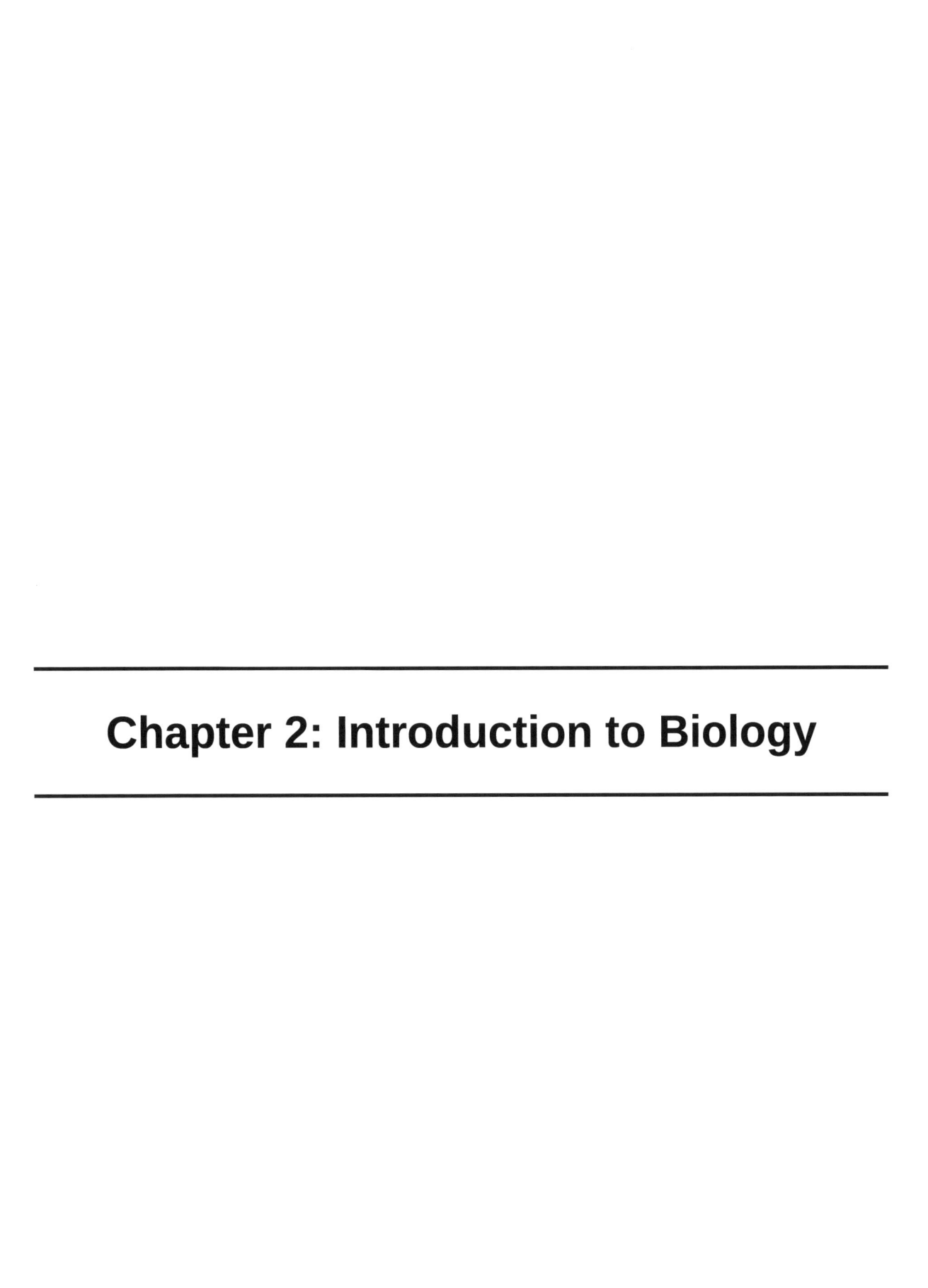

# Themes and Concepts of Biology

### Learning Objective

By the end of this section, you will be able to:
- Identify and describe the properties of life
- Describe the levels of organization among living things
- List examples of different sub disciplines in biology

Biology is the science that studies life. What exactly is life? This may sound like a silly question with an obvious answer, but it is not easy to define life. For example, a branch of biology called virology studies viruses, which exhibit some of the characteristics of living entities but lack others. It turns out that although viruses can attack living organisms, cause diseases, and even reproduce, they do not meet the criteria that biologists use to define life.

From its earliest beginnings, biology has wrestled with four questions: What are the shared properties that make something "alive"? How do those various living things function? When faced with the remarkable diversity of life, how do we organize the different kinds of organisms so that we can better understand them? And, finally—what biologists ultimately seek to understand—how did this diversity arise and how is it continuing? As new organisms are discovered every day, biologists continue to seek answers to these and other questions.

## PROPERTIES OF LIFE

All groups of living organisms share several key characteristics or functions: order, sensitivity or response to stimuli, reproduction, adaptation, growth and development, regulation, homeostasis, and energy processing. When viewed together, these eight characteristics serve to define life.

### Order

Organisms are highly organized structures that consist of one or more cells. Even very simple, single-celled organisms are remarkably complex. Inside each cell, atoms make up molecules. These in turn make up cell components or organelles. Multicellular organisms, which may consist of millions of individual cells, have an advantage over single-celled organisms in that their cells can be specialized to perform specific functions, and even sacrificed in certain situations for the good of the organism as a whole. How these specialized cells come together to form organs such as the heart, lung, or skin in organisms like the toad shown in [Figure 1] will be discussed later.

### Sensitivity or Response to Stimuli

Organisms respond to diverse stimuli. For example, plants can bend toward a source of light or respond to touch ([Figure 2]). Even tiny bacteria can move toward or away from chemicals (a process called chemotaxis) or light (phototaxis). Movement toward a stimulus is considered a positive response, while movement away from a stimulus is considered a negative response.

38  Introduction to Biology

*Figure 1: A toad represents a highly organized structure consisting of cells, tissues, organs, and organ systems. (credit: "Ivengo(RUS)"/Wikimedia Commons)*

*Figure 2: The leaves of this sensitive plant (Mimosa pudica) will instantly droop and fold when touched. After a few minutes, the plant returns to its normal state. (credit: Alex Lomas)*

Watch this video to see how the sensitive plant responds to a touch stimulus: https://commons.wikimedia.org/wiki/File:Mimosa_pudica_leaves_folding_when_touched_3.ogv.

## Reproduction

Single-celled organisms reproduce by first duplicating their DNA, which is the genetic material, and then dividing it equally as the cell prepares to divide to form two new cells. Many multicellular organisms (those made up of more than one cell) produce specialized reproductive cells that will form new individuals. When reproduction occurs, DNA containing genes is passed along to an organism's offspring. These genes are the reason that the offspring will belong to the same species and will have characteristics similar to the parent, such as fur color and blood type.

*Figure 3: Although no two look alike, these kittens have inherited genes from both parents and share many of the same characteristics. (credit: Pieter & Renée Lanser)*

*Adaptation*

All living organisms exhibit a "fit" to their environment. Biologists refer to this fit as adaptation and it is a consequence of evolution by natural selection, which operates in every lineage of reproducing organisms. Examples of adaptations are diverse and unique, from heat-resistant Archaea that live in boiling hot springs to the tongue length of a nectar-feeding moth that matches the size of the flower from which it feeds. All adaptations enhance the reproductive potential of the individual exhibiting them, including their ability to survive to reproduce. Adaptations are not constant. As an environment changes, natural selection causes the characteristics of the individuals in a population to track those changes.

*Growth and Development*

Organisms grow and develop according to specific instructions coded for by their genes. These genes provide instructions that will direct cellular growth and development, ensuring that a species' young (Figure 3) will grow up to exhibit many of the same characteristics as its parents.

*Regulation*

Even the smallest organisms are complex and require multiple regulatory mechanisms to coordinate internal functions, such as the transport of nutrients, response to stimuli, and coping with environmental stresses. For example, organ systems such as the digestive or circulatory systems perform specific functions like carrying oxygen throughout the body, removing wastes, delivering nutrients to every cell, and cooling the body.

*Homeostasis*

Living Things Must Maintain Homeostasis. **Homeostasis** means **"steady state"**. Homeostasis is the tendency of an organism or cell to **maintain a constant internal environment**. Living things constantly adjust to internal and external changes. Homeostasis means to maintain **dynamic equilibrium** in the body. It is dynamic because it is constantly adjusting to the changes that the body's systems encounter. It is equilibrium because body functions are kept within specific ranges or normal limits.

Maintaining homeostasis requires that the body continuously monitor its internal conditions. From body temperature to blood pressure to levels of certain nutrients, each physiological condition has a particular set point. A **set point** is the physiological value around which the normal range fluctuates. A **normal range** is the restricted set of values that is optimally healthful and stable. For example, the set point for normal human body temperature is approximately 37°C (98.6°F) Physiological parameters, such as body temperature and blood pressure, tend to fluctuate within a normal range a few degrees above and below that point. Even an animal that is apparently inactive is maintaining this homeostatic equilibrium. An inability to maintain homeostasis may lead to death or a disease, a condition known as **homeostatic imbalance.**

Homeostasis, in a general sense, refers to stability, balance, or equilibrium. Physiologically, it is the body's attempt to maintain a constant and balanced internal environment, which requires persistent monitoring and adjustments as conditions change.

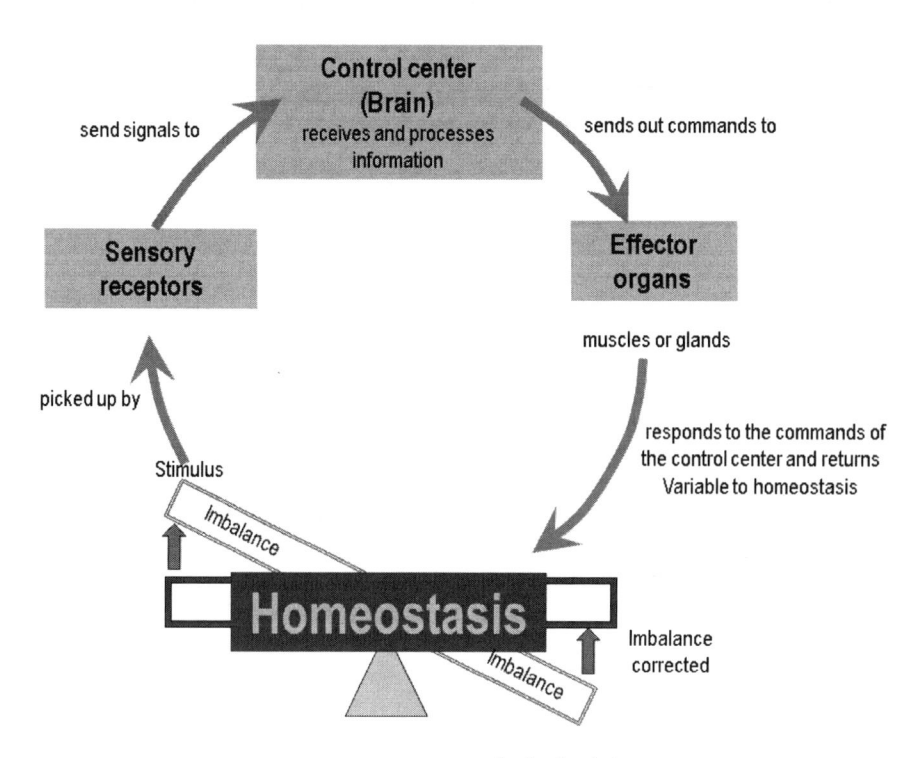

*Figure 2: Components of a feedback loop*

Adjustment of physiological systems within the body is called **homeostatic regulation**, which involves **three parts or mechanisms:**

1. **Sensory receptor**
2. **Control center – Brain**
3. **Effector organ**

- The **sensory receptor** receives information that something in the environment is changing. A change in the internal or external environment is called a **stimulus** and is detected by a receptor. sensory receptor monitors a physiological value. This value is reported to the control center.

- The **control center or integration center (Brain)** receives and processes information from the receptor.

- The **effector organ** responds to the commands of the control center by either opposing or enhancing the stimulus. The effector is a **muscle** (that contracts or relaxes) or a **gland** that secretes.

## Maintenance of Homeostasis by Feedback loops

When a change occurs in an animal's environment, an adjustment must be made. The receptor senses the change in the environment, then sends a signal to the control center (in most cases, the brain) which in turn generates a response that is signaled to an effector. Homeostasis is controlled by the nervous and endocrine system of mammals. **Homeostatsis is maintained by negative feedback loops.** Positive feedback loops actually push the organism further out of homeostasis, but may be necessary for life to occur.

### Types of homeostatic feedback systems / mechanisms:

- Negative Feedback Mechanism
- Positive Feedback Mechanism

Table 1: Examples for Homeostasis

| Negative Feedback Mechanism maintains | Positive Feedback Mechanism |
|---|---|
| Body Temperature | Childbirth |
| Blood pressure | Blood Clotting |
| Blood Sugar | Ovulation |
| Blood pH | Lactation |

### Negative Feedback Mechanisms

Any homeostatic process that changes the direction of the stimulus is a **negative feedback loop**. It may either increase or decrease the stimulus, but the stimulus is not allowed to continue as it did before the receptor sensed it. In other words, if a level is too high, the body does something to bring it down, and conversely, if a level is too low, the body does something to make it go up. Hence the term negative feedback.

### Thermoregulation:

Thermoregulation is a process that allows your body to maintain its core internal temperature. The set point for normal human body temperature is approximately 37°C (98.6°F). Body temperature affects body activities. Body proteins, including enzymes, begin to denature and lose their function with high heat (around 50°C for mammals). Enzyme activity will decrease by half for every ten degree centigrade drop in temperature, to the point of freezing, with a few exceptions.

During body temperature regulation, **temperature receptors in the skin (sensory receptor)** communicate information to the **brain (the control center)** which signals the **blood vessels and sweat glands in the skin (effectors)**. As the internal and external environment of the body are constantly changing, adjustments must be made continuously to stay at or near a specific value: the **set point**. (approximately 37°C / 98.6°F)

### When body temperature exceeds its normal range:

When the brain's temperature regulation center receives data from the sensors indicating that the body's temperature exceeds its normal range, the brain (the control center) which signals the effectors: blood vessels and sweat glands in the skin. This stimulation has three major effects:

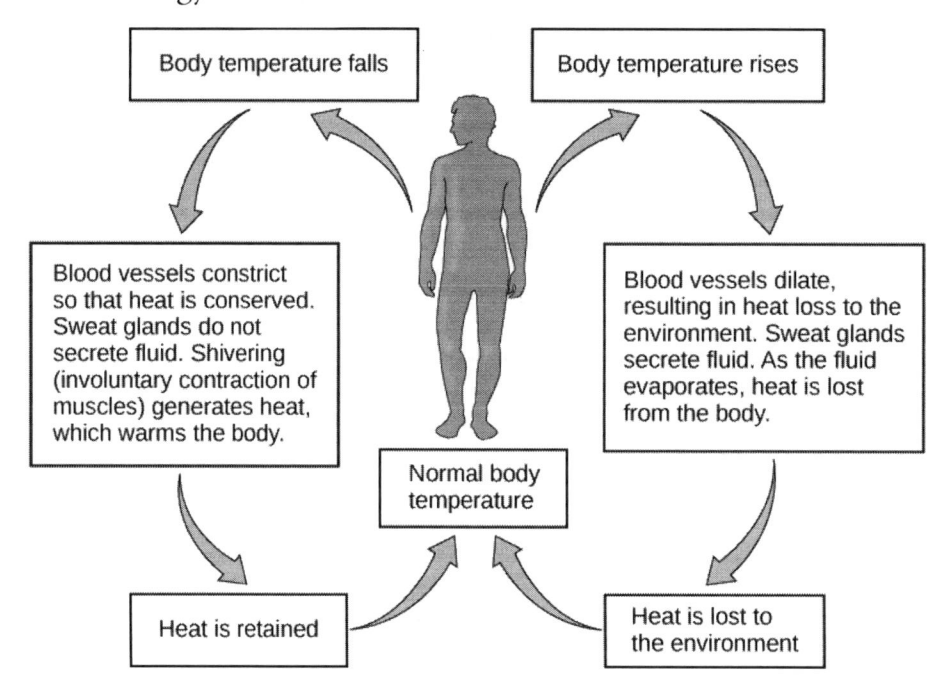

```
                   Body temperature falls        Body temperature rises

   ┌─────────────────────────────┐                    ┌─────────────────────────────┐
   │ Blood vessels constrict     │                    │ Blood vessels dilate,       │
   │ so that heat is conserved.  │                    │ resulting in heat loss to the│
   │ Sweat glands do not         │                    │ environment. Sweat glands   │
   │ secrete fluid. Shivering    │                    │ secrete fluid. As the fluid │
   │ (involuntary contraction of │                    │ evaporates, heat is lost    │
   │ muscles) generates heat,    │                    │ from the body.              │
   │ which warms the body.       │                    └─────────────────────────────┘
   └─────────────────────────────┘
                                       Normal body
                                       temperature

              Heat is retained                         Heat is lost to
                                                       the environment
```

*Figure 3: Flow chart shows how normal body temperature is maintained.*

- Blood vessels in the skin begin to dilate allowing more blood from the body core to flow to the surface of the skin allowing the heat to radiate into the environment.

- As blood flow to the skin increases, sweat glands are activated to increase their output. As the sweat evaporates from the skin surface into the surrounding air, it takes heat with it.

- The depth of respiration increases, and a person may breathe through an open mouth instead of through the nasal passageways. This further increases heat loss from the lungs.

**When body temperature reduces below its normal range:**

In contrast, activation of the brain's heat-gain center by exposure to cold

- reduces blood flow to the skin, and blood returning from the limbs is diverted into a network of deep veins. This arrangement traps heat closer to the body core and restricts heat loss.

- If heat loss is severe, the brain triggers an increase in random signals to skeletal muscles, causing them to contract and producing shivering. The muscle contractions of shivering release heat while using up ATP.

- The brain triggers the thyroid gland in the endocrine system to release thyroid hormone, which increases metabolic activity and heat production in cells throughout the body. The brain also signals the adrenal glands to release epinephrine (adrenaline), a hormone that causes the breakdown of glycogen into glucose, which can be used as an energy source. The breakdown of glycogen into glucose also results in increased metabolism and heat production.

*Positive Feedback Mechanism*

**Positive feedback** intensifies a change in the body's physiological condition rather than reversing it, maintains the direction of the stimulus, possibly accelerating it. A deviation from the normal range results in more change, and the system moves farther away from the normal range. Positive feedback in the body is normal only when there is a definite end point. Childbirth and the body's response to blood loss are two examples of positive feedback loops that are normal but are activated only when needed.

Themes and Concepts of Biology 43

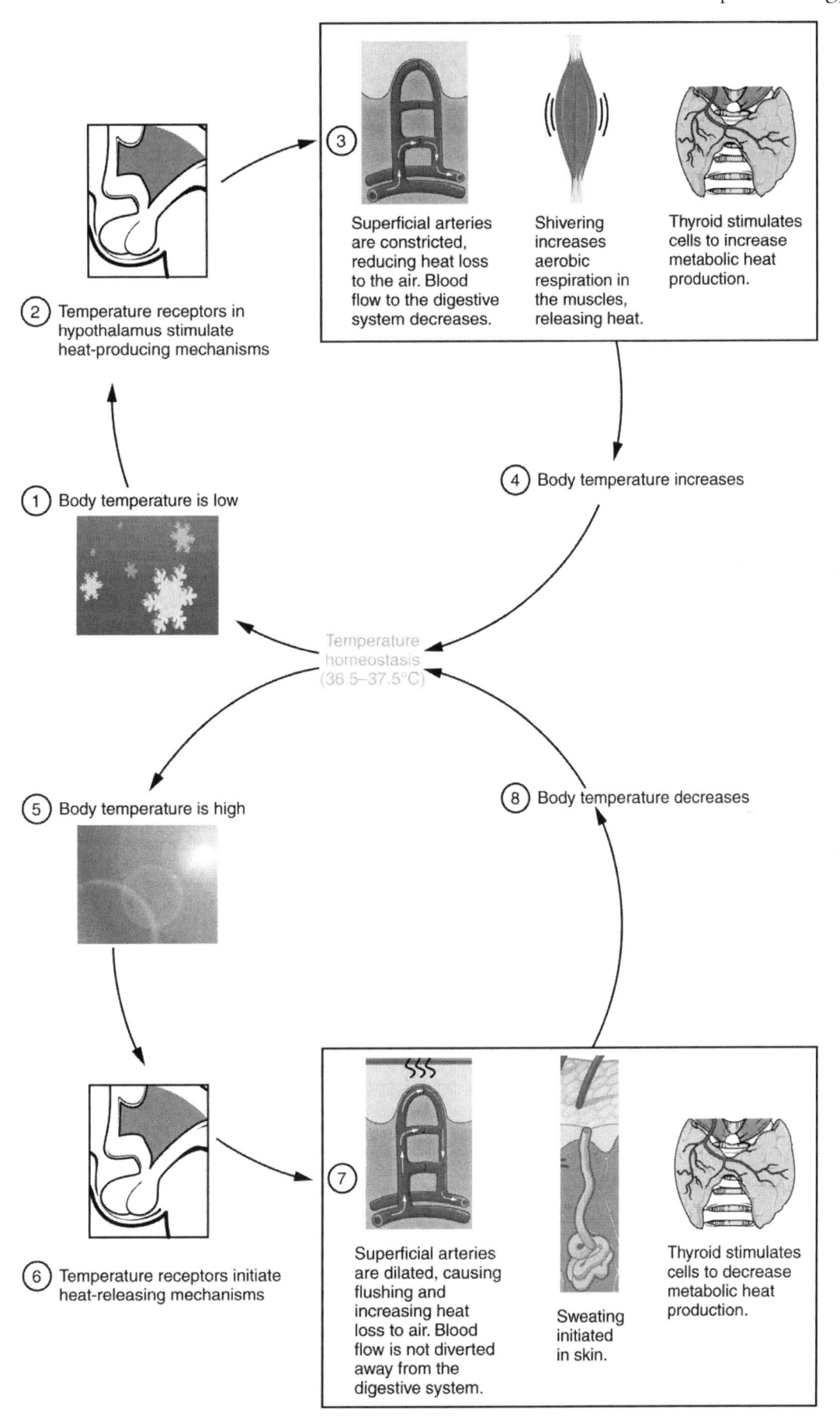

1. Body temperature is low

2. Temperature receptors in hypothalamus stimulate heat-producing mechanisms

3. Superficial arteries are constricted, reducing heat loss to the air. Blood flow to the digestive system decreases. Shivering increases aerobic respiration in the muscles, releasing heat. Thyroid stimulates cells to increase metabolic heat production.

4. Body temperature increases

Temperature homeostasis (36.5–37.5°C)

5. Body temperature is high

6. Temperature receptors initiate heat-releasing mechanisms

7. Superficial arteries are dilated, causing flushing and increasing heat loss to air. Blood flow is not diverted away from the digestive system. Sweating initiated in skin. Thyroid stimulates cells to decrease metabolic heat production.

8. Body temperature decreases

*Figure 4: The hypothalamus in Brain controls thermoregulation.*

### Example 1: Labor and Delivery During Childbirth:

The extreme muscular work of labor and delivery are the result of a positive feedback system (Figure 5). The first contractions of labor (the stimulus) push the baby toward the cervix (the lowest part of the uterus). The cervix contains **stretch-sensitive nerve cells (the sensory receptors)** that monitor the degree of stretching. These nerve cells send messages to the brain, which in turn causes the **pituitary gland at the base of the brain (control center)** to release the hormone **oxytocin** into the bloodstream. Oxytocin causes stronger contractions of the smooth muscles in of the **uterus (the effectors),** pushing the baby further down the birth canal. This causes even greater stretching of the cervix. The cycle of stretching, oxytocin release, and increasingly more forceful contractions stops only when the baby comes out of the uterus. At this point, the stretching of the cervix halts, stopping the release of oxytocin.

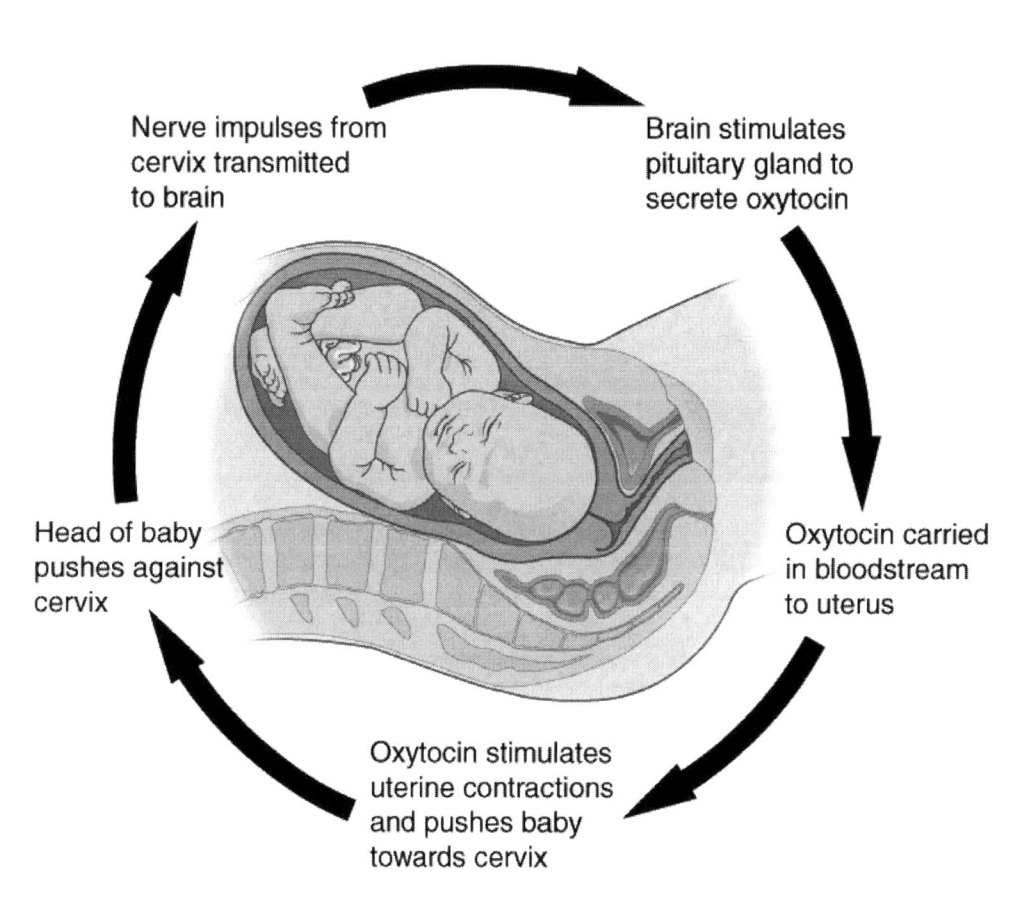

*Figure 5. Positive Feedback Loop. Normal childbirth is driven by a positive feedback loop. A positive feedback loop results in a change in the body's status, rather than a return to homeostasis.*

### Example 2: Steps in Blood Clotting

A second example of positive feedback centers on reversing extreme damage to the body. Following a penetrating wound, the most immediate threat is excessive blood loss. Less blood circulating means reduced blood pressure and reduced perfusion (penetration of blood) to the brain and other vital organs. If perfusion is severely reduced, vital organs will shut down and the person will die. The body responds to this potential catastrophe by releasing substances in the injured blood vessel wall that begin the process of blood clotting. **As each step of clotting occurs, it stimulates the release of more clotting substances.** This accelerates the processes of clotting and sealing off the damaged area. Clotting is contained in a local area based on the tightly controlled availability of clotting proteins. This is an adaptive, life-saving cascade of events.

*Energy Processing*

All organisms (such as the California condor shown in (Figure 5) use a source of energy for their metabolic activities. Some organisms capture energy from the Sun and convert it into chemical energy in food; others use chemical energy from molecules they take in.

## LEVELS OF ORGANIZATION OF LIVING THINGS

Living things are highly organized and structured, following a hierarchy on a scale from small to large. The atom is the smallest and most fundamental unit of matter. It consists of a nucleus surrounded by electrons. Atoms form molecules. A molecule is a chemical structure consisting of at least two atoms held together by a chemical bond. Many molecules that are biologically important are macromolecules, large molecules that are typically formed by combining smaller units called monomers. An example of a macromolecule is deoxyribonucleic acid (DNA) (Figure 6), which contains the instructions for the functioning of the organism that contains it.

Some cells contain aggregates of macromolecules surrounded by membranes; these are called organelles. Organelles are small structures that exist within cells and perform specialized functions. All living things are made of cells; the cell itself is the smallest fundamental unit of structure and function in living organisms. (This requirement is why viruses are not considered living: they are not made of cells. To make new viruses, they have to invade and hijack a living cell; only then can they obtain the materials they need to reproduce.) Some organisms consist of a single cell and others are multicellular. Cells are classified as prokaryotic or eukaryotic. Prokaryotes are single-celled organisms that lack organelles surrounded by a membrane and do not have nuclei surrounded by nuclear membranes; in contrast, the cells of eukaryotes do have membrane-bound organelles and nuclei.

In most multicellular organisms, cells combine to make tissues, which are groups of similar cells carrying out the same function. Organs are collections of tissues grouped together based on a common function. Organs are present not only in animals but also in plants. An organ system is a higher level of organization that consists of functionally related organs. For example vertebrate animals have many organ systems, such as the circulatory system that transports blood throughout the body and to and from the lungs; it includes organs such as the heart and blood vessels. Organisms are individual living entities. For example, each tree in a forest

*Figure 5: A lot of energy is required for a California condor to fly. Chemical energy derived from food is used to power flight. California condors are an endangered species; scientists have strived to place a wing tag on each bird to help them identify and locate each individual bird. (credit: Pacific Southwest Region U.S. Fish and Wildlife)*

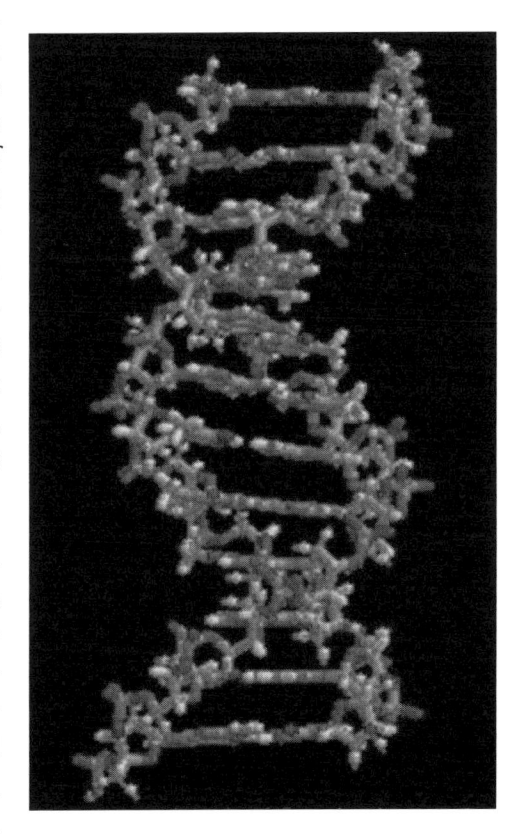

*Figure 6: A molecule, like this large DNA molecule, is composed of atoms. (credit: "Brian0918"/Wikimedia Commons)*

### Art Connection

**Atom:** A basic unit of matter that consists of a dense central nucleus surrounded by a cloud of negatively charged electrons.

**Molecule:** A phospholipid, composed of many atoms.

**Organelles:** Structures that perform functions within a cell. Highlighted in blue are a Golgi apparatus and a nucleus.

**Cells:** Human blood cells.

**Tissue:** Human skin tissue.

**Organs and organ systems:** Organs such as the stomach and intestine make up part of the human digestive system.

**Organisms, populations, and communities:** In a park, each person is an organism. Together, all the people make up a population. All the plant and animal species in the park comprise a community.

**Ecosystem:** The ecosystem of Central Park in New York includes living organisms and the environment in which they live.

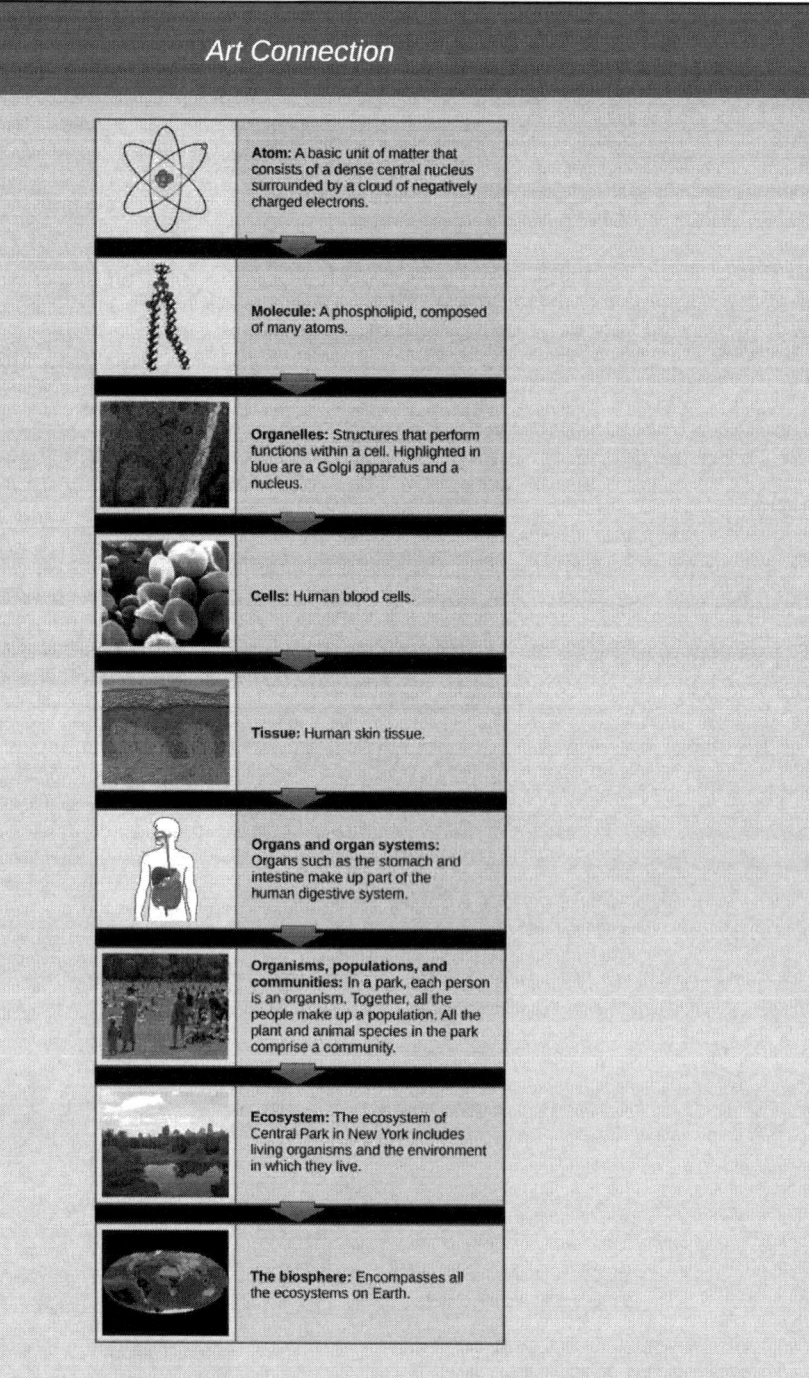

**The biosphere:** Encompasses all the ecosystems on Earth.

*Figure 7: From an atom to the entire Earth, biology examines all aspects of life. (credit "molecule": modification of work by Jane Whitney; credit "organelles": modification of work by Louisa Howard; credit "cells": modification of work by Bruce Wetzel, Harry Schaefer, National Cancer Institute; credit "tissue": modification of work by "Kilbad"/Wikimedia Commons; credit "organs": modification of work by Mariana Ruiz Villareal, Joaquim Alves Gaspar; credit "organisms": modification of work by Peter Dutton; credit "ecosystem": modification of work by "gigi4791"/Flickr; credit "biosphere": modification of work by NASA)*

Which of the following statements is false?
1. Tissues exist within organs which exist within organ systems.
2. Communities exist within populations which exist within ecosystems.
3. Organelles exist within cells which exist within tissues.
4. Communities exist within ecosystems which exist in the biosphere.

is an organism. Single-celled prokaryotes and single-celled eukaryotes are also considered organisms and are typically referred to as microorganisms.

All the individuals of a species living within a specific area are collectively called a population. For example, a forest may include many white pine trees. All of these pine trees represent the population of white pine trees in this forest. Different populations may live in the same specific area. For example, the forest with the pine trees includes populations of flowering plants and also insects and microbial populations. A community is the set of populations inhabiting a particular area. For instance, all of the trees, flowers, insects, and other populations in a forest form the forest's community. The forest itself is an ecosystem. An ecosystem consists of all the living things in a particular area together with the abiotic, or non-living, parts of that environment such as nitrogen in the soil or rainwater. At the highest level of organization (Figure 7), the biosphere is the collection of all ecosystems, and it represents the zones of life on Earth. It includes land, water, and portions of the atmosphere.

## THE DIVERSITY OF LIFE

The science of biology is very broad in scope because there is a tremendous diversity of life on Earth. The source of this diversity is evolution, the process of gradual change during which new species arise from older species. Evolutionary biologists study the evolution of living things in everything from the microscopic world to ecosystems.

In the 18th century, a scientist named Carl Linnaeus first proposed organizing the known species of organisms into a hierarchical taxonomy. In this system, species that are most similar to each other are put together within a grouping known as a genus. Furthermore, similar genera (the plural of genus) are put together within a family. This grouping continues until all organisms are collected together into groups at the highest level. The current taxonomic system now has eight levels in its hierarchy, from lowest to highest, they are: species, genus, family, order, class, phylum, kingdom, domain. Thus species are grouped within genera, genera are grouped within families, families are grouped within orders, and so on ([Figure 8]).

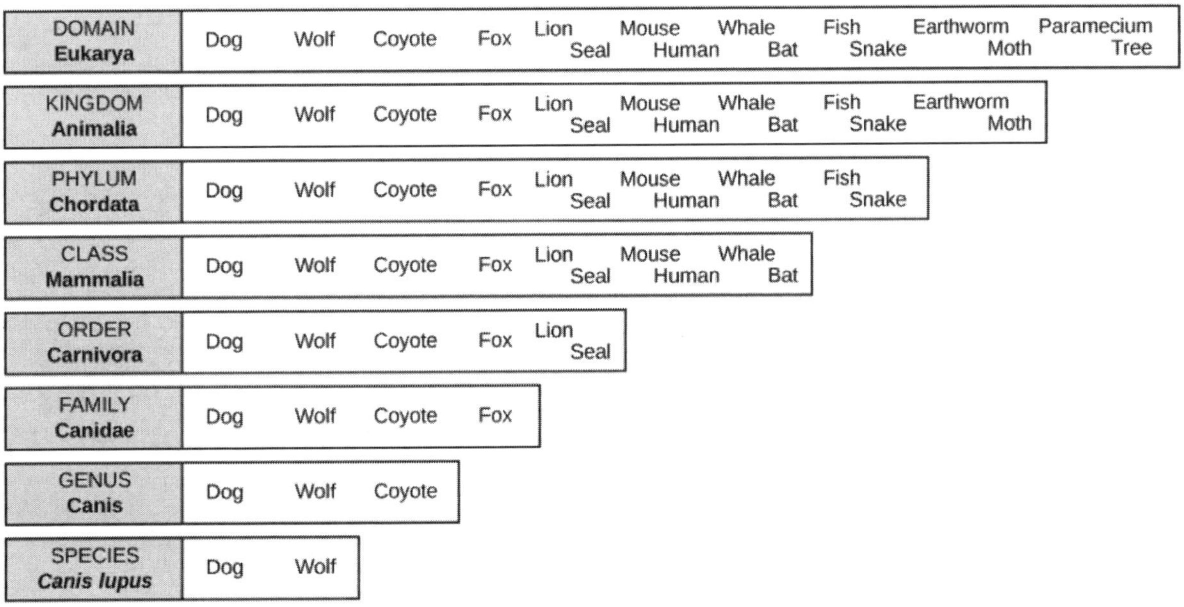

*Figure 8: This diagram shows the levels of taxonomic hierarchy for a dog, from the broadest category—domain—to the most specific—species.*

# 48   Introduction to Biology

The highest level, domain, is a relatively new addition to the system since the 1990s. Scientists now recognize three domains of life, the Eukarya, the Archaea, and the Bacteria. The domain Eukarya contains organisms that have cells with nuclei. It includes the kingdoms of fungi, plants, animals, and several kingdoms of protists. The Archaea, are single-celled organisms without nuclei and include many extremophiles that live in harsh environments like hot springs. The Bacteria are another quite different group of single-celled organisms without nuclei ([Figure 9]). Both the Archaea and the Bacteria are prokaryotes, an informal name for cells without nuclei. The recognition in the 1990s that certain "bacteria," now known as the Archaea, were as different genetically and biochemically from other bacterial cells as they were from eukaryotes, motivated the recommendation to divide life into three domains. This dramatic change in our knowledge of the tree of life demonstrates that classifications are not permanent and will change when new information becomes available.

In addition to the hierarchical taxonomic system, Linnaeus was the first to name organisms using two unique names, now called the binomial naming system. Before Linnaeus, the use of common names to refer to organisms caused confusion because there were regional differences in these common names. Binomial names consist of the genus name (which is capitalized) and the species name (all lower-case). Both names are set in italics when they are printed. Every species is given a unique binomial which is recognized the world over, so that a scientist in any location can know which organism is being referred to. For example, the North American blue jay is known uniquely as *Cyanocitta cristata*. Our own species is *Homo sapiens*.

## Branches of Biological Study

The scope of biology is broad and therefore contains many branches and sub disciplines. Biologists may pursue one of those sub disciplines and work in a more focused field. For instance, molecular biology studies biological processes at the molecular level, including interactions among molecules such as DNA, RNA, and proteins, as well as the way they are regulated. Microbiology is the study of the structure and function of microorganisms. It is quite a broad branch itself, and depending on the subject of study, there are also microbial physiologists, ecologists, and geneticists, among others.

Another field of biological study, neurobiology, studies the biology of the nervous system, and although it is considered a branch of biology, it is also recognized as an interdisciplinary field of study known as neuroscience. Because of its interdisciplinary nature, this sub discipline studies different functions of the nervous system using molecular, cellular, developmental, medical, and computational approaches.

*These images represent different domains. The scanning electron micrograph shows (a) bacterial cells belong to the domain Bacteria, while the (b) extremophiles, seen all together as colored mats in this hot spring, belong to domain Archaea. Both the (c) sunflower and (d) lion are part of domain Eukarya. (credit a: modification of work by Rocky Mountain Laboratories, NIAID, NIH; credit b: modification of work by Steve Jurvetson; credit c: modification of work by Michael Arrighi; credit d: modification of work by Frank Vassen)*

## Key Takeaways

Evolution in Action

Carl Woese and the Phylogenetic TreeThe evolutionary relationships of various life forms on Earth can be summarized in a phylogenetic tree. A phylogenetic tree is a diagram showing the evolutionary relationships among biological species based on similarities and differences in genetic or physical traits or both. A phylogenetic tree is composed of branch points, or nodes, and branches. The internal nodes represent ancestors and are points in evolution when, based on scientific evidence, an ancestor is thought to have diverged to form two new species. The length of each branch can be considered as estimates of relative time.

In the past, biologists grouped living organisms into five kingdoms: animals, plants, fungi, protists, and bacteria. The pioneering work of American microbiologist Carl Woese in the early 1970s has shown, however, that life on Earth has evolved along three lineages, now called domains—Bacteria, Archaea, and Eukarya. Woese proposed the domain as a new taxonomic level and Archaea as a new domain, to reflect the new phylogenetic tree ([Figure 10]). Many organisms belonging to the Archaea domain live under extreme conditions and are called extremophiles. To construct his tree, Woese used genetic relationships rather than similarities based on morphology (shape). Various genes were used in phylogenetic studies. Woese's tree was constructed from comparative sequencing of the genes that are universally distributed, found in some slightly altered form in every organism, conserved (meaning that these genes have remained only slightly changed throughout evolution), and of an appropriate length.

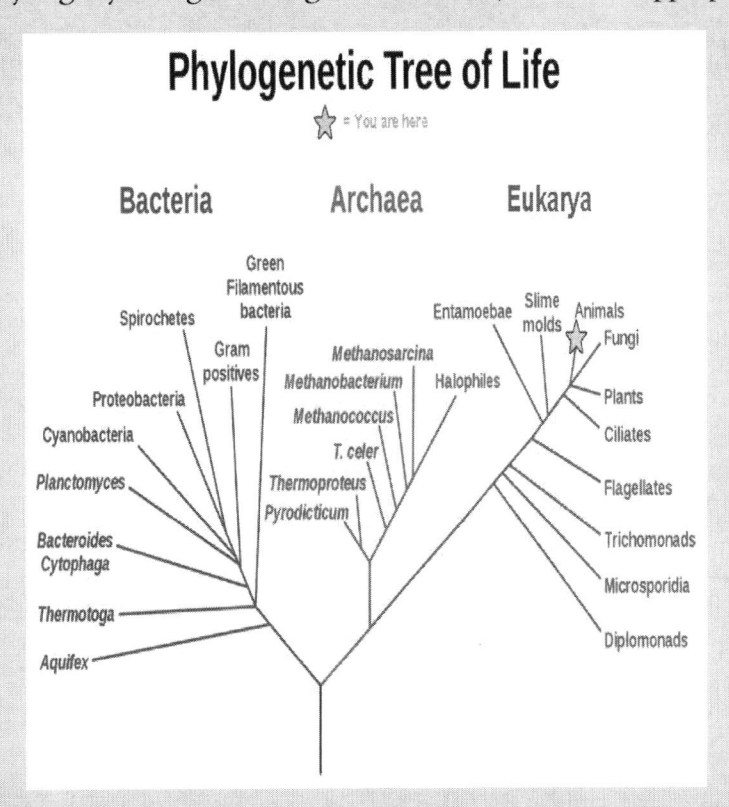

*Figure 10: This phylogenetic tree was constructed by microbiologist Carl Woese using genetic relationships. The tree shows the separation of living organisms into three domains: Bacteria, Archaea, and Eukarya. Bacteria and Archaea are organisms without a nucleus or other organelles surrounded by a membrane and, therefore, are prokaryotes. (credit: modification of work by Eric Gaba)*

*Figure 11: Researchers work on excavating dinosaur fossils at a site in Castellón, Spain. (credit: Mario Modesto)*

Paleontology, another branch of biology, uses fossils to study life's history ([Figure 11]). Zoology and botany are the study of animals and plants, respectively. Biologists can also specialize as biotechnologists, ecologists, or physiologists, to name just a few areas. Biotechnologists apply the knowledge of biology to create useful products. Ecologists study the interactions of organisms in their environments. Physiologists study the workings of cells, tissues and organs. This is just a small sample of the many fields that biologists can pursue. From our own bodies to the world we live in, discoveries in biology can affect us in very direct and important ways. We depend on these discoveries for our health, our food sources, and the benefits provided by our ecosystem. Because of this, knowledge of biology can benefit us in making decisions in our day-to-day lives.

The development of technology in the twentieth century that continues today, particularly the technology to describe and manipulate the genetic material, DNA, has transformed biology. This transformation will allow biologists to continue to understand the history of life in greater detail, how the human body works, our human origins, and how humans can survive as a species on this planet despite the stresses caused by our increasing numbers. Biologists continue to decipher huge mysteries about life suggesting that we have only begun to understand life on the planet, its history, and our relationship to it. For this and other reasons, the knowledge of biology gained through this textbook and other printed and electronic media should be a benefit in whichever field you enter.

*Forensic Science*

Forensic science is the application of science to answer questions related to the law. Biologists as well as chemists and biochemists can be forensic scientists. Forensic scientists provide scientific evidence for use in courts, and their job involves examining trace material associated with crimes. Interest in forensic science has increased in the last few years, possibly because of popular television shows that feature forensic scientists on the job. Also, the development of molecular techniques and the establishment of DNA databases have updated the types of work that forensic scientists can do. Their job activities are primarily related to crimes against people such as murder, rape, and assault. Their work involves analyzing samples such as hair, blood, and other body fluids and also processing DNA ([Figure 12]) found in many different environments and materials. Forensic scientists also analyze other biological evidence left at crime scenes, such as insect parts or pollen grains. Students who want to pursue careers in forensic science will most likely be required to take chemistry and biology courses as well as some intensive math courses.

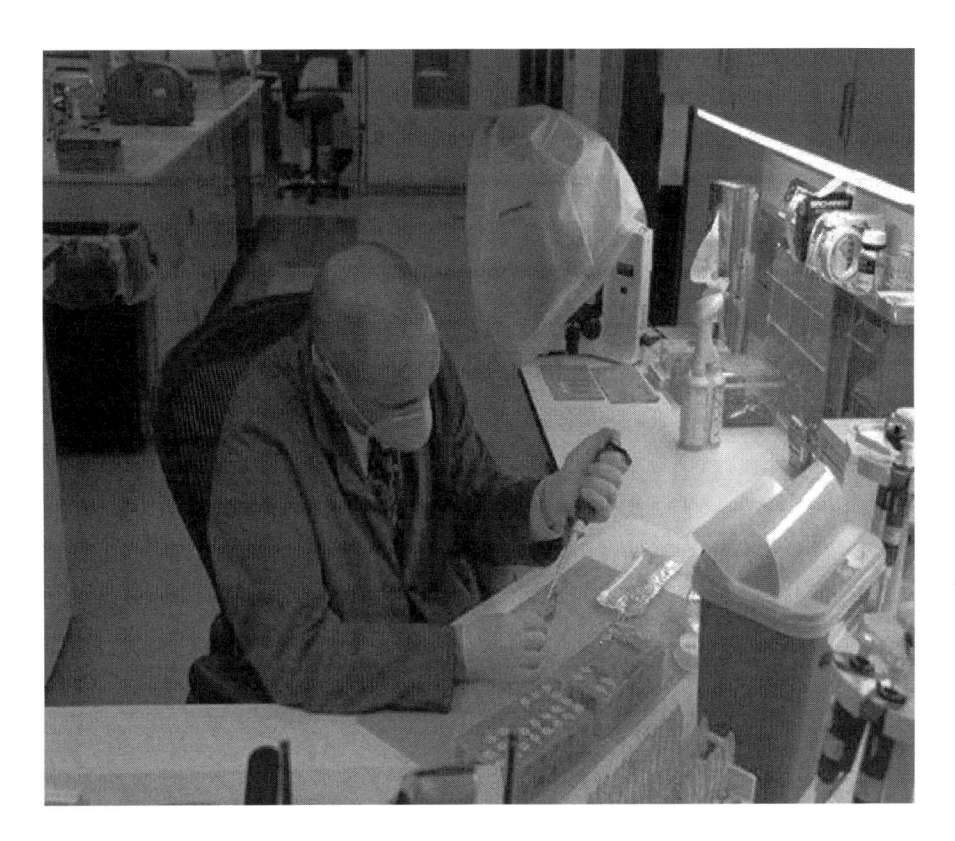

*Figure 12: This forensic scientist works in a DNA extraction room at the U.S. Army Criminal Investigation Laboratory. (credit: U.S. Army CID Command Public Affairs)*

## SECTION SUMMARY

Biology is the science of life. All living organisms share several key properties such as order, sensitivity or response to stimuli, reproduction, adaptation, growth and development, regulation, homeostasis, and energy processing. Living things are highly organized following a hierarchy that includes atoms, molecules, organelles, cells, tissues, organs, and organ systems. Organisms, in turn, are grouped as populations, communities, ecosystems, and the biosphere. Evolution is the source of the tremendous biological diversity on Earth today. A diagram called a phylogenetic tree can be used to show evolutionary relationships among organisms. Biology is very broad and includes many branches and sub disciplines. Examples include molecular biology, microbiology, neurobiology, zoology, and botany, among others.

## Glossary

**atom:** a basic unit of matter that cannot be broken down by normal chemical reactions

**biology:** the study of living organisms and their interactions with one another and their environments

**biosphere:** a collection of all ecosystems on Earth

**cell:** the smallest fundamental unit of structure and function in living things

**community:** a set of populations inhabiting a particular area

**ecosystem:** all living things in a particular area together with the abiotic, nonliving parts of that environment

**eukaryote:** an organism with cells that have nuclei and membrane-bound organelles

**evolution:** the process of gradual change in a population that can also lead to new species arising from older species

**homeostasis:** the ability of an organism to maintain constant internal conditions

**macromolecule:** a large molecule typically formed by the joining of smaller molecules

**molecule:** a chemical structure consisting of at least two atoms held together by a chemical bond

**organ:** a structure formed of tissues operating together to perform a common function

**organ system:** the higher level of organization that consists of functionally related organs

**organelle:** a membrane-bound compartment or sac within a cell

**organism:** an individual living entity

**phylogenetic tree:** a diagram showing the evolutionary relationships among biological species based on similarities and differences in genetic or physical traits or both

**population:** all individuals within a species living within a specific area

**prokaryote:** a unicellular organism that lacks a nucleus or any other membrane-bound organelle

**tissue:** a group of similar cells carrying out the same function

# The Process of Science

### Learning Objective

By the end of this section, you will be able to:

- Identify the shared characteristics of the natural sciences
- Understand the process of scientific inquiry
- Compare inductive reasoning with deductive reasoning
- Describe the goals of basic science and applied science

*Figure 1: Formerly called blue-green algae, the (a) cyanobacteria seen through a light microscope are some of Earth's oldest life forms. These (b) stromatolites along the shores of Lake Thetis in Western Australia are ancient structures formed by the layering of cyanobacteria in shallow waters. (credit a: modification of work by NASA; scale-bar data from Matt Russell; credit b: modification of work by Ruth Ellison)*

Like geology, physics, and chemistry, biology is a science that gathers knowledge about the natural world. Specifically, biology is the study of life. The discoveries of biology are made by a community of researchers who work individually and together using agreed-on methods. In this sense, biology, like all sciences is a social enterprise like politics or the arts. The methods of science include careful observation, record keeping, logical and mathematical reasoning, experimentation, and submitting conclusions to the scrutiny of others. Science also requires considerable imagination and creativity; a well-designed experiment is commonly described as elegant, or beautiful. Like politics, science has considerable practical implications and some science is dedicated to practical applications, such as the prevention of disease (see [Figure 2]). Other science proceeds largely motivated by curiosity. Whatever its goal, there is no doubt that science, including biology, has transformed human existence and will continue to do so.

*Figure 2: Biologists may choose to study Escherichia coli (E. coli), a bacterium that is a normal resident of our digestive tracts but which is also sometimes responsible for disease outbreaks. In this micrograph, the bacterium is visualized using a scanning electron microscope and digital colorization. (credit: Eric Erbe; digital colorization by Christopher Pooley, USDA-ARS)*

## THE NATURE OF SCIENCE

Biology is a science, but what exactly is science? What does the study of biology share with other scientific disciplines? Science (from the Latin *scientia,* meaning "knowledge") can be defined as knowledge about the natural world.

Science is a very specific way of learning, or knowing, about the world. The history of the past 500 years demonstrates that science is a very powerful way of knowing about the world; it is largely responsible for the technological revolutions that have taken place during this time. There are however, areas of knowledge and human experience that the methods of science cannot be applied to. These include such things as answering purely moral questions, aesthetic questions, or what can be generally categorized as spiritual questions. Science has cannot investigate these areas because they are outside the realm of material phenomena, the phenomena of matter and energy, and cannot be observed and measured.

The scientific method is a method of research with defined steps that include experiments and careful observation. The steps of the scientific method will be examined in detail later, but one of the most important aspects of this method is the testing of hypotheses. A hypothesis is a suggested explanation for an event, which can be tested. Hypotheses, or tentative explanations, are generally produced within the context of a scientific theory. A scientific theory is a generally accepted, thoroughly tested and confirmed explanation for a set of observations or phenomena. Scientific theory is the foundation of scientific knowledge. In addition, in many scientific disciplines (less so in biology) there are scientific laws, often expressed in mathematical formulas, which describe how elements of nature will behave under certain specific conditions. There is not an evolution of hypotheses through theories to laws as if they represented some increase in certainty about the world. Hypotheses are the day-to-day material that scientists work with and they are developed within the context of theories. Laws are concise descriptions of parts of the world that are amenable to formulaic or mathematical description.

## Natural Sciences

What would you expect to see in a museum of natural sciences? Frogs? Plants? Dinosaur skeletons? Exhibits about how the brain functions? A planetarium? Gems and minerals? Or maybe all of the above? Science includes such diverse fields as astronomy, biology, computer sciences, geology, logic, physics, chemistry, and mathematics ([Figure 3]). However, those fields of science related to the physical world and its phenomena and processes are considered natural sciences. Thus, a museum of natural sciences might contain any of the items listed above.

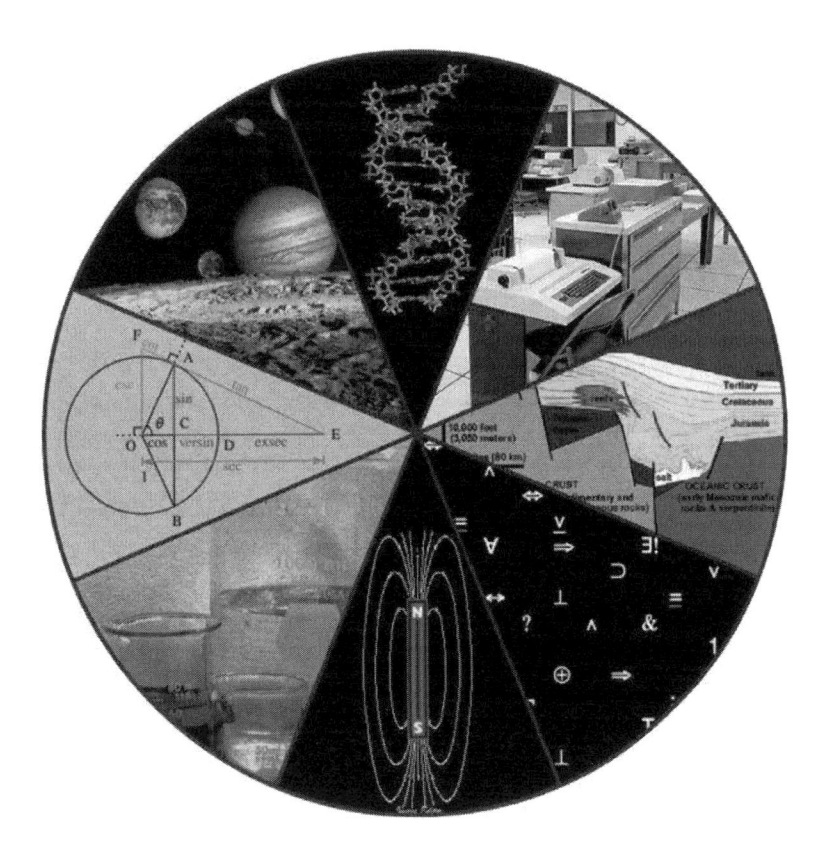

*Figure 3: Some fields of science include astronomy, biology, computer science, geology, logic, physics, chemistry, and mathematics. (credit: "Image Editor"/Flickr)*

There is no complete agreement when it comes to defining what the natural sciences include. For some experts, the natural sciences are astronomy, biology, chemistry, earth science, and physics. Other scholars choose to divide natural sciences into life sciences, which study living things and include biology, and physical sciences, which study nonliving matter and include astronomy, physics, and chemistry. Some disciplines such as biophysics and biochemistry build on two sciences and are interdisciplinary.

### Scientific Inquiry

One thing is common to all forms of science: an ultimate goal "to know." Curiosity and inquiry are the driving forces for the development of science. Scientists seek to understand the world and the way it operates. Two methods of logical thinking are used: inductive reasoning and deductive reasoning.

Inductive reasoning is a form of logical thinking that uses related observations to arrive at a general conclusion. This type of reasoning is common in descriptive science. A life scientist such as a biologist makes observations and records them. These data can be qualitative (descriptive) or quantitative (consisting of numbers), and the raw data can be supplemented with drawings, pictures, photos, or videos.

From many observations, the scientist can infer conclusions (inductions) based on evidence. Inductive reasoning involves formulating generalizations inferred from careful observation and the analysis of a large amount of data. Brain studies often work this way. Many brains are observed while people are doing a task. The part of the brain that lights up, indicating activity, is then demonstrated to be the part controlling the response to that task.

Deductive reasoning or deduction is the type of logic used in hypothesis-based science. In deductive reasoning, the pattern of thinking moves in the opposite direction as compared to inductive reasoning. Deductive reasoning is a form of logical thinking that uses a general principle or law to forecast specific results. From those general principles, a scientist can extrapolate and predict the specific results that would be valid as long as the general principles are valid. For example, a prediction would be that if the climate is becoming warmer in a region, the distribution of plants and animals should change. Comparisons have been made between distributions in the past and the present, and the many changes that have been found are consistent with a warming climate. Finding the change in distribution is evidence that the climate change conclusion is a valid one.

Both types of logical thinking are related to the two main pathways of scientific study: descriptive science and hypothesis-based science. Descriptive (or discovery) science aims to observe, explore, and discover, while hypothesis-based science begins with a specific question or problem and a potential answer or solution that can be tested. The boundary between these two forms of study is often blurred, because most scientific endeavors combine both approaches. Observations lead to questions, questions lead to forming a hypothesis as a possible answer to those questions, and then the hypothesis is tested. Thus, descriptive science and hypothesis-based science are in continuous dialogue.

## HYPOTHESIS TESTING

Biologists study the living world by posing questions about it and seeking science-based responses. This approach is common to other sciences as well and is often referred to as the scientific method. The scientific method was used even in ancient times, but it was first documented by England's Sir Francis Bacon (1561–1626) ([Figure 4]), who set up inductive methods for scientific inquiry. The scientific method is not exclusively used by biologists but can be applied to almost anything as a logical problem-solving method.

The scientific process typically starts with an observation (often a problem to be solved) that leads to a question. Let's think about a simple problem that starts with an observation and apply the scientific method to solve the problem. One Monday morning, a student arrives at class and quickly discovers that the classroom is too warm. That is an observation that also describes a problem: the classroom is too warm. The student then asks a question: "Why is the classroom so warm?"

Recall that a hypothesis is a suggested explanation that can be tested. To solve a problem, several hypotheses may be proposed. For example, one hypothesis might be, "The classroom is warm because no one turned on

*Figure 4: Sir Francis Bacon is credited with being the first to document the scientific method.*

the air conditioning." But there could be other responses to the question, and therefore other hypotheses may be proposed. A second hypothesis might be, "The classroom is warm because there is a power failure, and so the air conditioning doesn't work."

Once a hypothesis has been selected, a prediction may be made. A prediction is similar to a hypothesis but it typically has the format "If . . . then . . . ." For example, the prediction for the first hypothesis might be, "*If* the student turns on the air conditioning, *then* the classroom will no longer be too warm."

A hypothesis must be testable to ensure that it is valid. For example, a hypothesis that depends on what a bear thinks is not testable, because it can never be known what a bear thinks. It should also be falsifiable, meaning that it can be disproven by experimental results. An example of an unfalsifiable hypothesis is "Botticelli's *Birth of Venus* is beautiful." There is no experiment that might show this statement to be false. To test a hypothesis, a researcher will conduct one or more experiments designed to eliminate one or more of the hypotheses. This is important. A hypothesis can be disproven, or eliminated, but it can never be proven. Science does not deal in proofs like mathematics. If an experiment fails to disprove a hypothesis, then we find support for that explanation, but this is not to say that down the road a better explanation will not be found, or a more carefully designed experiment will be found to falsify the hypothesis.

Each experiment will have one or more variables and one or more controls. A variable is any part of the experiment that can vary or change during the experiment. A control is a part of the experiment that does not change. Look for the variables and controls in the example that follows. As a simple example, an experiment might be conducted to test the hypothesis that phosphate limits the growth of algae in freshwater ponds. A series of artificial ponds are filled with water and half of them are treated by adding phosphate each week, while the other half are treated by adding a salt that is known not to be used by algae. The variable here is the phosphate (or lack of phosphate), the experimental or treatment cases are the ponds with added phosphate and the control ponds are those with something inert added, such as the salt. Just adding something is also a control against the possibility that adding extra matter to the pond has an effect. If the treated ponds show lesser growth of algae, then we have found support for our hypothesis. If they do not, then we reject our hypothesis. Be aware that rejecting one hypothesis does not determine whether or not the other hypotheses can be accepted; it simply eliminates one hypothesis that is not valid ([Figure 5]). Using the scientific method, the hypotheses that are inconsistent with experimental data are rejected.

## *BASIC AND APPLIED SCIENCE*

The scientific community has been debating for the last few decades about the value of different types of science. Is it valuable to pursue science for the sake of simply gaining knowledge, or does scientific knowledge only have worth if we can apply it to solving a specific problem or bettering our lives? This question focuses on the differences between two types of science: basic science and applied science.

Basic science or "pure" science seeks to expand knowledge regardless of the short-term application of that knowledge. It is not focused on developing a product or a service of immediate public or commercial value. The immediate goal of basic science is knowledge for knowledge's sake, though this does not mean that in the end it may not result in an application.

In contrast, applied science or "technology," aims to use science to solve real-world problems, making it possible, for example, to improve a crop yield, find a cure for a particular disease, or save animals threatened by a natural disaster. In applied science, the problem is usually defined for the researcher.

Some individuals may perceive applied science as "useful" and basic science as "useless." A question these people might pose to a scientist advocating knowledge acquisition would be, "What for?"

## Scientific Method

A flow chart shows the steps in the scientific method. The steps are: Make an observation → Ask a question → Form a hypothesis that answers the question → Make a prediction based on the hypothesis → Do an experiment to test the prediction → Analyze the results → Hypothesis is SUPPORTED or Hypothesis is NOT SUPPORTED → Report results. If the hypothesis is not supported, try again by forming a new hypothesis.

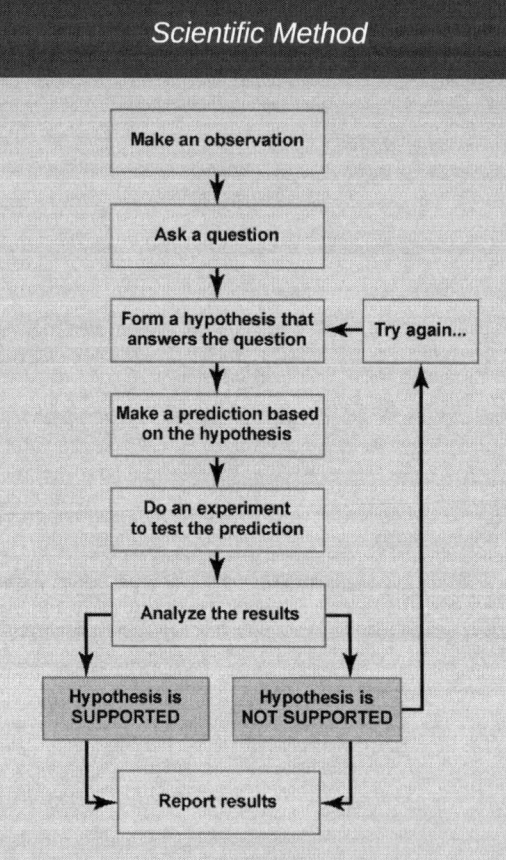

A flow chart shows the steps in the scientific method. In step 1, an observation is made. In step 2, a question is asked about the observation. In step 3, an answer to the question, called a hypothesis, is proposed. In step 4, a prediction is made based on the hypothesis. In step 5, an experiment is done to test the prediction. In step 6, the results are analyzed to determine whether or not the hypothesis is supported. If the hypothesis is not supported, another hypothesis is made. In either case, the results are reported.

In the example below, the scientific method is used to solve an everyday problem. Which part in the example below is the hypothesis? Which is the prediction? Based on the results of the experiment, is the hypothesis supported? If it is not supported, propose some alternative hypotheses.

1. My toaster doesn't toast my bread.
2. Why doesn't my toaster work?
3. There is something wrong with the electrical outlet.
4. If something is wrong with the outlet, my coffeemaker also won't work when plugged into it.
5. I plug my coffeemaker into the outlet.
6. My coffeemaker works.

In practice, the scientific method is not as rigid and structured as it might at first appear. Sometimes an experiment leads to conclusions that favor a change in approach; often, an experiment brings entirely new scientific questions to the puzzle. Many times, science does not operate in a linear fashion; instead, scientists continually draw inferences and make generalizations, finding patterns as their research proceeds. Scientific reasoning is more complex than the scientific method alone suggests.

A careful look at the history of science, however, reveals that basic knowledge has resulted in many remarkable applications of great value. Many scientists think that a basic understanding of science is necessary before an application is developed; therefore, applied science relies on the results generated through basic science. Other scientists think that it is time to move on from basic science and instead to find solutions to actual problems. Both approaches are valid. It is true that there are problems that demand immediate attention; however, few solutions would be found without the help of the knowledge generated through basic science.

One example of how basic and applied science can work together to solve practical problems occurred after the discovery of DNA structure led to an understanding of the molecular mechanisms governing DNA replication. Strands of DNA, unique in every human, are found in our cells, where they provide the instructions necessary for life. During DNA replication, new copies of DNA are made, shortly before a cell divides to form new cells. Understanding the mechanisms of DNA replication enabled scientists to develop laboratory techniques that are now used to identify genetic diseases, pinpoint individuals who were at a crime scene, and determine paternity. Without basic science, it is unlikely that applied science would exist.

Another example of the link between basic and applied research is the Human Genome Project, a study in which each human chromosome was analyzed and mapped to determine the precise sequence of DNA subunits and the exact location of each gene. (The gene is the basic unit of heredity; an individual's complete collection of genes is his or her genome.) Other organisms have also been studied as part of this project to gain a better understanding of human chromosomes. The Human Genome Project ([Figure 6]) relied on basic research carried out with non-human organisms and, later, with the human genome. An important end goal eventually became using the data for applied research seeking cures for genetically related diseases.

*Figure 6: The Human Genome Project was a 13-year collaborative effort among researchers working in several different fields of science. The project was completed in 2003. (credit: the U.S. Department of Energy Genome Programs)*

## REPORTING SCIENTIFIC WORK

Whether scientific research is basic science or applied science, scientists must share their findings for other researchers to expand and build upon their discoveries. Communication and collaboration within and between sub disciplines of science are key to the advancement of knowledge in science. For this reason, an important aspect of a scientist's work is disseminating results and communicating with peers. Scientists can share results by presenting them at a scientific meeting or conference, but this approach can reach only the limited few who are present. Instead, most scientists present their results in peer-reviewed articles that are published in scientific journals. Peer-reviewed articles are scientific papers that are reviewed, usually anonymously by a scientist's colleagues, or peers. These colleagues are qualified individuals, often experts in the same research area, who judge whether or not the scientist's work is suitable for publication. The process of peer review helps to ensure that the research described in a scientific paper or grant proposal is original, significant, logical, and thorough. Grant proposals, which are requests for research funding, are also subject to peer review. Scientists publish their work so other scientists can reproduce their experiments under similar or different conditions to expand on the findings. The experimental results must be consistent with the findings of other scientists.

There are many journals and the popular press that do not use a peer-review system. A large number of online open-access journals, journals with articles available without cost, are now available many of which use rigorous peer-review systems, but some of which do not. Results of any studies published in these forums without peer review are not reliable and should not form the basis for other scientific work. In one exception, journals may allow a researcher to cite a personal communication from another researcher about unpublished results with the cited author's permission.

## SECTION SUMMARY

Biology is the science that studies living organisms and their interactions with one another and their environments. Science attempts to describe and understand the nature of the universe in whole or in part. Science has many fields; those fields related to the physical world and its phenomena are considered natural sciences.

A hypothesis is a tentative explanation for an observation. A scientific theory is a well-tested and consistently verified explanation for a set of observations or phenomena. A scientific law is a description, often in the form of a mathematical formula, of the behavior of an aspect of nature under certain circumstances. Two types of logical reasoning are used in science. Inductive reasoning uses results to produce general scientific principles. Deductive reasoning is a form of logical thinking that predicts results by applying general principles. The common thread throughout scientific research is the use of the scientific method. Scientists present their results in peer-reviewed scientific papers published in scientific journals.

Science can be basic or applied. The main goal of basic science is to expand knowledge without any expectation of short-term practical application of that knowledge. The primary goal of applied research, however, is to solve practical problems.

## Glossary

**applied science:** a form of science that solves real-world problems

**basic science:** science that seeks to expand knowledge regardless of the short-term application of that knowledge

**control:** a part of an experiment that does not change during the experiment

**deductive reasoning:** a form of logical thinking that uses a general statement to forecast specific results

**descriptive science:** a form of science that aims to observe, explore, and find things out

**falsifiable:** able to be disproven by experimental results

**hypothesis:** a suggested explanation for an event, which can be tested

**hypothesis-based science:** a form of science that begins with a specific explanation that is then tested

**inductive reasoning:** a form of logical thinking that uses related observations to arrive at a general conclusion

**life science:** a field of science, such as biology, that studies living things

**natural science:** a field of science that studies the physical world, its phenomena, and processes

**peer-reviewed article:** a scientific report that is reviewed by a scientist's colleagues before publication

**physical science:** a field of science, such as astronomy, physics, and chemistry, that studies nonliving matter

**science:** knowledge that covers general truths or the operation of general laws, especially when acquired and tested by the scientific method

**scientific law:** a description, often in the form of a mathematical formula, for the behavior of some aspect of nature under certain specific conditions

**scientific method:** a method of research with defined steps that include experiments and careful observation

**scientific theory:** a thoroughly tested and confirmed explanation for observations or phenomena

**variable:** a part of an experiment that can vary or change

# Exercises

## PART 1

1. The smallest unit of biological structure that meets the functional requirements of "living" is the _____.

    a. organ

    b. organelle

    c. cell

    d. macromolecule

2. Which of the following sequences represents the hierarchy of biological organization from the most complex to the least complex level?

    a. organelle, tissue, biosphere, ecosystem, population

    b. organ, organism, tissue, organelle, molecule

    c. organism, community, biosphere, molecule, tissue, organ

    d. biosphere, ecosystem, community, population, organism

3. Using examples, explain how biology can be studied from a microscopic approach to a global approach:

## PART 2

1. A suggested and testable explanation for an event is called a _____.

    a. hypothesis

    b. variable

    c. theory

    d. control

2. The type of logical thinking that uses related observations to arrive at a general conclusion is called _____.

    a. deductive reasoning

    b. the scientific method

    c. hypothesis-based science

    d. inductive reasoning

3. Give an example of how applied science has had a direct effect on your daily life:

# Check Your Knowledge: Self-Test

1. List the 8 properties of life and define them:

2. Imagine you are investigating if something is alive. After some laboratory tests, you detect that it is processing energy, it reproduces and maintains a relatively stable internal environment. What would be your conclusion? Justify your answer.

3. Imagine you are investigating if something is alive. After some laboratory observations, you notice that it is capable of moving, but there seems to be no energy processing, no ability to reproduce, no regulation and the internal environment is not stable. What would be your conclusion? Justify your answer:

4. The maintenance a relatively stable internal environment is called:

    a. regulation

    b. homeostasis

    c. development

    d. energy processing

5. Organisms possess traits that have been shaped by natural selection and they enhance the individual's ability to reproduce and to survive in their environment. What are such traits called?

    a. reproduction

    b. adaptation

    c. positive traits

    d. sensitivity

6. How would you classify a feedback loop that enhances and amplifies change in the internal environment? Justify your answer:

7. How would you classify a feedback loop that reverses change in the internal environment? Justify your answer:

8. List the steps of the scientific method in order:

9. New knowledge can be generated by extensive observation of the natural world, leading to general conclusions. This procedure is characterized by _____ reasoning.

10. New knowledge can be generated by testing hypothesis in scientific experiments. This procedure is characterized by _____ reasoning.

11. What is a hypothesis?

    a. It is the predicted results of an experiment.

    b. It is a type of inductive reasoning.

    c. It is a type of scientific question.

    d. It is a tentative answer to a scientific question.

12. What are the differences between basic and applied science?

# Chapter 3: Chemistry of Life

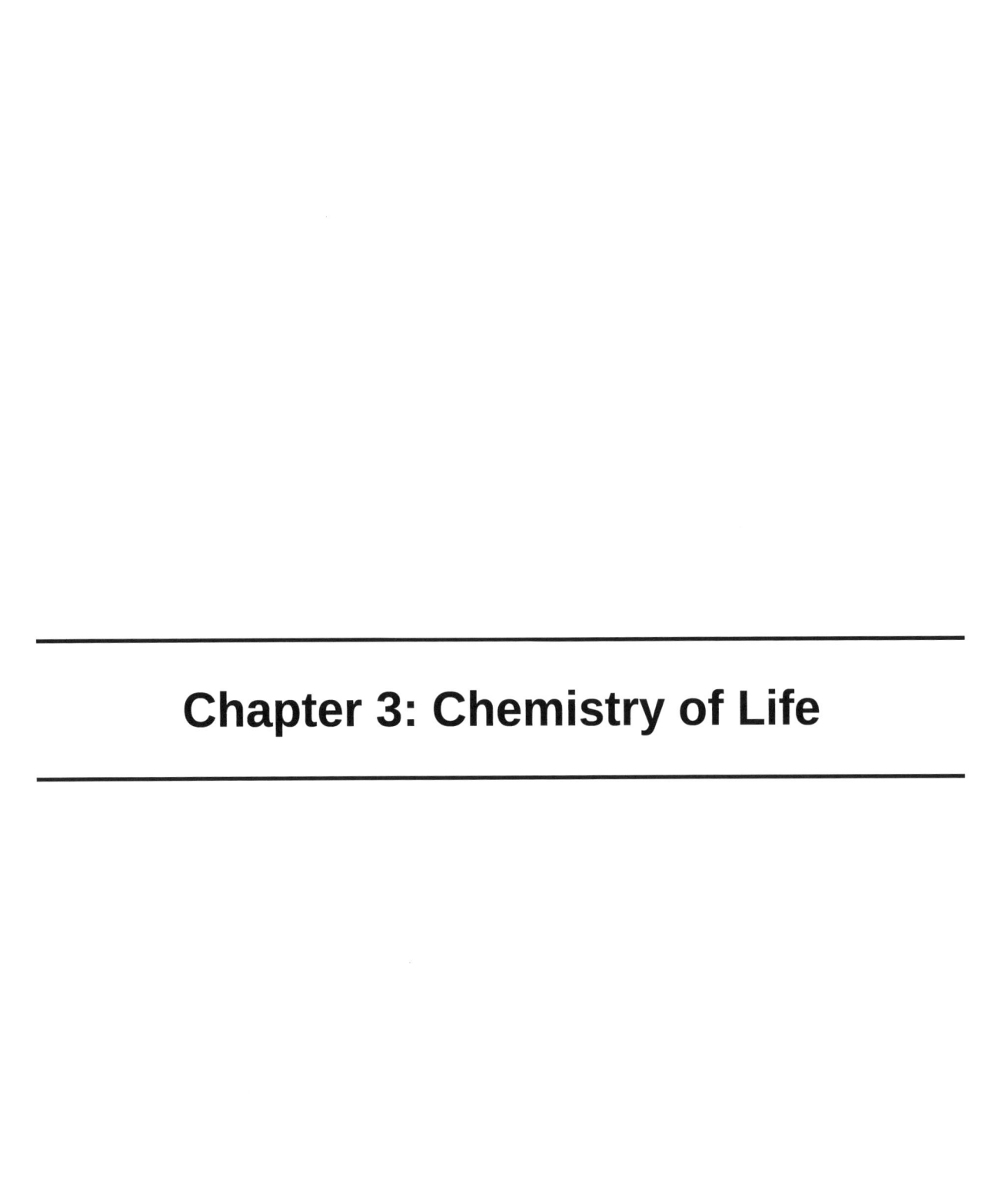

# Phases and Classification of Matter

### Learning Objectives

By the end of this module, you will be able to:

- Describe the basic properties of each physical state of matter: solid, liquid, and gas
- Define and give examples of atoms and molecules
- Classify matter as an element, compound, homogeneous mixture, or heterogeneous mixture with regard to its physical state and composition
- Distinguish between mass and weight
- Apply the law of conservation of matter

**Matter** is defined as anything that occupies space and has mass, and it is all around us. Solids and liquids are more obviously matter: We can see that they take up space, and their weight tells us that they have mass. Gases are also matter; if gases did not take up space, a balloon would stay collapsed rather than inflate when filled with gas.

Solids, liquids, and gases are the three states of matter commonly found on earth (Figure 1). A **solid** is rigid and possesses a definite shape. A **liquid** flows and takes the shape of a container, except that it forms a flat or slightly curved upper surface when acted upon by gravity. (In zero gravity, liquids assume a spherical shape.) Both liquid and solid samples have volumes that are very nearly independent of pressure. A **gas** takes both the shape and volume of its container.

Solid — Has fixed shape and volume

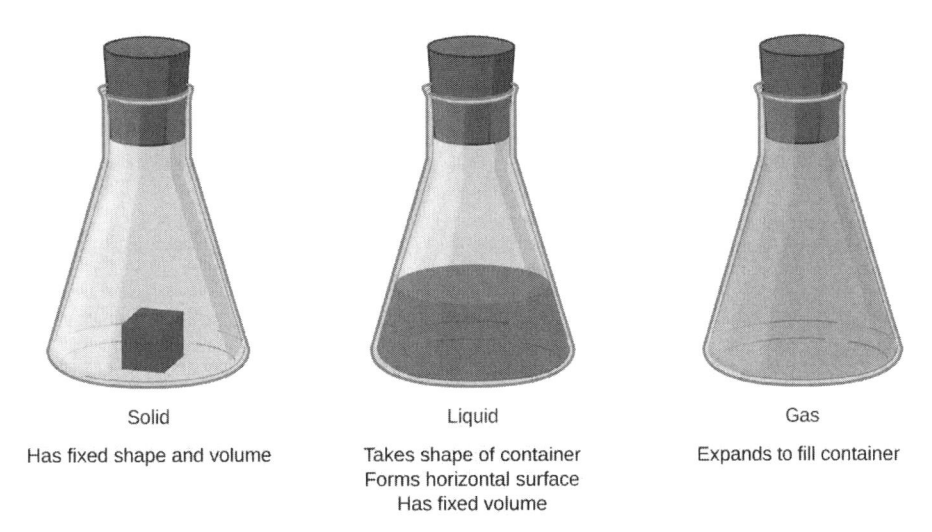

Liquid — Takes shape of container, Forms horizontal surface, Has fixed volume

Gas — Expands to fill container

*Figure 1. The three most common states or phases of matter are solid, liquid, and gas.*

A fourth state of matter, plasma, occurs naturally in the interiors of stars. A **plasma** is a gaseous state of matter that contains appreciable numbers of electrically charged particles (Figure 2). The presence of these charged particles imparts unique properties to plasmas that justify their classification as a state of matter distinct from gases. In addition to stars, plasmas are found in some other high-temperature environments (both natural and man-made), such as lightning strikes, certain television screens, and specialized analytical instruments used to detect trace amounts of metals.

*Figure 2. A plasma torch can be used to cut metal. (credit: "Hypertherm"/Wikimedia Commons)*

> In a tiny cell in a plasma television, the plasma emits ultraviolet light, which in turn causes the display at that location to appear a specific color. The composite of these tiny dots of color makes up the image that you see. Watch this video to learn more about plasma and the places you encounter it.

Some samples of matter appear to have properties of solids, liquids, and/or gases at the same time. This can occur when the sample is composed of many small pieces. For example, we can pour sand as if it were a liquid because it is composed of many small grains of solid sand. Matter can also have properties of more than one state when it is a mixture, such as with clouds. Clouds appear to behave somewhat like gases, but they are actually mixtures of air (gas) and tiny particles of water (liquid or solid).

The **mass** of an object is a measure of the amount of matter in it. One way to measure an object's mass is to measure the force it takes to accelerate the object. It takes much more force to accelerate a car than a bicycle because the car has much more mass. A more common way to determine the mass of an object is to use a balance to compare its mass with a standard mass.

Although weight is related to mass, it is not the same thing. **Weight** refers to the force that gravity exerts on an object. This force is directly proportional to the mass of the object. The weight of an object changes as the force of gravity changes, but its mass does not. An astronaut's mass does not change just because she goes to the moon. But her weight on the moon is only one-sixth her earth-bound weight because the moon's gravity is only one-sixth that of the earth's. She may feel "weightless" during her trip when she experiences negligible external forces (gravitational or any other), although she is, of course, never "massless."

The **law of conservation of matter** summarizes many scientific observations about matter: It states that *there is no detectable change in the total quantity of matter present when matter converts from one type to another (a chemical change) or changes among solid, liquid, or gaseous states (a physical change)*. Brewing beer and the operation of batteries provide examples of the conservation of matter (Figure 3). During the brewing of beer, the ingredients (water, yeast, grains, malt, hops, and sugar) are converted into beer (water, alcohol, carbonation, and flavoring substances) with no actual loss of substance. This is most clearly seen during the bottling process, when glucose turns into ethanol and carbon dioxide, and the total mass of the substances does not change. This can also be seen in a lead-acid car battery: The original substances (lead, lead oxide, and sulfuric acid), which are capable of producing electricity, are changed into other substances (lead sulfate and water) that do not produce electricity, with no change in the actual amount of matter.

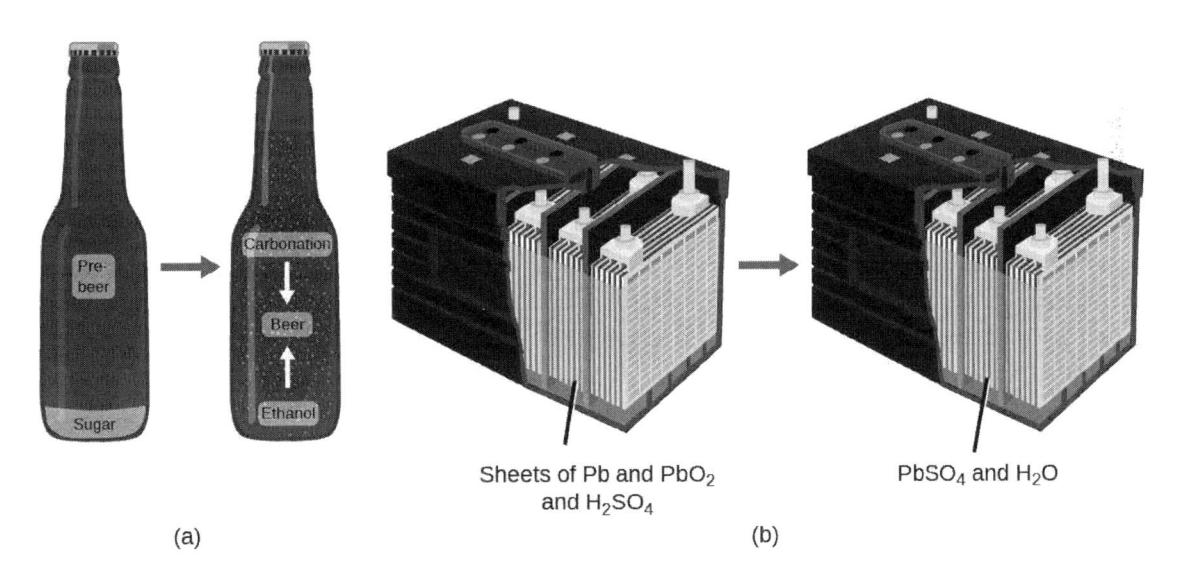

*Figure 3. (a) The mass of beer precursor materials is the same as the mass of beer produced: Sugar has become alcohol and carbonation. (b) The mass of the lead, lead oxide plates, and sulfuric acid that goes into the production of electricity is exactly equal to the mass of lead sulfate and water that is formed.*

Although this conservation law holds true for all conversions of matter, convincing examples are few and far between because, outside of the controlled conditions in a laboratory, we seldom collect all of the material that is produced during a particular conversion. For example, when you eat, digest, and assimilate food, all of the matter in the original food is preserved. But because some of the matter is incorporated into your body, and much is excreted as various types of waste, it is challenging to verify by measurement.

## ATOMS AND MOLECULES

An **atom** is the smallest particle of an element that has the properties of that element and can enter into a chemical combination.

Consider the element gold, for example. Imagine cutting a gold nugget in half, then cutting one of the halves in half, and repeating this process until a piece of gold remained that was so small that it could not be cut in half (regardless of how tiny your knife may be). This minimally sized piece of gold is an atom (from the Greek *atomos*, meaning "indivisible") (Figure 4). This atom would no longer be gold if it were divided any further.

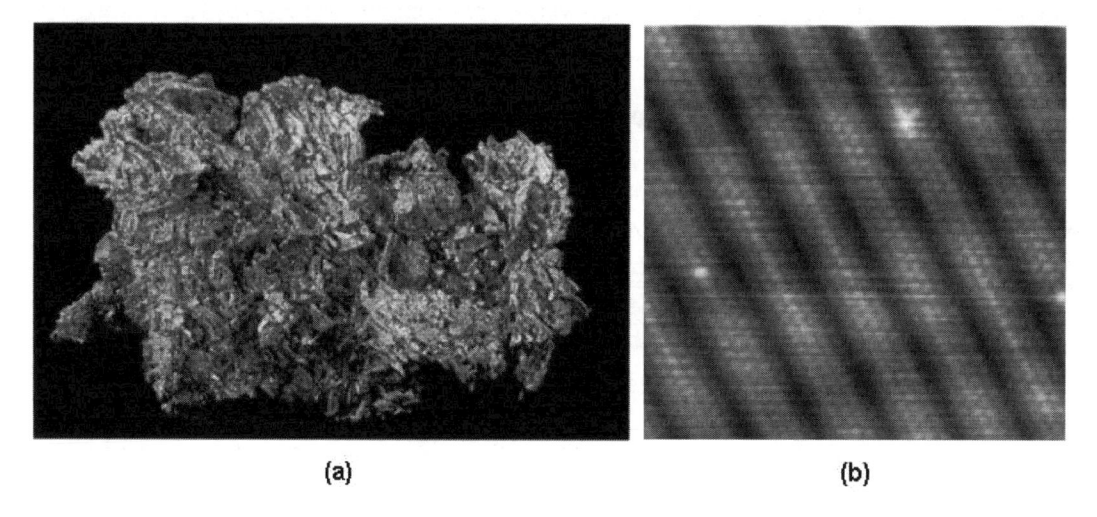

*Figure 4. (a) This photograph shows a gold nugget. uniform stripes of light and dark gold, as seen through microscope (b) A scanning-tunneling microscope (STM) can generate views of the surfaces of solids, such as this image of a gold crystal. Each sphere represents one gold atom. (credit a: modification of work by United States Geological Survey; credit b: modification of work by "Erwinrossen"/Wikimedia Commons)*

The first suggestion that matter is composed of atoms is attributed to the Greek philosophers Leucippus and Democritus, who developed their ideas in the 5th century BCE. However, it was not until the early nineteenth century that John **Dalton** (1766–1844), a British schoolteacher with a keen interest in science, supported this hypothesis with quantitative measurements. Since that time, repeated experiments have confirmed many aspects of this hypothesis, and it has become one of the central theories of chemistry. Other aspects of Dalton's atomic theory are still used but with minor revisions (details of Dalton's theory are provided in the chapter on atoms and molecules).

An atom is so small that its size is difficult to imagine. One of the smallest things we can see with our unaided eye is a single thread of a spider web: These strands are about 1/10,000 of a centimeter (0.00001 cm) in diameter. Although the cross-section of one strand is almost impossible to see without a microscope, it is huge on an atomic scale. A single carbon atom in the web has a diameter of about 0.000000015 centimeter, and it would take about 7000 carbon atoms to span the diameter of the strand. To put this in perspective, if a carbon atom were the size of a dime, the cross-section of one strand would be larger than a football field, which would require about 150 million carbon atom "dimes" to cover it. (Figure 5) shows increasingly close microscopic and atomic-level views of ordinary cotton.

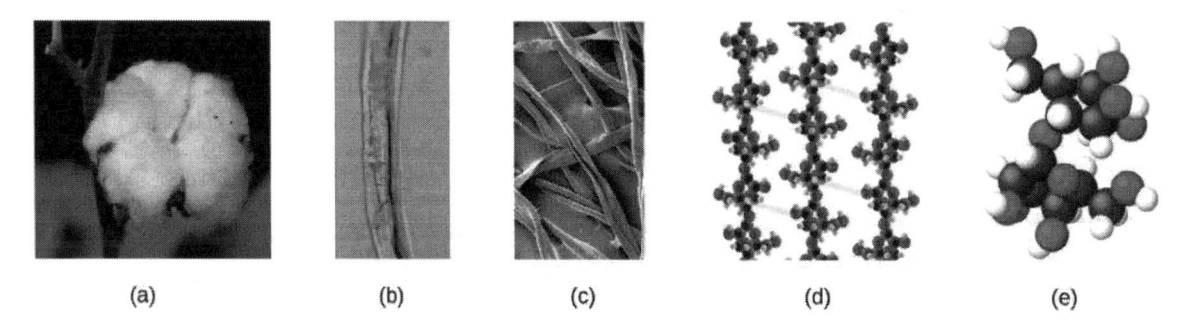

*Figure 5. These images provide an increasingly closer view: (a) a cotton boll, (b) a single cotton fiber viewed under an optical microscope (magnified 40 times), (c) an image of a cotton fiber obtained with an electron microscope (much higher magnification than with the optical microscope); and (d and e) atomic-level models of the fiber (spheres of different colors represent atoms of different elements). (credit c: modification of work by "Featheredtar"/Wikimedia Commons)*

An atom is so light that its mass is also difficult to imagine. A billion lead atoms (1,000,000,000 atoms) weigh about $3 \times 10^{-13}$ grams, a mass that is far too light to be weighed on even the world's most sensitive balances. It would require over 300,000,000,000,000 lead atoms (300 trillion, or $3 \times 10^{14}$) to be weighed, and they would weigh only 0.0000001 gram.

It is rare to find collections of individual atoms. Only a few elements, such as the gases helium, neon, and argon, consist of a collection of individual atoms that move about independently of one another. Other elements, such as the gases hydrogen, nitrogen, oxygen, and chlorine, are composed of units that consist of pairs of atoms (Figure 6). One form of the element phosphorus consists of units composed of four phosphorus atoms. The element sulfur exists in various forms, one of which consists of units composed of eight sulfur atoms. These units are called molecules. **A molecule** consists of two or more atoms joined by strong forces called chemical bonds. The atoms in a molecule move around as a unit, much like the cans of soda in a six-pack or a bunch of keys joined together on a single key ring. A molecule may consist of two or more identical atoms, as in the molecules found in the elements hydrogen, oxygen, and sulfur, or it may consist of two or more different atoms, as in the molecules found in water. Each water molecule is a unit that contains two hydrogen atoms and one oxygen atom. Each glucose molecule is a unit that contains 6 carbon atoms, 12 hydrogen atoms, and 6 oxygen atoms. Like atoms, molecules are incredibly small and light. If an ordinary glass of water were enlarged to the size of the earth, the water molecules inside it would be about the size of golf balls.

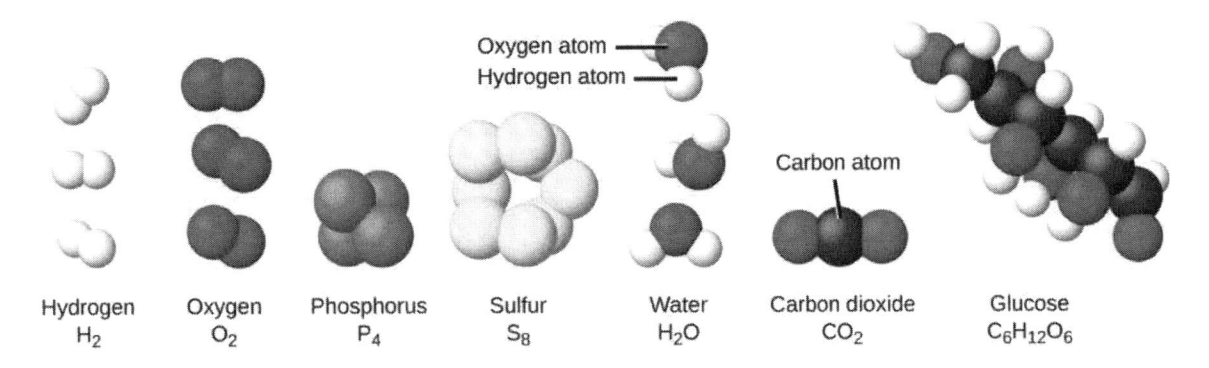

*Figure 6. The elements hydrogen, oxygen, phosphorus, and sulfur form molecules consisting of two or more atoms of the same element. The compounds water, carbon dioxide, and glucose consist of combinations of atoms of different elements.*

## CLASSIFYING MATTER

We can classify matter into several categories. Two broad categories are mixtures and pure substances. **A pure substance** has a constant composition. All specimens of a pure substance have exactly the same makeup and properties. Any sample of sucrose (table sugar) consists of 42.1% carbon, 6.5% hydrogen, and 51.4% oxygen by mass. Any sample of sucrose also has the same physical properties, such as melting point, color, and sweetness, regardless of the source from which it is isolated.

We can divide pure substances into two classes: elements and compounds. Pure substances that cannot be broken down into simpler substances by chemical changes are called **elements**. Iron, silver, gold, aluminum, sulfur, oxygen, and copper are familiar examples of the more than 100 known elements, of which about 90 occur naturally on the earth, and two dozen or so have been created in laboratories.

Pure substances that can be broken down by chemical changes are called **compounds**. This breakdown may produce either elements or other compounds, or both. Mercury(II) oxide, an orange, crystalline solid, can be broken down by heat into the elements mercury and oxygen (Figure 7). When heated in the absence of air, the compound sucrose is broken down into the element carbon and the compound water. (The initial stage of this process, when the sugar is turning brown, is known as caramelization—this is what imparts the characteristic sweet and nutty flavor to caramel apples, caramelized onions, and caramel). Silver(I) chloride is a white solid that can be broken down into its elements, silver and chlorine, by absorption of light. This property is the basis for the use of this compound in photographic films and photochromic eyeglasses (those with lenses that darken when exposed to light).

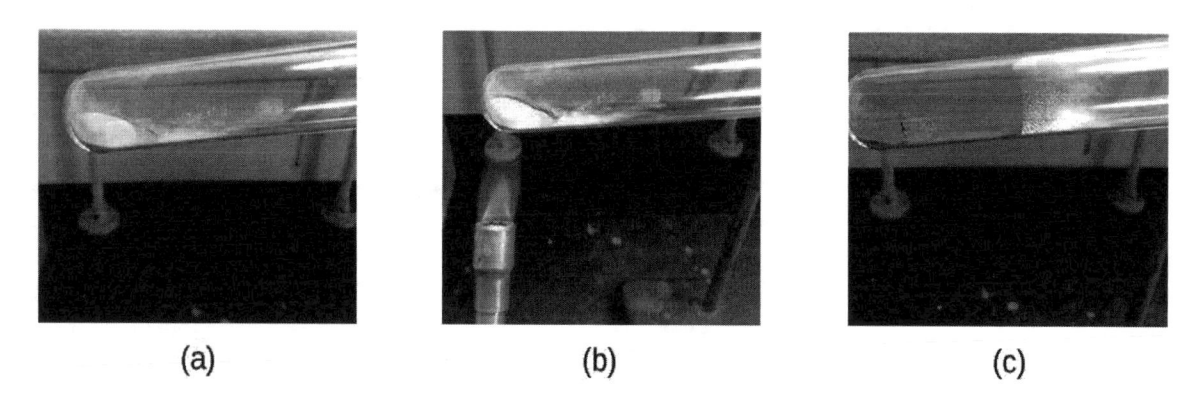

*Figure 7. (a)The compound mercury(II) oxide, (b)when heated, (c) decomposes into silvery droplets of liquid mercury and invisible oxygen gas. (credit: modification of work by Paul Flowers)*

Many compounds break down when heated. This video shows the breakdown of mercury oxide, HgO.

https://youtu.be/_Y1alDuXm6A

You can also view an example of the photochemical decomposition of silver chloride (AgCl), the basis of early photography.

https://youtu.be/ZLEYyzW427I

The properties of combined elements are different from those in the free, or uncombined, state. For example, white crystalline sugar (sucrose) is a compound resulting from the chemical combination of the element carbon, which is a black solid in one of its uncombined forms, and the two elements hydrogen and oxygen, which are colorless gases when uncombined. Free sodium, an element that is a soft, shiny, metallic solid, and free chlorine, an element that is a yellow-green gas, combine to form sodium chloride (table salt), a compound that is a white, crystalline solid.

A **mixture** is composed of two or more types of matter that can be present in varying amounts and can be separated by physical changes, such as evaporation (you will learn more about this later). A mixture with a composition that varies from point to point is called a **heterogeneous mixture**. Italian dressing is an example of a heterogeneous mixture (Figure 8). Its composition can vary because we can make it from varying amounts of oil, vinegar, and herbs. It is not the same from point to point throughout the mixture—one drop may be mostly vinegar, whereas a different drop may be mostly oil or herbs because the oil and vinegar separate and the herbs settle. Other examples of heterogeneous mixtures are chocolate chip cookies (we can see the separate bits of chocolate, nuts, and cookie dough) and granite (we can see the quartz, mica, feldspar, and more).

A **homogeneous mixture**, also called a **solution**, exhibits a uniform composition and appears visually the same throughout. An example of a solution is a sports drink, consisting of water, sugar, coloring, flavoring, and electrolytes mixed together uniformly (Figure 8). Each drop of a sports drink tastes the same because each drop contains the same amounts of water, sugar, and other components. Note that the composition of a sports drink can vary—it could be made with somewhat more or less sugar, flavoring, or other components, and still be a sports drink. Other examples of homogeneous mixtures include air, maple syrup, gasoline, and a solution of salt in water.

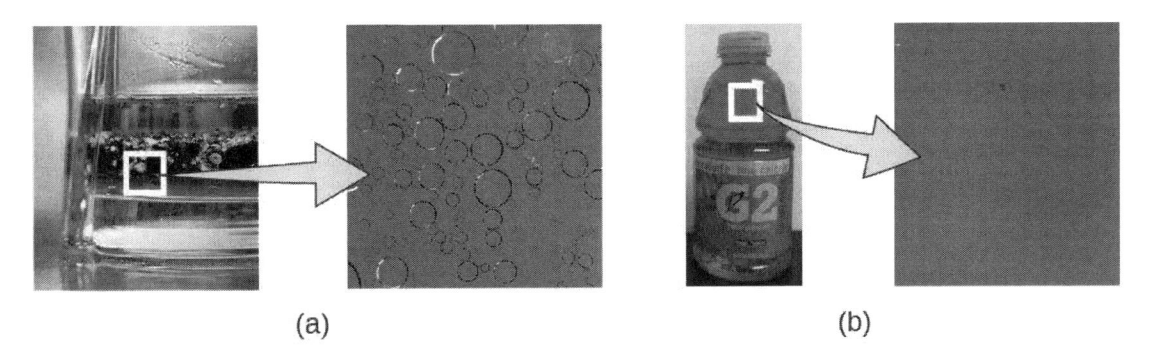

*Figure 8. (a) Oil and vinegar salad dressing is a heterogeneous mixture because its composition is not uniform throughout. (b) A commercial sports drink is a homogeneous mixture because its composition is uniform throughout. (credit a "left": modification of work by John Mayer; credit a "right": modification of work by Umberto Salvagnin; credit b "left: modification of work by Jeff Bedford)*

Although there are just over 100 elements, tens of millions of chemical compounds result from different combinations of these elements. Each compound has a specific composition and possesses definite chemical and physical properties by which we can distinguish it from all other compounds. And, of course, there are innumerable ways to combine elements and compounds to form different mixtures. A summary of how to distinguish between the various major classifications of matter is shown in (Figure 9).

## Chemistry of Life

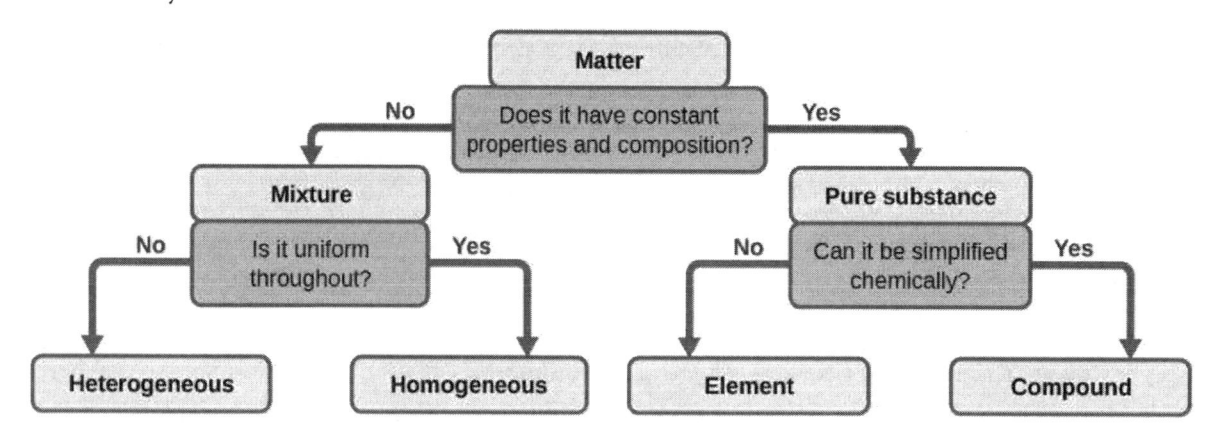

*Figure 9. Depending on its properties, a given substance can be classified as a homogeneous mixture, a heterogeneous mixture, a compound, or an element.*

Eleven elements make up about 99% of the earth's crust and atmosphere (Table 1). Oxygen constitutes nearly one-half and silicon about one-quarter of the total quantity of these elements. A majority of elements on earth are found in chemical combinations with other elements; about one-quarter of the elements are also found in the free state.

*Table 1. Elemental Composition of Earth*

| Element | Symbol | Percent Mass | Element | Symbol | Percent Mass |
|---|---|---|---|---|---|
| oxygen | O | 49.20 | chlorine | Cl | 0.19 |
| silicon | Si | 25.67 | phosphorus | P | 0.11 |
| aluminum | Al | 7.50 | manganese | Mn | 0.09 |
| iron | Fe | 4.71 | carbon | C | 0.08 |
| calcium | Ca | 3.39 | sulfur | S | 0.06 |
| sodium | Na | 2.63 | barium | Ba | 0.04 |
| potassium | K | 2.40 | nitrogen | N | 0.03 |
| magnesium | Mg | 1.93 | fluorine | F | 0.03 |
| hydrogen | H | 0.87 | strontium | Sr | 0.02 |
| titanium | Ti | 0.58 | all others | - | 0.47 |

## Decomposition of Water / Production of Hydrogen

Water consists of the elements hydrogen and oxygen combined in a 2 to 1 ratio. Water can be broken down into hydrogen and oxygen gases by the addition of energy. One way to do this is with a battery or power supply, as shown in (Figure 10).

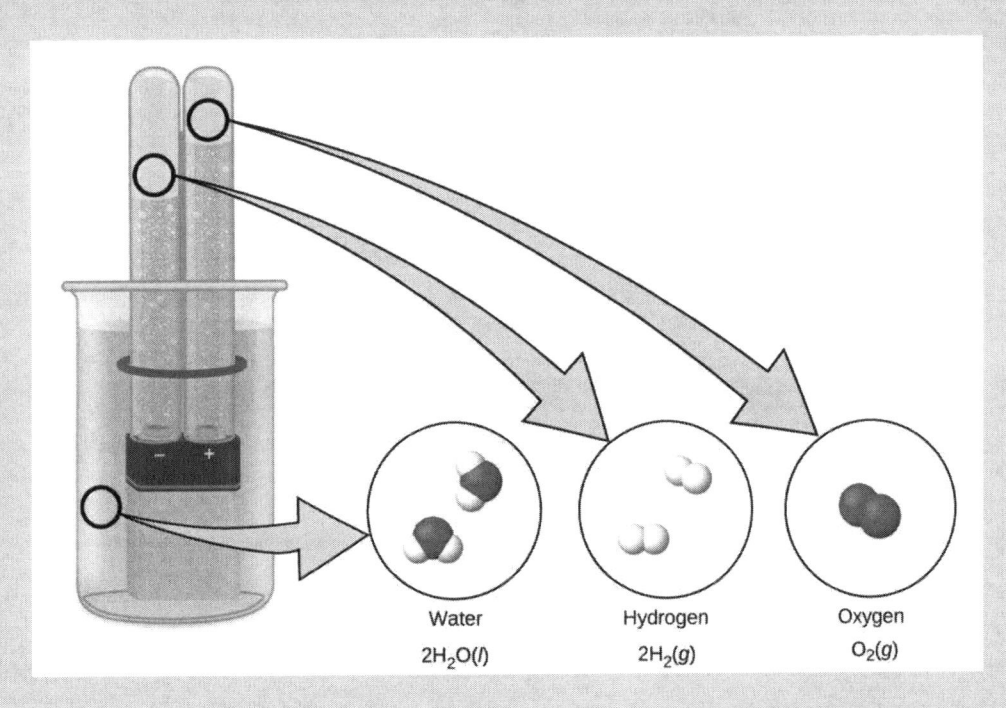

Water  
$2H_2O(l)$

Hydrogen  
$2H_2(g)$

Oxygen  
$O_2(g)$

*Figure 10. The decomposition of water is shown at the macroscopic, microscopic, and symbolic levels. The battery provides an electric current (microscopic) that decomposes water. At the macroscopic level, the liquid separates into the gases hydrogen (on the left) and oxygen (on the right). Symbolically, this change is presented by showing how liquid $H_2O$ separates into $H_2$ and $O_2$ gases.*

The breakdown of water involves a rearrangement of the atoms in water molecules into different molecules, each composed of two hydrogen atoms and two oxygen atoms, respectively. Two water molecules form one oxygen molecule and two hydrogen molecules. The representation for what occurs, $2H_2O(l) \rightarrow 2H_2(g) + O_2(g)$, will be explored in more depth in later chapters. The two gases produced have distinctly different properties. Oxygen is not flammable but is required for combustion of a fuel, and hydrogen is highly flammable and a potent energy source. How might this knowledge be applied in our world? One application involves research into more fuel-efficient transportation. Fuel-cell vehicles (FCV) run on hydrogen instead of gasoline (Figure 11). They are more efficient than vehicles with internal combustion engines, are nonpolluting, and reduce greenhouse gas emissions, making us less dependent on fossil fuels. FCVs are not yet economically viable, however, and current hydrogen production depends on natural gas. If we can develop a process to economically decompose water, or produce hydrogen in another environmentally sound way, FCVs may be the way of the future.

76  Chemistry of Life

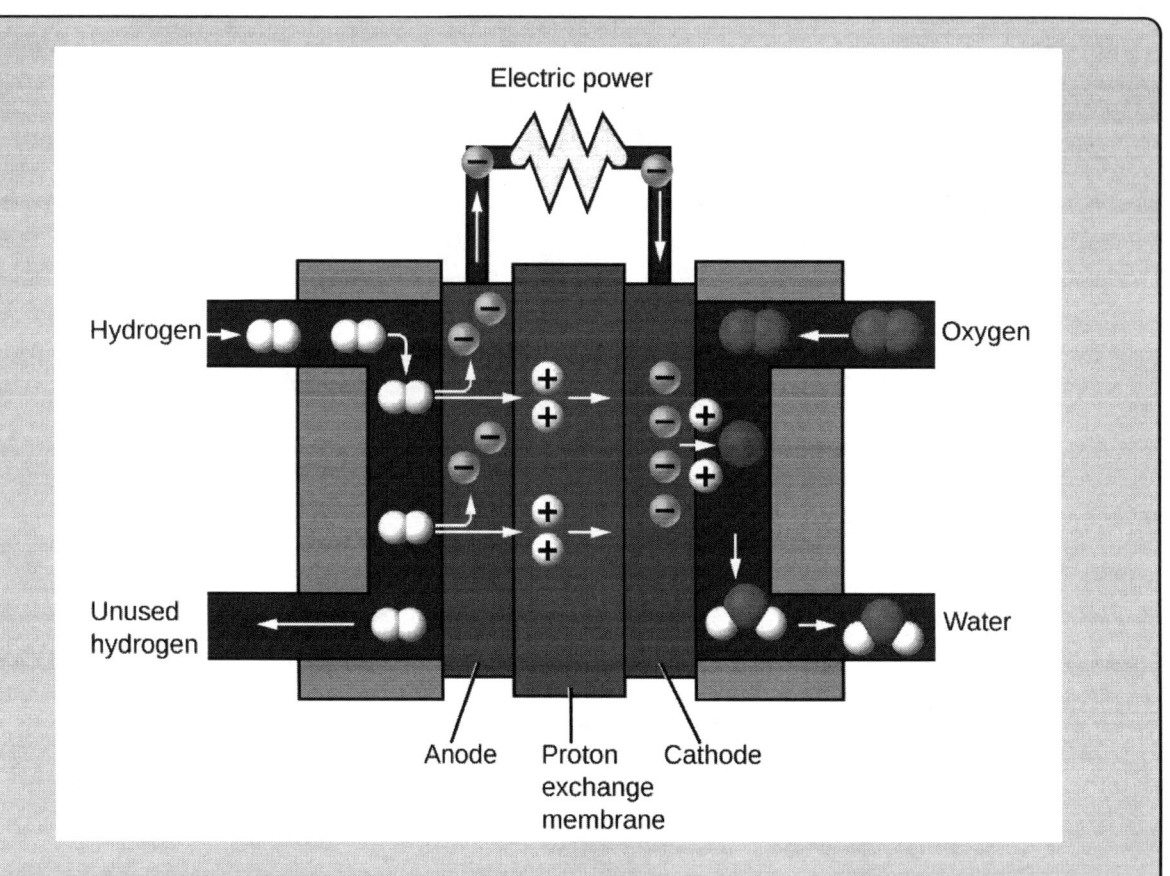

*Figure 11. A fuel cell generates electrical energy from hydrogen and oxygen via an electrochemical process and produces only water as the waste product.*

### Chemistry in Everyday Life: Chemistry of Cell Phones

Imagine how different your life would be without cell phones (Figure 12) and other smart devices. Cell phones are made from numerous chemical substances, which are extracted, refined, purified, and assembled using an extensive and in-depth understanding of chemical principles. About 30% of the elements that are found in nature are found within a typical smart phone. The case/body/frame consists of a combination of sturdy, durable polymers comprised primarily of carbon, hydrogen, oxygen, and nitrogen [acrylonitrile butadiene styrene (ABS) and polycarbonate thermoplastics], and light, strong, structural metals, such as aluminum, magnesium, and iron. The display screen is made from a specially toughened glass (silica glass strengthened by the addition of aluminum, sodium, and potassium) and coated with a material to make it conductive (such as indium tin oxide). The circuit board uses a semiconductor material (usually silicon); commonly used metals like copper, tin, silver, and gold; and more unfamiliar elements such as yttrium, praseodymium, and gadolinium. The battery relies upon lithium ions and a variety of other materials, including iron, cobalt, copper, polyethylene oxide, and polyacrylonitrile.

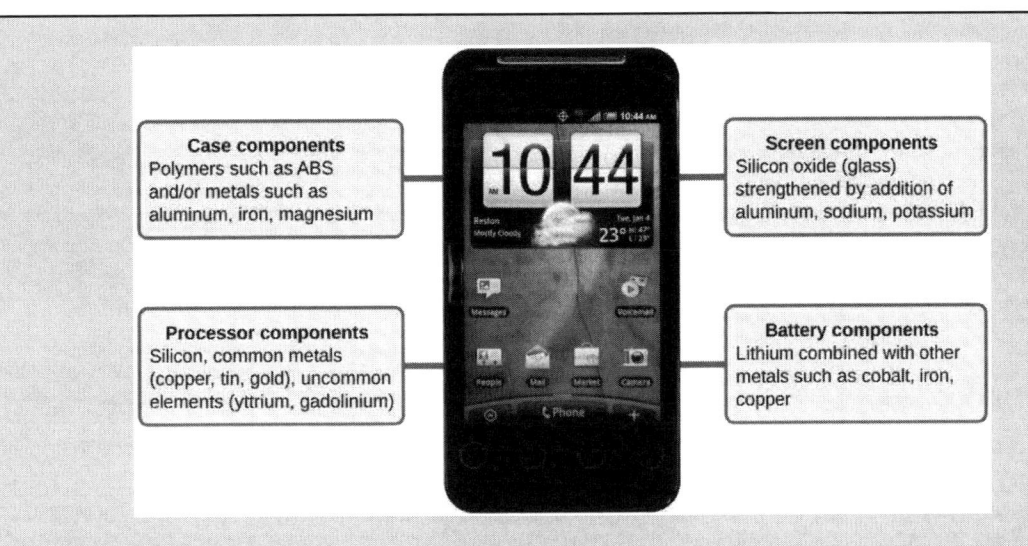

*Figure 12. Almost one-third of naturally occurring elements are used to make a cell phone. (credit: modification of work by John Taylor)*

## Key Concepts and Summary

Matter is anything that occupies space and has mass. The basic building block of matter is the atom, the smallest unit of an element that can enter into combinations with atoms of the same or other elements. In many substances, atoms are combined into molecules. On earth, matter commonly exists in three states: solids, of fixed shape and volume; liquids, of variable shape but fixed volume; and gases, of variable shape and volume. Under high-temperature conditions, matter also can exist as a plasma. Most matter is a mixture: It is composed of two or more types of matter that can be present in varying amounts and can be separated by physical means. Heterogeneous mixtures vary in composition from point to point; homogeneous mixtures have the same composition from point to point. Pure substances consist of only one type of matter. A pure substance can be an element, which consists of only one type of atom and cannot be broken down by a chemical change, or a compound, which consists of two or more types of atoms.

## Glossary

**atom:** smallest particle of an element that can enter into a chemical combination

**compound:** pure substance that can be decomposed into two or more elements

**element:** substance that is composed of a single type of atom; a substance that cannot be decomposed by a chemical change

**gas:** state in which matter has neither definite volume nor shape

**heterogeneous mixture:** combination of substances with a composition that varies from point to point

**homogeneous mixture:** (also, solution) combination of substances with a composition that is uniform throughout

**liquid:** state of matter that has a definite volume but indefinite shape

**law of conservation of matter:** when matter converts from one type to another or changes form, there is no detectable change in the total amount of matter present

**mass:** fundamental property indicating amount of matter

**matter:** anything that occupies space and has mass

**mixture:** matter that can be separated into its components by physical means

**molecule:** bonded collection of two or more atoms of the same or different elements

**plasma:** gaseous state of matter containing a large number of electrically charged atoms and/or molecules

**pure substance:** homogeneous substance that has a constant composition

**solid:** state of matter that is rigid, has a definite shape, and has a fairly constant volume

**weight:** force that gravity exerts on an object

# Physical and Chemical Properties

### Learning Objectives

By the end of this section, you will be able to:
- Identify properties of and changes in matter as physical or chemical
- Identify properties of matter as extensive or intensive

The characteristics that enable us to distinguish one substance from another are called properties. A **physical property** is a characteristic of matter that is not associated with a change in its chemical composition. Familiar examples of physical properties include density, color, hardness, melting and boiling points, and electrical conductivity. We can observe some physical properties, such as density and color, without changing the physical state of the matter observed. Other physical properties, such as the melting temperature of iron or the freezing temperature of water, can only be observed as matter undergoes a physical change. A **physical change** is a change in the state or properties of matter without any accompanying change in its chemical composition (the identities of the substances contained in the matter). We observe a physical change when wax melts, when sugar dissolves in coffee, and when steam condenses into liquid water (Figure 1). Other examples of physical changes include magnetizing and demagnetizing metals (as is done with common antitheft security tags) and grinding solids into powders (which can sometimes yield noticeable changes in color). In each of these examples, there is a change in the physical state, form, or properties of the substance, but no change in its chemical composition.

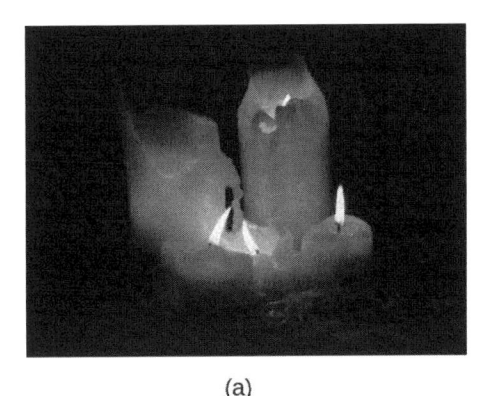

*Figure 1. (a) Wax undergoes a physical change when solid wax is heated and forms liquid wax. (b) Steam condensing inside a cooking pot is a physical change, as water vapor is changed into liquid water. (credit a: modification of work by "95jb14"/ Wikimedia Commons; credit b: modification of work by "mjneuby"/Flickr)*

The change of one type of matter into another type (or the inability to change) is a **chemical property**. Examples of chemical properties include flammability, toxicity, acidity, reactivity (many types), and heat of combustion. Iron, for example, combines with oxygen in the presence of water to form rust; chromium does not oxidize (Figure 2). Nitroglycerin is very dangerous because it explodes easily; neon poses almost no hazard because it is very unreactive.

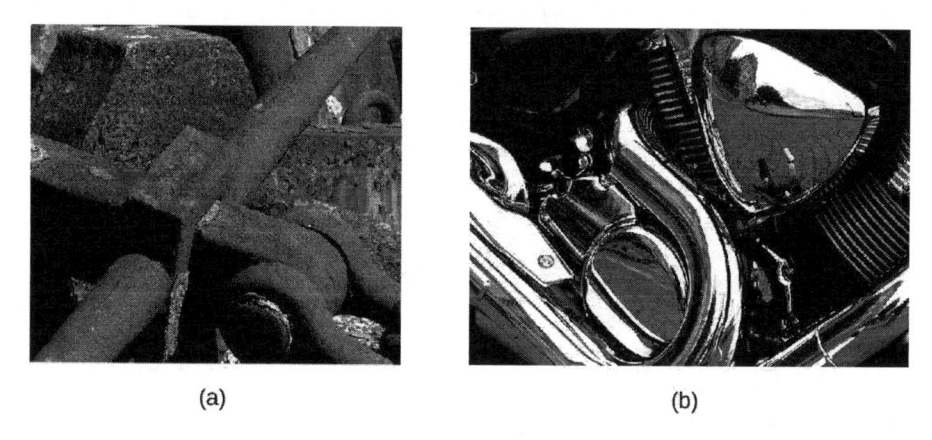

*Figure 2. (a) One of the chemical properties of iron is that it rusts; (b) one of the chemical properties of chromium is that it does not. (credit a: modification of work by Tony Hisgett; credit b: modification of work by "Atoma"/Wikimedia Commons)*

To identify a chemical property, we look for a chemical change. **A chemical change** always produces one or more types of matter that differ from the matter present before the change. The formation of rust is a chemical change because rust is a different kind of matter than the iron, oxygen, and water present before the rust formed. The explosion of nitroglycerin is a chemical change because the gases produced are very different kinds of matter from the original substance. Other examples of chemical changes include reactions that are performed in a lab (such as copper reacting with nitric acid), all forms of combustion (burning), and food being cooked, digested, or rotting (Figure 3).

*Figure 3. (a) Copper and nitric acid undergo a chemical change to form copper nitrate and brown, gaseous nitrogen dioxide. (b) During the combustion of a match, cellulose in the match and oxygen from the air undergo a chemical change to form carbon dioxide and water vapor. (c) Cooking red meat causes a number of chemical changes, including the oxidation of iron in myoglobin that results in the familiar red-to-brown color change. (d) A banana turning brown is a chemical change as new, darker (and less tasty) substances form. (credit b: modification of work by Jeff Turner; credit c: modification of work by Gloria Cabada-Leman; credit d: modification of work by Roberto Verzo)*

Properties of matter fall into one of two categories. If the property depends on the amount of matter present, it is an **extensive property**. The mass and volume of a substance are examples of extensive properties; for instance, a gallon of milk has a larger mass and volume than a cup of milk. The value of an extensive property is directly proportional to the amount of matter in question. If the property of a sample of matter does not depend on the amount of matter present, it is an **intensive property**. Temperature is an example of an intensive property. If the gallon and cup of milk are each at 20 °C (room temperature), when they are combined, the temperature remains at 20 °C. As another example, consider the distinct but related properties of heat and temperature. A drop of hot cooking oil spattered on your arm causes brief, minor discomfort, whereas a pot of hot oil yields severe burns. Both the drop and the pot of oil are at the same temperature (an intensive property), but the pot clearly contains much more heat (extensive property).

### Hazard Diamond

You may have seen the symbol shown in Figure 4 on containers of chemicals in a laboratory or workplace. Sometimes called a "fire diamond" or "hazard diamond," this chemical hazard diamond provides valuable information that briefly summarizes the various dangers of which to be aware when working with a particular substance.

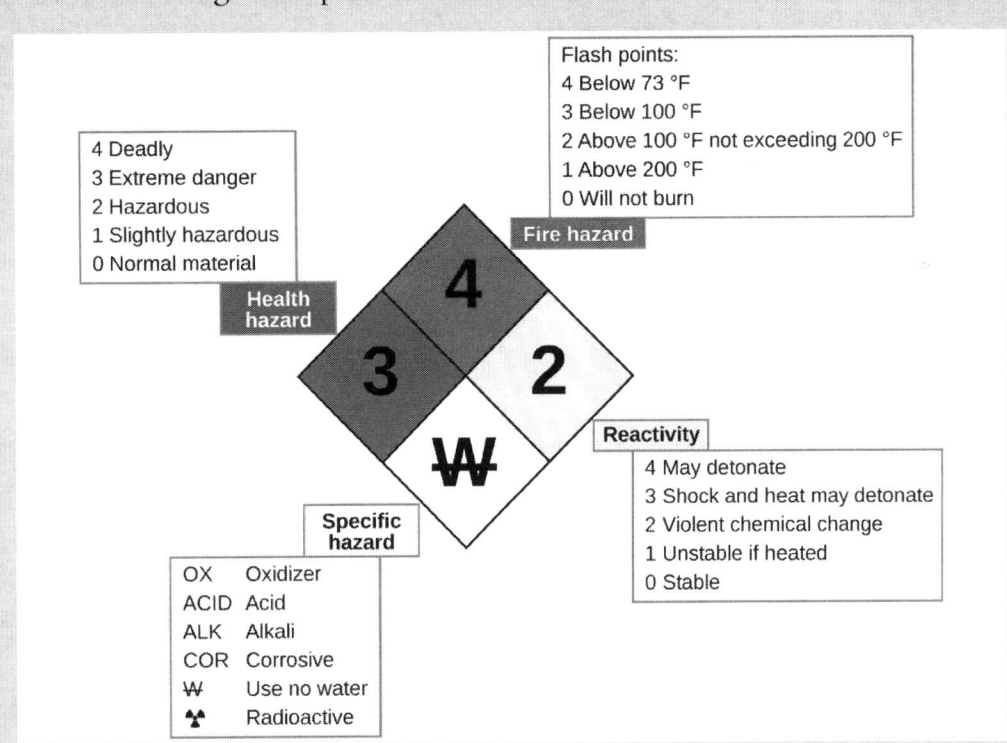

Health hazard
4 Deadly
3 Extreme danger
2 Hazardous
1 Slightly hazardous
0 Normal material

Fire hazard — Flash points:
4 Below 73 °F
3 Below 100 °F
2 Above 100 °F not exceeding 200 °F
1 Above 200 °F
0 Will not burn

Reactivity
4 May detonate
3 Shock and heat may detonate
2 Violent chemical change
1 Unstable if heated
0 Stable

Specific hazard
OX    Oxidizer
ACID  Acid
ALK   Alkali
COR   Corrosive
W̵     Use no water
☢     Radioactive

*Figure 4. The National Fire Protection Agency (NFPA) hazard diamond summarizes the major hazards of a chemical substance.*

The National Fire Protection Agency (NFPA) 704 Hazard Identification System was developed by NFPA to provide safety information about certain substances. The system details flammability, reactivity, health, and other hazards. Within the overall diamond symbol, the top (red) diamond specifies the level of fire hazard (temperature range for flash point). The blue (left) diamond indicates the level of health hazard. The yellow (right) diamond describes reactivity hazards, such as how readily the substance will undergo detonation or a violent chemical change. The white (bottom) diamond points out special hazards, such as if it is an oxidizer (which allows the substance to burn in the absence of air/oxygen), undergoes an unusual or dangerous reaction with water, is corrosive, acidic, alkaline, a biological hazard, radioactive, and so on. Each hazard is rated on a scale from 0 to 4, with 0 being no hazard and 4 being extremely hazardous.

While many elements differ dramatically in their chemical and physical properties, some elements have similar properties. We can identify sets of elements that exhibit common behaviors. For example, many elements conduct heat and electricity well, whereas others are poor conductors. These properties can be used to sort the elements into three classes: metals (elements that conduct well), nonmetals (elements that conduct poorly), and metalloids (elements that have properties of both metals and nonmetals).

The periodic table is a table of elements that places elements with similar properties close together (Figure 5). You will learn more about the periodic table as you continue your study of chemistry.

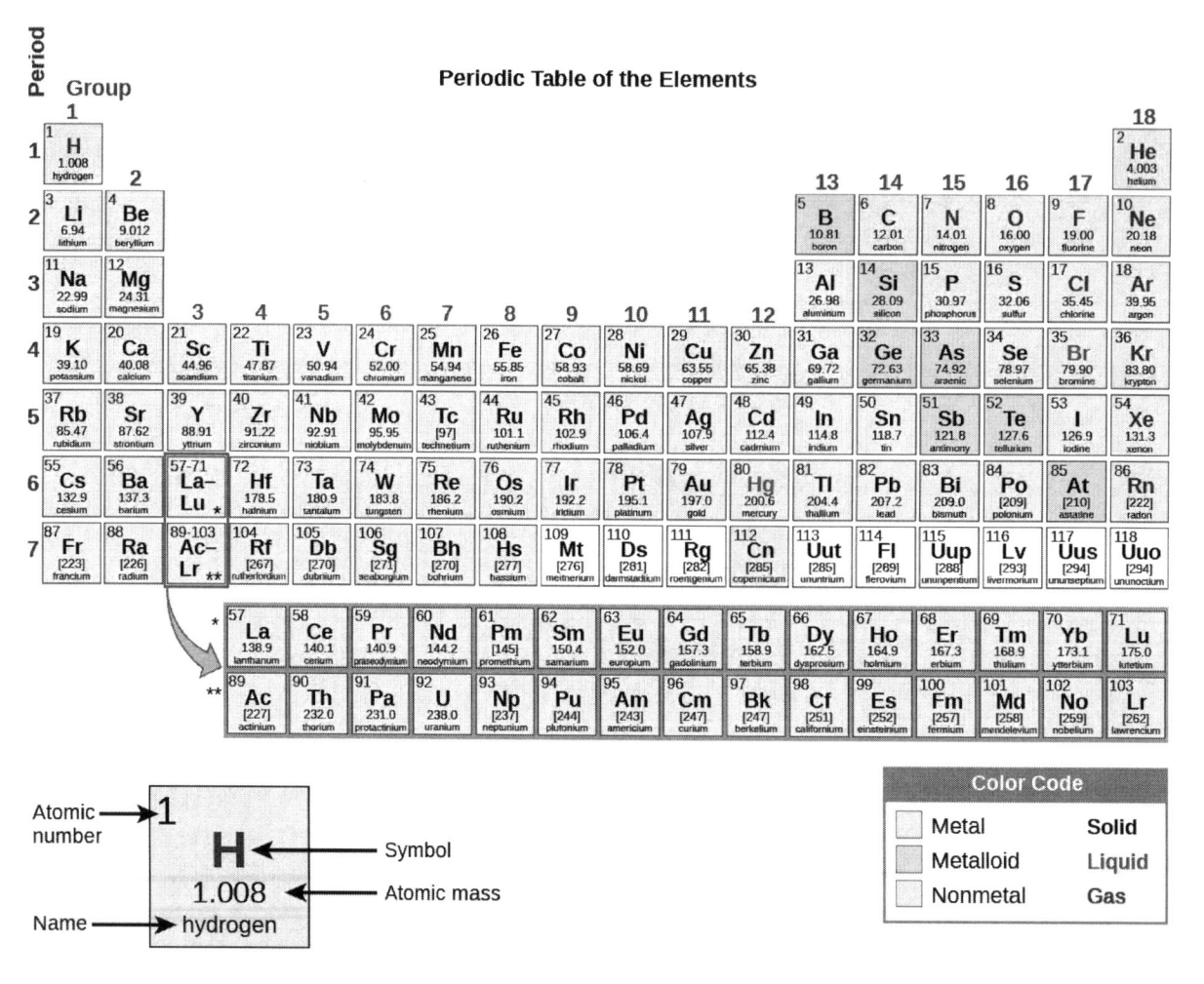

*Figure 5. The periodic table shows how elements may be grouped according to certain similar properties. Note the background color denotes whether an element is a metal, metalloid, or nonmetal, whereas the element symbol color indicates whether it is a solid, liquid, or gas.*

### Key Concepts and Summary

All substances have distinct physical and chemical properties, and may undergo physical or chemical changes. Physical properties, such as hardness and boiling point, and physical changes, such as melting or freezing, do not involve a change in the composition of matter. Chemical properties, such flammability and acidity, and chemical changes, such as rusting, involve production of matter that differs from that present beforehand.

Measurable properties fall into one of two categories. Extensive properties depend on the amount of matter present, for example, the mass of gold. Intensive properties do not depend on the amount of matter present, for example, the density of gold. Heat is an example of an extensive property, and temperature is an example of an intensive property.

### Glossary

**chemical change:** change producing a different kind of matter from the original kind of matter

**chemical property:** behavior that is related to the change of one kind of matter into another kind of matter

**extensive property:** property of a substance that depends on the amount of the substance

**intensive property:** property of a substance that is independent of the amount of the substance

**physical change:** change in the state or properties of matter that does not involve a change in its chemical composition

# Atoms, Isotopes, Ions, and Molecules: The Building Blocks

## Learning Objectives

By the end of this section, you will be able to:

- Define matter and elements
- Describe the interrelationship between protons, neutrons, and electrons
- Compare the ways in which electrons can be donated or shared between atoms
- Explain the ways in which naturally occurring elements combine to create molecules, cells, tissues, organ systems, and organisms

At its most fundamental level, life is made up of matter. Matter is any substance that occupies space and has mass. Elements are unique forms of matter with specific chemical and physical properties that cannot be broken down into smaller substances by ordinary chemical reactions. There are 118 elements, but only 92 occur naturally. The remaining elements are synthesized in laboratories and are unstable.

Each element is designated by its chemical symbol, which is a single capital letter or, when the first letter is already "taken" by another element, a combination of two letters. Some elements follow the English term for the element, such as C for carbon and Ca for calcium. Other elements' chemical symbols derive from their Latin names; for example, the symbol for sodium is Na, referring to *natrium*, the Latin word for sodium.

The four elements common to all living organisms are oxygen (O), carbon (C), hydrogen (H), and nitrogen (N). In the non-living world, elements are found in different proportions, and some elements common to living organisms are relatively rare on the earth as a whole, as shown in [Figure 1]. For example, the atmosphere is rich in nitrogen and oxygen but contains little carbon and hydrogen, while the earth's crust, although it contains oxygen and a small amount of hydrogen, has little nitrogen and carbon. In spite of their differences in abundance, all elements and the chemical reactions between them obey the same chemical and physical laws regardless of whether they are a part of the living or non-living world.

*Approximate Percentage of Elements in Living Organisms (Humans) Compared to the Non-living World*

| Element | Life (Humans) | Atmosphere | Earth's Crust |
|---|---|---|---|
| Oxygen (O) | 65% | 21% | 46% |
| Carbon (C) | 18% | trace | trace |
| Hydrogen (H) | 10% | trace | 0.1% |
| Nitrogen (N) | 3% | 78% | trace |

*Elements and Compounds*

All matter in the natural world is composed of one or more of the 92 fundamental substances called elements. An **element** is a pure substance that is distinguished from all other matter by the fact that it cannot be created or broken down by ordinary chemical means. While your body can assemble many of the chemical compounds needed for life from their constituent elements, it cannot make elements. They must come from the environment. A familiar example of an element that you must take in is calcium ($Ca^{++}$). Calcium is essential to the human body; it is absorbed and used for a number of processes, including strengthening bones. When you consume dairy products your digestive system breaks down the food into components small enough to cross into the bloodstream. Among these is calcium, which, because it is an element, cannot be broken down further. The elemental calcium in cheese, therefore, is the same as the calcium that forms your bones. Some other elements you might be familiar with are oxygen, sodium, and iron. The elements in the human body are shown in Table 1, beginning with the most abundant: oxygen (O), carbon (C), hydrogen (H), and nitrogen (N). Each element's name can be replaced by a one- or two-letter symbol; you will become familiar with some of these during this course. All the elements in your body are derived from the foods you eat and the air you breathe.

*Table 1. Elements of the Human Body. The main elements that compose the human body are shown from most abundant to least abundant.*

| Element | Symbol | Percentage in Body | At a Look |
|---|---|---|---|
| Oxygen | O | 65.0 | |
| Carbon | C | 18.5 | |
| Hydrogen | H | 9.5 | |
| Nitrogen | N | 3.2 | |
| Calcium | Ca | 1.5 | |
| Phosphorus | P | 1.0 | |
| Potassium | K | 0.4 | |
| Sulfur | S | 0.3 | |
| Sodium | Na | 0.2 | |
| Chlorine | Cl | 0.2 | |
| Magnesium | Mg | 0.1 | |
| Trace elements include boron (B), chromium (Cr), cobalt (Co), copper (Cu), fluorine (F), iodine (I), iron (Fe), manganese (Mn), molybdenum (Mo), selenium (Se), silicon (Si), tin (Sn), vanadium (V), and zinc (Zn) | | less than 1.0 | |

In nature, elements rarely occur alone. Instead, they combine to form compounds. A *compound* is a substance composed of two or more elements joined by chemical bonds. For example, the compound glucose is an important body fuel. It is always composed of the same three elements: carbon, hydrogen, and oxygen. Moreover, the elements that make up any given compound always occur in the same relative amounts. In glucose, there are always six carbon and six oxygen units for every twelve hydrogen units. But what, exactly, are these "units" of elements?

## THE STRUCTURE OF THE ATOM

To understand how elements come together, we must first discuss the smallest component or building block of an element, the atom. An atom is the smallest unit of matter that retains all of the chemical properties of an element. For example, one gold atom has all of the properties of gold in that it is a solid metal at room temperature. A gold coin is simply a very large number of gold atoms molded into the shape of a coin and containing small amounts of other elements known as impurities. Gold atoms cannot be broken down into anything smaller while still retaining the properties of gold.

An atom is composed of two regions: the nucleus, which is in the center of the atom and contains protons and neutrons, and the outermost region of the atom which holds its electrons in orbit around the nucleus, as illustrated in [Figure 1]. Atoms contain protons, electrons, and neutrons, among other subatomic particles. The only exception is hydrogen (H), which is made of one proton and one electron with no neutrons.

Protons and neutrons have approximately the same mass, about $1.67 \times 10^{-24}$ grams. Scientists arbitrarily define this amount of mass as one atomic mass unit (amu) or one Dalton, as shown in [Figure 2]. Although similar in mass, protons and neutrons differ in their electric charge. A proton is positively charged whereas a neutron is uncharged. Therefore, the number of neutrons in an atom contributes significantly to its mass, but not to its charge. Electrons are much smaller in mass than protons, weighing only $9.11 \times 10^{-28}$ grams, or about 1/1800 of an atomic mass unit. Hence, they do not contribute much to an element's overall atomic mass. Therefore, when considering atomic mass, it is customary to ignore the mass of any electrons and calculate the atom's mass based on the number of protons and neutrons alone. Although not significant contributors to mass, electrons do contribute greatly to the atom's charge, as each electron has a negative charge equal to the positive charge of a proton. In uncharged, neutral atoms, the number of electrons orbiting the nucleus is equal to the number of protons inside the nucleus. In these atoms, the positive and negative charges cancel each other out, leading to an atom with no net charge.

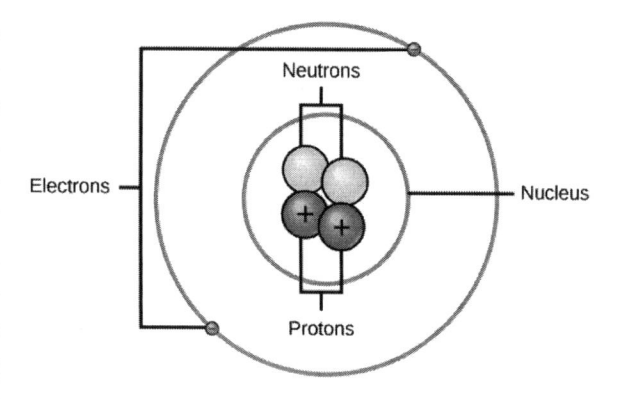

*Figure 1. Elements, such as helium, depicted here, are made up of atoms. Atoms are made up of protons and neutrons located within the nucleus, with electrons in orbitals surrounding the nucleus.*

Accounting for the sizes of protons, neutrons, and electrons, most of the volume of an atom—greater than 99 percent—is, in fact, empty space. With all this empty space, one might ask why so-called solid objects do not just pass through one another. The reason they do not is that the electrons that surround all atoms are negatively charged and negative charges repel each other.

Protons, Neutrons, and Electrons

|  | Charge | Mass (amu) | Location |
|---|---|---|---|
| Proton | +1 | 1 | nucleus |
| Neutron | 0 | 1 | nucleus |
| Electron | −1 | 0 | orbitals |

*Atomic Number and Mass Number*

An atom of carbon is unique to carbon, but a proton of carbon is not. One proton is the same as another, whether it is found in an atom of carbon, sodium (Na), or iron (Fe). The same is true for neutrons and electrons. So, what gives an element its distinctive properties—what makes carbon so different from sodium or iron? The answer is the unique quantity of protons each contains. Carbon by definition is an element whose atoms contain six protons. No other element has exactly six protons in its atoms. Moreover, *all* atoms of carbon, whether found in your liver or in a lump of coal, contain six protons. Thus, the ***atomic number***, which is the number of protons in the nucleus of the atom, identifies the element. Because an atom usually has the same number of electrons as protons, the atomic number identifies the usual number of electrons as well.

In their most common form, many elements also contain the same number of neutrons as protons. The most common form of carbon, for example, has six neutrons as well as six protons, for a total of 12 subatomic particles in its nucleus. An element's **mass number** is the sum of the number of protons and neutrons in its nucleus. So the most common form of carbon's mass number is 12. (Electrons have so little mass that they do not appreciably contribute to the mass of an atom.) Carbon is a relatively light element. Uranium (U), in contrast, has a mass number of 238 and is referred to as a heavy metal. Its atomic number is 92 (it has 92 protons) but it contains 146 neutrons; it has the most mass of all the naturally occurring elements.

*Periodic Table Of Elements*

By the twentieth century, it became apparent that the periodic relationship involved atomic numbers rather than atomic masses. The modern statement of this relationship, the **periodic law**, is as follows: *the properties of the elements are periodic functions of their atomic numbers*. A modern **periodic table** arranges the elements in increasing order of their atomic numbers and groups atoms with similar properties in the same vertical column (Figure 2). Each box represents an element and contains its atomic number, symbol, average atomic mass, and (sometimes) name. The elements are arranged in seven horizontal rows, called **periods** or **series**, and 18 vertical columns, called **groups**. Groups are labeled at the top of each column. In the United States, the labels traditionally were numerals with capital letters. However, IUPAC recommends that the numbers 1 through 18 be used, and these labels are more common. For the table to fit on a single page, parts of two of the rows, a total of 14 columns, are usually written below the main body of the table.

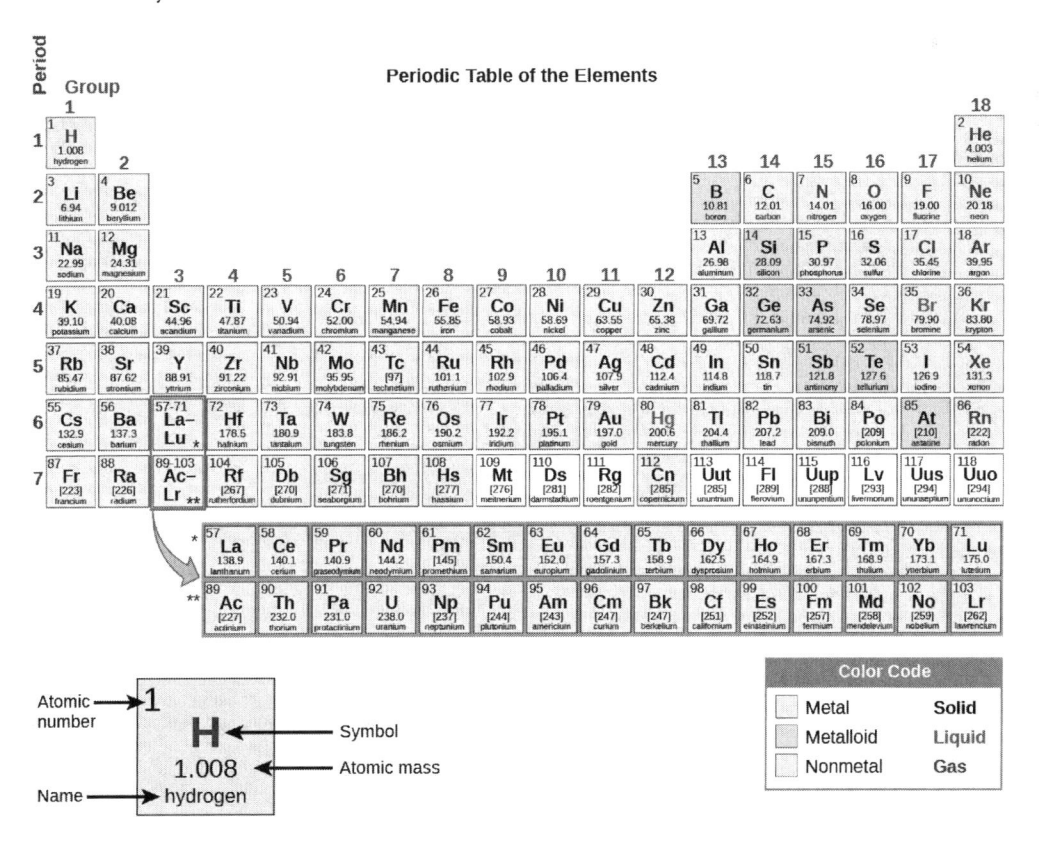

*Figure 2. Elements in the periodic table are organized according to their properties.*

Many elements differ dramatically in their chemical and physical properties, but some elements are similar in their behaviors. For example, many elements appear shiny, are malleable (able to be deformed without breaking) and ductile (can be drawn into wires), and conduct heat and electricity well. Other elements are not shiny, malleable, or ductile, and are poor conductors of heat and electricity. We can sort the elements into large classes with common properties: **metals** (elements that are shiny, malleable, good conductors of heat and electricity—shaded yellow); **nonmetals** (elements that appear dull, poor conductors of heat and electricity—shaded green); and **metalloids** (elements that conduct heat and electricity moderately well, and possess some properties of metals and some properties of nonmetals—shaded purple).

The elements can also be classified into the **main-group elements** (or **representative elements**) in the columns labeled 1, 2, and 13–18; the **transition metals** in the columns labeled 3–12; and **inner transition metals** in the two rows at the bottom of the table (the top-row elements are called **lanthanides** and the bottom-row elements are **actinides**; Figure 3). The elements can be subdivided further by more specific properties, such as the composition of the compounds they form. For example, the elements in group 1 (the first column) form compounds that consist of one atom of the element and one atom of hydrogen. These elements (except hydrogen) are known as **alkali metals**, and they all have similar chemical properties. The elements in group 2 (the second column) form compounds consisting of one atom of the element and two atoms of hydrogen: These are called **alkaline earth metals**, with similar properties among members of that group. Other groups with specific names are the **pnictogens** (group 15), **chalcogens** (group 16), **halogens** (group 17), and the **noble gases** (group 18, also known as **inert gases**). The groups can also be referred to by the first element of the group: For example, the chalcogens can be called the oxygen group or oxygen family. Hydrogen is a unique, nonmetallic element with properties similar to both group 1A and group 7A elements. For that reason, hydrogen may be shown at the top of both groups, or by itself.

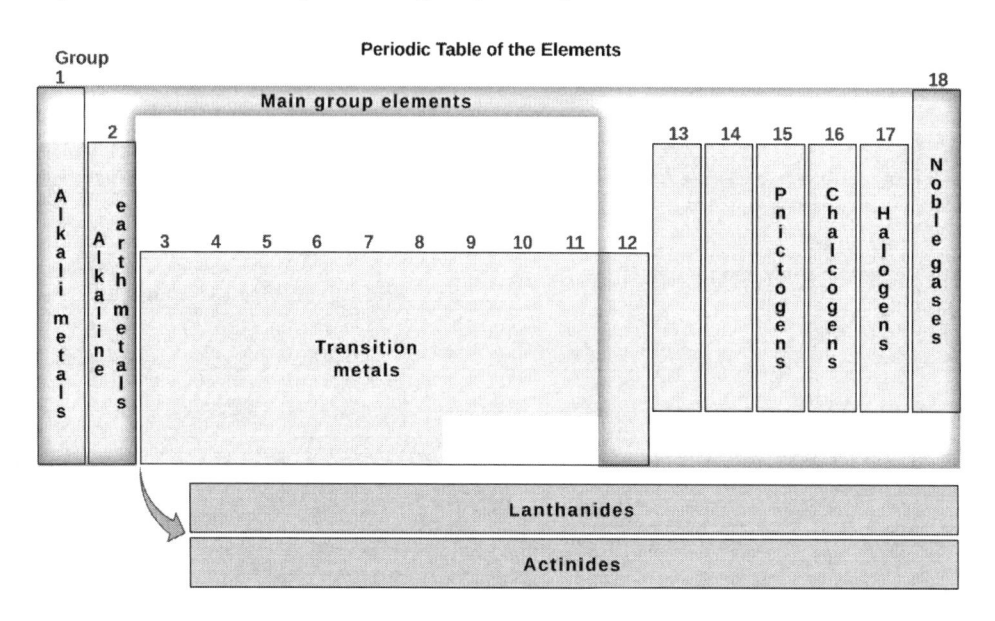

*Figure 3. The periodic table organizes elements with similar properties into groups.*

Click on this link to the Royal Society of Chemistry for an interactive periodic table, which you can use to explore the properties of the elements (includes podcasts and videos of each element). You may also want to try this one from PeriodicTable.com that shows photos of all the elements.

In studying the periodic table, you might have noticed something about the atomic masses of some of the elements. Element 43 (technetium), element 61 (promethium), and most of the elements with atomic number 84 (polonium) and higher have their atomic mass given in square brackets. This is done for elements that consist entirely of unstable, radioactive isotopes (you will learn more about radioactivity in the nuclear chemistry chapter). An average atomic weight cannot be determined for these elements because their radioisotopes may vary significantly in relative abundance, depending on the source, or may not even exist in nature. The number in square brackets is the atomic mass number (and approximate atomic mass) of the most stable isotope of that element.

> **Visit this website to view the periodic table.** In the periodic table of the elements, elements in a single row have the same number of electrons that can participate in a chemical reaction. These electrons are known as "valence electrons." For example, the elements in the first row all have a single valence electron, an electron that can be "donated" in a chemical reaction with another atom. What is the meaning of a mass number shown in parentheses?

## Isotopes

Although each element has a unique number of protons, it can exist as different isotopes. An *isotope* is one of the different forms of an element, distinguished from one another by different numbers of neutrons. The standard isotope of carbon is $^{12}C$, commonly called carbon twelve. $^{12}C$ has six protons and six neutrons, for a mass number of twelve. All of the isotopes of carbon have the same number of protons; therefore, $^{13}C$ has seven neutrons, and $^{14}C$ has eight neutrons. The different isotopes of an element can also be indicated with the mass number hyphenated (for example, C-12 instead of $^{12}C$). Hydrogen has three common isotopes, shown in Figure 3.

Protium ($^1H$)    Deuterium ($^2H$)    Tritium ($^3H$)

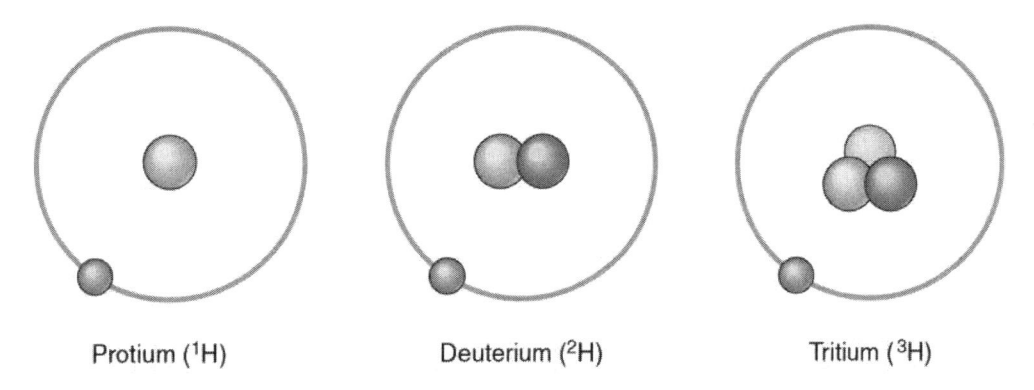

*Figure 3. Isotopes of Hydrogen.* Protium, designated $^1H$, has one proton and no neutrons. It is by far the most abundant isotope of hydrogen in nature. Deuterium, designated $^2H$, has one proton and one neutron. Tritium, designated $^3H$, has two neutrons.

An isotope that contains more than the usual number of neutrons is referred to as a heavy isotope. An example is $^{14}C$. Heavy isotopes tend to be unstable, and unstable isotopes are radioactive. A *radioactive isotope* is an isotope whose nucleus readily decays, giving off subatomic particles and electromagnetic energy. Different radioactive isotopes (also called radioisotopes) differ in their half-life, the time it takes for half of any size sample of an isotope to decay. For example, the half-life of tritium—a radioisotope of hydrogen—is about 12 years, indicating it takes 12 years for half of the tritium nuclei in a sample to decay. Excessive exposure to radioactive isotopes can damage human cells and even cause cancer and birth defects, but when exposure is controlled, some radioactive isotopes can be useful in medicine. For more information, see the Career Connections.

## Career Connection: Interventional Radiologist

The controlled use of radioisotopes has advanced medical diagnosis and treatment of disease. Interventional radiologists are physicians who treat disease by using minimally invasive techniques involving radiation. Many conditions that could once only be treated with a lengthy and traumatic operation can now be treated non-surgically, reducing the cost, pain, length of hospital stay, and recovery time for patients. For example, in the past, the only options for a patient with one or more tumors in the liver were surgery and chemotherapy (the administration of drugs to treat cancer). Some liver tumors, however, are difficult to access surgically, and others could require the surgeon to remove too much of the liver. Moreover, chemotherapy is highly toxic to the liver, and certain tumors do not respond well to it anyway. In some such cases, an interventional radiologist can treat the tumors by disrupting their blood supply, which they need if they are to continue to grow. In this procedure, called radioembolization, the radiologist accesses the liver with a fine needle, threaded through one of the patient's blood vessels. The radiologist then inserts tiny radioactive "seeds" into the blood vessels that supply the tumors. In the days and weeks following the procedure, the radiation emitted from the seeds destroys the vessels and directly kills the tumor cells in the vicinity of the treatment.

Radioisotopes emit subatomic particles that can be detected and tracked by imaging technologies. One of the most advanced uses of radioisotopes in medicine is the positron emission tomography (PET) scanner, which detects the activity in the body of a very small injection of radioactive glucose, the simple sugar that cells use for energy. The PET camera reveals to the medical team which of the patient's tissues are taking up the most glucose. Thus, the most metabolically active tissues show up as bright "hot spots" on the images (Figure 4). PET can reveal some cancerous masses because cancer cells consume glucose at a high rate to fuel their rapid reproduction.

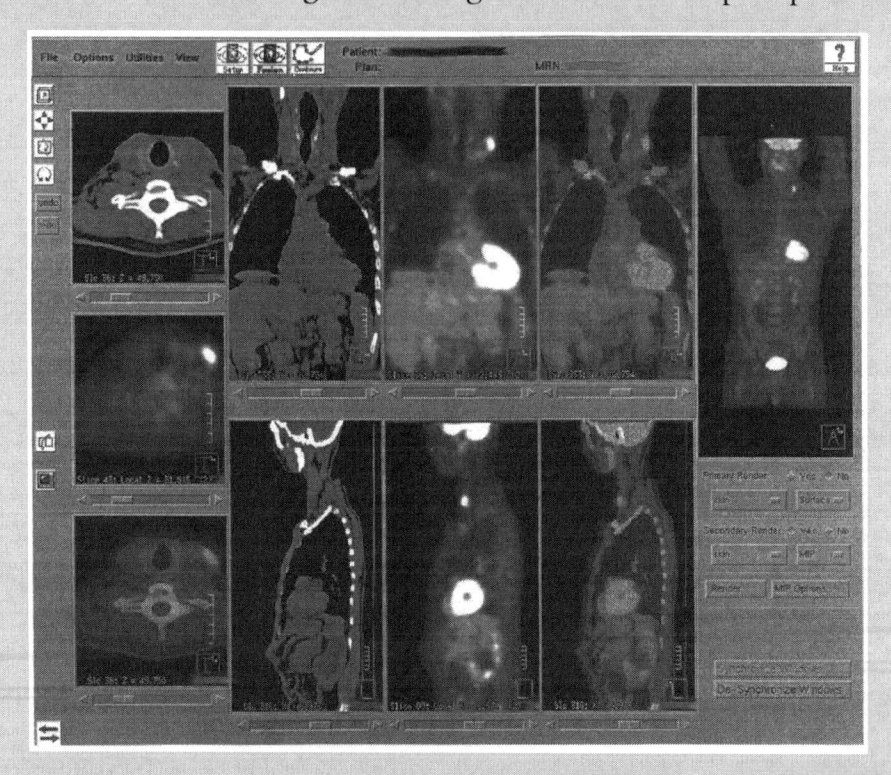

*Figure 4. PET Scan. PET highlights areas in the body where there is relatively high glucose use, which is characteristic of cancerous tissue. This PET scan shows sites of the spread of a large primary tumor to other sites.*

Evolution Connection

*Carbon Dating*

Carbon is normally present in the atmosphere in the form of gaseous compounds like carbon dioxide and methane. Carbon-14 ($^{14}C$) is a naturally occurring radioisotope that is created in the atmosphere from atmospheric $^{14}N$ (nitrogen) by the addition of a neutron and the loss of a proton because of cosmic rays. This is a continuous process, so more $^{14}C$ is always being created. As a living organism incorporates $^{14}C$ initially as carbon dioxide fixed in the process of photosynthesis, the relative amount of $^{14}C$ in its body is equal to the concentration of $^{14}C$ in the atmosphere. When an organism dies, it is no longer ingesting $^{14}C$, so the ratio between $^{14}C$ and $^{12}C$ will decline as $^{14}C$ decays gradually to $^{14}N$ by a process called beta decay—the emission of electrons or positrons. This decay gives off energy in a slow process.

After approximately 5,730 years, half of the starting concentration of $^{14}C$ will have been converted back to $^{14}N$. The time it takes for half of the original concentration of an isotope to decay back to its more stable form is called its half-life. Because the half-life of $^{14}C$ is long, it is used to date formerly living objects such as old bones or wood. Comparing the ratio of the $^{14}C$ concentration found in an object to the amount of $^{14}C$ detected in the atmosphere, the amount of the isotope that has not yet decayed can be determined. On the basis of this amount, the age of the material, such as the pygmy mammoth shown in [Figure 3], can be calculated with accuracy if it is not much older than about 50,000 years. Other elements have isotopes with different half lives. For example, $^{40}K$ (potassium-40) has a half-life of 1.25 billion years, and $^{235}U$ (Uranium 235) has a half-life of about 700 million years. Through the use of radiometric dating, scientists can study the age of fossils or other remains of extinct organisms to understand how organisms have evolved from earlier species.

*Figure 3. The age of carbon-containing remains less than about 50,000 years old, such as this pygmy mammoth, can be determined using carbon dating. (credit: Bill Faulkner, NPS)*

To learn more about atoms, isotopes, and how to tell one isotope from another, run the simulation.

# ELECTRON SHELLS AND THE BOHR MODEL

## The Behavior of Electrons

In the human body, atoms do not exist as independent entities. Rather, they are constantly reacting with other atoms to form and to break down more complex substances. To fully understand anatomy and physiology you must grasp how atoms participate in such reactions. The key is understanding the behavior of electrons.

Although electrons do not follow rigid orbits a set distance away from the atom's nucleus, they do tend to stay within certain regions of space called electron shells. An *electron shell* is a layer of electrons that encircle the nucleus at a distinct energy level.

The atoms of the elements found in the human body have from one to five electron shells, and all electron shells hold eight electrons except the first shell, which can only hold two. This configuration of electron shells is the same for all atoms. The precise number of shells depends on the number of electrons in the atom. Hydrogen and helium have just one and two electrons, respectively. If you take a look at the periodic table of the elements, you will notice that hydrogen and helium are placed alone on either sides of the top row; they are the only elements that have just one electron shell (Figure 5). A second shell is necessary to hold the electrons in all elements larger than hydrogen and helium.

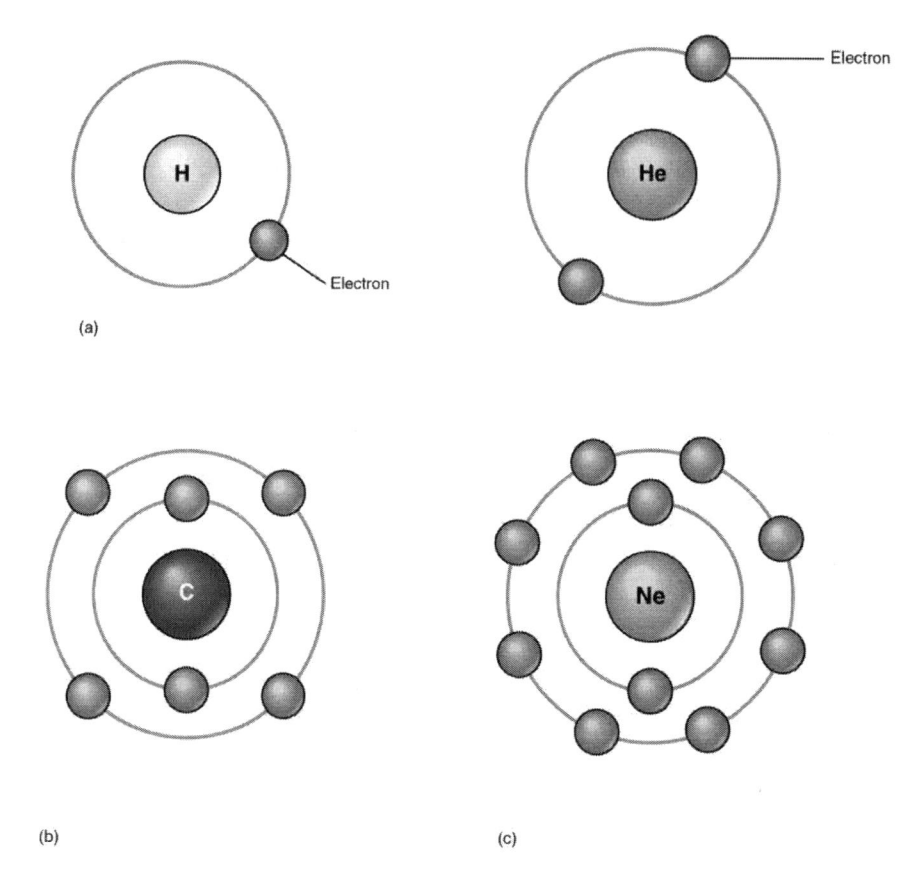

*Figure 5. Electron Shells.* Electrons orbit the atomic nucleus at distinct levels of energy called electron shells. (a) With one electron, hydrogen only half-fills its electron shell. Helium also has a single shell, but its two electrons completely fill it. (b) The electrons of carbon completely fill its first electron shell, but only half-fills its second. (c) Neon, an element that does not occur in the body, has 10 electrons, filling both of its electron shells.

Lithium (Li), whose atomic number is 3, has three electrons. Two of these fill the first electron shell, and the third spills over into a second shell. The second electron shell can accommodate as many as eight electrons. Carbon, with its six electrons, entirely fills its first shell, and half-fills its second. With ten electrons, neon (Ne) entirely fills its two electron shells. Again, a look at the periodic table reveals that all of the elements in the second row, from lithium to neon, have just two electron shells. Atoms with more than ten electrons require more than two shells. These elements occupy the third and subsequent rows of the periodic table.

The factor that most strongly governs the tendency of an atom to participate in chemical reactions is the number of electrons in its valence shell. A *valence shell* is an atom's outermost electron shell. If the valence shell is full, the atom is stable; meaning its electrons are unlikely to be pulled away from the nucleus by the electrical charge of other atoms. If the valence shell is not full, the atom is reactive; meaning it will tend to react with other atoms in ways that make the valence shell full. Consider hydrogen, with its one electron only half-filling its valence shell. This single electron is likely to be drawn into relationships with the atoms of other elements, so that hydrogen's single valence shell can be stabilized.

All atoms (except hydrogen and helium with their single electron shells) are most stable when there are exactly eight electrons in their valence shell. This principle is referred to as the octet rule, and it states that an atom will give up, gain, or share electrons with another atom so that it ends up with eight electrons in its own valence shell. For example, oxygen, with six electrons in its valence shell, is likely to react with other atoms in a way that results in the addition of two electrons to oxygen's valence shell, bringing the number to eight. When two hydrogen atoms each share their single electron with oxygen, covalent bonds are formed, resulting in a molecule of water, $H_2O$.

In nature, atoms of one element tend to join with atoms of other elements in characteristic ways. For example, carbon commonly fills its valence shell by linking up with four atoms of hydrogen. In so doing, the two elements form the simplest of organic molecules, methane, which also is one of the most abundant and stable carbon-containing compounds on Earth. As stated above, another example is water; oxygen needs two electrons to fill its valence shell. It commonly interacts with two atoms of hydrogen, forming $H_2O$. Incidentally, the name "hydrogen" reflects its contribution to water (hydro- = "water"; -gen = "maker"). Thus, hydrogen is the "water maker."

It should be stressed that there is a connection between the number of protons in an element, the atomic number that distinguishes one element from another, and the number of electrons it has. In all electrically neutral atoms, the number of electrons is the same as the number of protons. Thus, each element, at least when electrically neutral, has a characteristic number of electrons equal to its atomic number.

Electrons fill orbitals in a consistent order: they first fill the orbitals closest to the nucleus, then they continue to fill orbitals of increasing energy further from the nucleus. If there are multiple orbitals of equal energy, they will be filled with one electron in each energy level before a second electron is added. The electrons of the outermost energy level determine the energetic stability of the atom and its tendency to form chemical bonds with other atoms to form molecules.

Under standard conditions, atoms fill the inner shells first, often resulting in a variable number of electrons in the outermost shell. The innermost shell has a maximum of two electrons but the next two electron shells can each have a maximum of eight electrons. This is known as the octet rule, which states, with the exception of the innermost shell, that atoms are more stable energetically when they have eight electrons in their valence shell, the outermost electron shell. Examples of some neutral atoms and their electron configurations are shown in [**Figure 6**]. Notice that in this [**Figure 6**], helium has a complete outer electron shell, with two electrons filling its first and only shell. Similarly, neon has a complete outer 2n shell containing eight electrons. In contrast, chlorine and sodium have seven and one in their outer shells, respectively, but theoretically they would be more energetically stable if they followed the octet rule and had eight.

## Chemistry of Life

### Art Connection

An atom may give, take, or share electrons with another atom to achieve a full valence shell, the most stable electron configuration. Looking at this figure, how many electrons do elements in group 1 need to lose in order to achieve a stable electron configuration? How many electrons do elements in groups 14 and 17 need to gain to achieve a stable configuration?

<!--<para> Elements in group 1 need to lose one electron to achieve a stable electron configuration. Elements in groups 14 and 17 need to gain four and one electrons, respectively, to achieve a stable configuration.-->

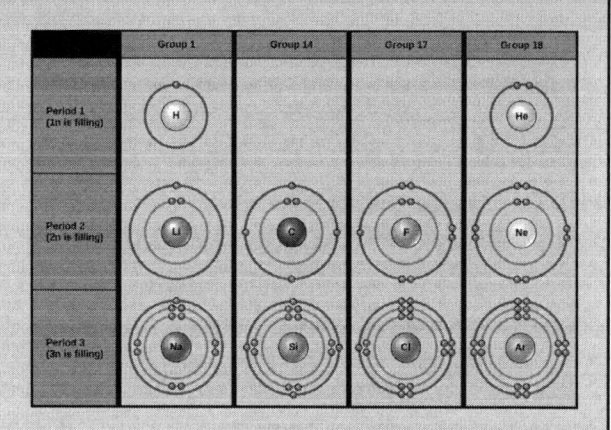

*Figure 6. Bohr diagrams indicate how many electrons fill each principal shell. Group 18 elements (helium, neon, and argon are shown) have a full outer, or valence, shell. A full valence shell is the most stable electron configuration. Elements in other groups have partially filled valence shells and gain or lose electrons to achieve a stable electron configuration.*

Understanding that the organization of the periodic table is based on the total number of protons (and electrons) helps us know how electrons are distributed among the outer shell. The periodic table is arranged in columns and rows based on the number of electrons and where these electrons are located. Take a closer look at the some of the elements in the table's far right column in [**Figure 4**]. The group 18 atoms helium (He), neon (Ne), and argon (Ar) all have filled outer electron shells, making it unnecessary for them to share electrons with other atoms to attain stability; they are highly stable as single atoms. Their non-reactivity has resulted in their being named the inert gases (or noble gases). Compare this to the group 1 elements in the left-hand column. These elements, including hydrogen (H), lithium (Li), and sodium (Na), all have one electron in their outermost shells. That means that they can achieve a stable configuration and a filled outer shell by donating or sharing one electron with another atom or a molecule such as water. Hydrogen will donate or share its electron to achieve this configuration, while lithium and sodium will donate their electron to become stable. As a result of losing a negatively charged electron, they become positively charged ions. Group 17 elements, including fluorine and chlorine, have seven electrons in their outmost shells, so they tend to fill this shell with an electron from other atoms or molecules, making them negatively charged ions. Group 14 elements, of which carbon is the most important to living systems, have four electrons in their outer shell allowing them to make several covalent bonds (discussed below) with other atoms. Thus, the columns of the periodic table represent the potential shared state of these elements' outer electron shells that is responsible for their similar chemical characteristics.

### CHEMICAL REACTIONS AND MOLECULES

All elements are most stable when their outermost shell is filled with electrons according to the octet rule. This is because it is energetically favorable for atoms to be in that configuration and it makes them stable. However, since not all elements have enough electrons to fill their outermost shells, atoms form chemical bonds with other atoms thereby obtaining the electrons they need to attain a stable electron configuration. When two or more atoms chemically bond with each other, the resultant chemical structure is a molecule. The familiar water molecule, $H_2O$, consists of two hydrogen atoms and one oxygen atom; these bond together to form water, as illustrated in [**Figure 8**]. Atoms can form molecules by donating, accepting, or sharing electrons to fill their outer shells.

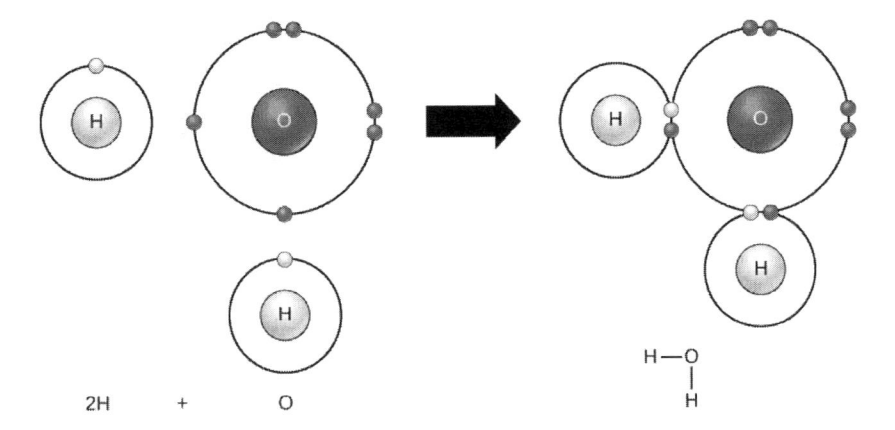

*Figure 8. Two or more atoms may bond with each other to form a molecule. When two hydrogens and an oxygen share electrons via covalent bonds, a water molecule is formed.*

Chemical reactions occur when two or more atoms bond together to form molecules or when bonded atoms are broken apart. The substances used in the beginning of a chemical reaction are called the reactants (usually found on the left side of a chemical equation), and the substances found at the end of the reaction are known as the products (usually found on the right side of a chemical equation). An arrow is typically drawn between the reactants and products to indicate the direction of the chemical reaction; this direction is not always a "one-way street." For the creation of the water molecule shown above, the chemical equation would be:

An example of a simple chemical reaction is the breaking down of hydrogen peroxide molecules, each of which consists of two hydrogen atoms bonded to two oxygen atoms ($H_2O_2$). The reactant hydrogen peroxide is broken down into water, containing one oxygen atom bound to two hydrogen atoms ($H_2O$), and oxygen, which consists of two bonded oxygen atoms ($O_2$). In the equation below, the reaction includes two hydrogen peroxide molecules and two water molecules. This is an example of a balanced chemical equation, wherein the number of atoms of each element is the same on each side of the equation. According to the law of conservation of matter, the number of atoms before and after a chemical reaction should be equal, such that no atoms are, under normal circumstances, created or destroyed.

Even though all of the reactants and products of this reaction are molecules (each atom remains bonded to at least one other atom), in this reaction only hydrogen peroxide and water are representatives of compounds: they contain atoms of more than one type of element. Molecular oxygen, on the other hand, as shown in [**Figure 9**], consists of two doubly bonded oxygen atoms and is not classified as a compound but as a mononuclear molecule.

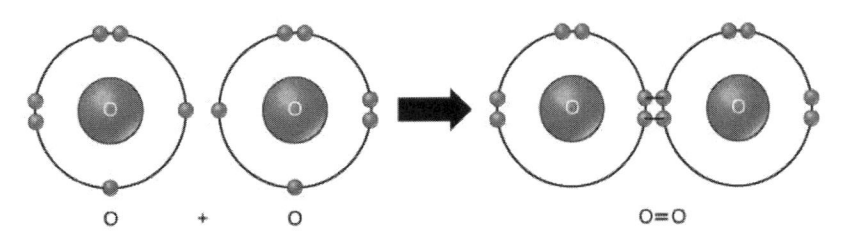

*Figure 9. The oxygen atoms in an $O_2$ molecule are joined by a double bond.*

Some chemical reactions, such as the one shown above, can proceed in one direction until the reactants are all used up. The equations that describe these reactions contain a unidirectional arrow and

# 96 Chemistry of Life

are irreversible. Reversible reactions are those that can go in either direction. In reversible reactions, reactants are turned into products, but when the concentration of product goes beyond a certain threshold (characteristic of the particular reaction), some of these products will be converted back into reactants; at this point, the designations of products and reactants are reversed. This back and forth continues until a certain relative balance between reactants and products occurs—a state called equilibrium. These situations of reversible reactions are often denoted by a chemical equation with a double headed arrow pointing towards both the reactants and products.

For example, in human blood, excess hydrogen ions (H$^+$) bind to bicarbonate ions (HCO$_3^-$) forming an equilibrium state with carbonic acid (H$_2$CO$_3$). If carbonic acid were added to this system, some of it would be converted to bicarbonate and hydrogen ions.

In biological reactions, however, equilibrium is rarely obtained because the concentrations of the reactants or products or both are constantly changing, often with a product of one reaction being a reactant for another. To return to the example of excess hydrogen ions in the blood, the formation of carbonic acid will be the major direction of the reaction. However, the carbonic acid can also leave the body as carbon dioxide gas (via exhalation) instead of being converted back to bicarbonate ion, thus driving the reaction to the right by the chemical law known as law of mass action. These reactions are important for maintaining the homeostasis of our blood.

## IONS AND IONIC BONDS

Some atoms are more stable when they gain or lose an electron (or possibly two) and form ions. This fills their outermost electron shell and makes them energetically more stable. Because the number of electrons does not equal the number of protons, each ion has a net charge. Cations are positive ions that are formed by losing electrons. Negative ions are formed by gaining electrons and are called anions. Anions are designated by their elemental name being altered to end in "-ide": the anion of chlorine is called chloride, and the anion of sulfur is called sulfide, for example.

This movement of electrons from one element to another is referred to as electron transfer. As **[Figure 10]** illustrates, sodium (Na) only has one electron in its outer electron shell. It takes less energy for sodium to donate that one electron than it does to accept seven more electrons to fill the outer shell. If sodium loses an electron, it now has 11 protons, 11 neutrons, and only 10 electrons, leaving it with an overall charge of +1. It is now referred to as a sodium ion. Chlorine (Cl) in its lowest energy state (called the ground state) has seven electrons in its outer shell. Again, it is more energy-efficient for chlorine to gain one electron than to lose seven. Therefore, it tends to gain an electron to create an ion with 17 protons, 17 neutrons, and 18 electrons, giving it a net negative (−1) charge. It is now referred to as a chloride ion. In this example, sodium will donate its one electron to empty its shell, and chlorine will accept that electron to fill its shell. Both ions now satisfy the octet rule and have complete outermost shells. Because the number of electrons is no longer equal to the number of protons, each is now an ion and has a +1 (sodium cation) or −1 (chloride anion) charge. Note that these transactions can normally only take place simultaneously: in order for a sodium atom to lose an electron, it must be in the presence of a suitable recipient like a chlorine atom.

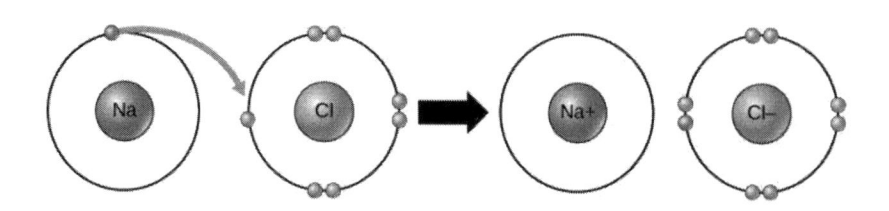

*Figure 10. In the formation of an ionic compound, metals lose electrons and nonmetals gain electrons to achieve an octet.*

Atoms, Isotopes, Ions, and Molecules: The Building Blocks 97

Ionic bonds are formed between ions with opposite charges. For instance, positively charged sodium ions and negatively charged chloride ions bond together to make crystals of sodium chloride, or table salt, creating a crystalline molecule with zero net charge.

Certain salts are referred to in physiology as electrolytes (including sodium, potassium, and calcium), ions necessary for nerve impulse conduction, muscle contractions and water balance. Many sports drinks and dietary supplements provide these ions to replace those lost from the body via sweating during exercise.

## COVALENT BONDS AND OTHER BONDS AND INTERACTIONS

Another way the octet rule can be satisfied is by the sharing of electrons between atoms to form covalent bonds. These bonds are stronger and much more common than ionic bonds in the molecules of living organisms. Covalent bonds are commonly found in carbon-based organic molecules, such as our DNA and proteins. Covalent bonds are also found in inorganic molecules like $H_2O$, $CO_2$, and $O_2$. One, two, or three pairs of electrons may be shared, making single, double, and triple bonds, respectively. The more covalent bonds between two atoms, the stronger their connection. Thus, triple bonds are the strongest.

The strength of different levels of covalent bonding is one of the main reasons living organisms have a difficult time in acquiring nitrogen for use in constructing their molecules, even though molecular nitrogen, $N_2$, is the most abundant gas in the atmosphere. Molecular nitrogen consists of two nitrogen atoms triple bonded to each other and, as with all molecules, the sharing of these three pairs of electrons between the two nitrogen atoms allows for the filling of their outer electron shells, making the molecule more stable than the individual nitrogen atoms. This strong triple bond makes it difficult for living systems to break apart this nitrogen in order to use it as constituents of proteins and DNA.

The formation of water molecules provides an example of covalent bonding. The hydrogen and oxygen atoms that combine to form water molecules are bound together by covalent bonds, as shown in [**Figure 8**]. The electron from the hydrogen splits its time between the incomplete outer shell of the hydrogen atoms and the incomplete outer shell of the oxygen atoms. To completely fill the outer shell of oxygen, which has six electrons in its outer shell but which would be more stable with eight, two electrons (one from each hydrogen atom) are needed: hence the well-known formula $H_2O$. The electrons are shared between the two elements to fill the outer shell of each, making both elements more stable.

View this short video to see an animation of ionic and covalent bonding.

### Polar Covalent Bonds

There are two types of covalent bonds: polar and nonpolar. In a polar covalent bond, shown in [**Figure 11**], the electrons are unequally shared by the atoms and are attracted more to one nucleus than the other. Because of the unequal distribution of electrons between the atoms of different elements, a slightly positive (δ+) or slightly negative (δ–) charge develops. This partial charge is an important property of water and accounts for many of its characteristics.

Water is a polar molecule, with the hydrogen atoms acquiring a partial positive charge and the oxygen a partial negative charge. This occurs because the nucleus of the oxygen atom is more attractive to the electrons of the hydrogen atoms than the hydrogen nucleus is to the oxygen's electrons. Thus oxygen has a higher electronegativity than hydrogen and the shared electrons spend more time in the vicinity of the oxygen nucleus than they do near the nucleus of the hydrogen atoms, giving the atoms of oxygen and hydrogen slightly negative and positive charges, respectively. Another way of stating this is that the probability of finding a shared electron near an oxygen nucleus is more likely than finding it near a hydrogen nucleus. Either way, the atom's relative electronegativity contributes to the development of partial charges whenever one element is significantly more electronegative than the other, and the charges generated by these polar bonds may then be used for the formation of hydrogen bonds based on the attraction of opposite partial charges. (Hydrogen bonds, which are discussed in detail below, are weak bonds between slightly positively charged hydrogen atoms to slightly negatively charged atoms in other molecules.) Since macromolecules often have atoms within them that differ in electronegativity, polar bonds are often present in organic molecules.

## Nonpolar Covalent Bonds

Nonpolar covalent bonds form between two atoms of the same element or between different elements that share electrons equally. For example, molecular oxygen ($O_2$) is nonpolar because the electrons will be equally distributed between the two oxygen atoms.

Another example of a nonpolar covalent bond is methane ($CH_4$), also shown in [**Figure 11**]. Carbon has four electrons in its outermost shell and needs four more to fill it. It gets these four from four hydrogen atoms, each atom providing one, making a stable outer shell of eight electrons. Carbon and hydrogen do not have the same electronegativity but are similar; thus, nonpolar bonds form. The hydrogen atoms each need one electron for their outermost shell, which is filled when it contains two electrons. These elements share the electrons equally among the carbons and the hydrogen atoms, creating a nonpolar covalent molecule.

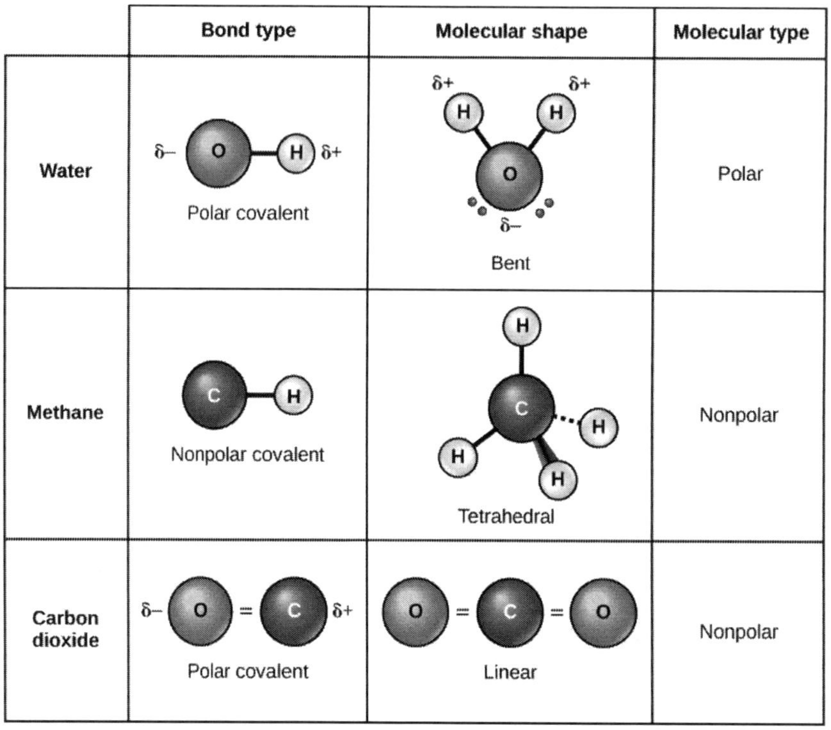

|  | Bond type | Molecular shape | Molecular type |
|---|---|---|---|
| Water | Polar covalent | Bent | Polar |
| Methane | Nonpolar covalent | Tetrahedral | Nonpolar |
| Carbon dioxide | Polar covalent | Linear | Nonpolar |

*Figure 11. Whether a molecule is polar or nonpolar depends both on bond type and molecular shape. Both water and carbon dioxide have polar covalent bonds, but carbon dioxide is linear, so the partial charges on the molecule cancel each other out.*

## Hydrogen Bonds and Van Der Waals Interactions

Ionic and covalent bonds between elements require energy to break. Ionic bonds are not as strong as covalent, which determines their behavior in biological systems. However, not all bonds are ionic or covalent bonds. Weaker bonds can also form between molecules. Two weak bonds that occur frequently are hydrogen bonds and van der Waals interactions. Without these two types of bonds, life as we know it would not exist. Hydrogen bonds provide many of the critical, life-sustaining properties of water and also stabilize the structures of proteins and DNA, the building block of cells.

When polar covalent bonds containing hydrogen form, the hydrogen in that bond has a slightly positive charge because hydrogen's electron is pulled more strongly toward the other element and away from the hydrogen. Because the hydrogen is slightly positive, it will be attracted to neighboring negative charges. When this happens, a weak interaction occurs between the $\delta^+$ of the hydrogen from one molecule and the $\delta-$ charge on the more electronegative atoms of another molecule, usually oxygen or nitrogen, or within the same molecule. This interaction is called a hydrogen bond. This type of bond is common and occurs regularly between water molecules. Individual hydrogen bonds are weak and easily broken; however, they occur in very large numbers in water and in organic polymers, creating a major force in combination. Hydrogen bonds are also responsible for zipping together the DNA double helix.

Like hydrogen bonds, van der Waals interactions are weak attractions or interactions between molecules. Van der Waals attractions can occur between any two or more molecules and are dependent on slight fluctuations of the electron densities, which are not always symmetrical around an atom. For these attractions to happen, the molecules need to be very close to one another. These bonds—along with ionic, covalent, and hydrogen bonds—contribute to the three-dimensional structure of the proteins in our cells that is necessary for their proper function.

Career Connection

*Pharmaceutical Chemist*

Pharmaceutical chemists are responsible for the development of new drugs and trying to determine the mode of action of both old and new drugs. They are involved in every step of the drug development process. Drugs can be found in the natural environment or can be synthesized in the laboratory. In many cases, potential drugs found in nature are changed chemically in the laboratory to make them safer and more effective, and sometimes synthetic versions of drugs substitute for the version found in nature.

After the initial discovery or synthesis of a drug, the chemist then develops the drug, perhaps chemically altering it, testing it to see if the drug is toxic, and then designing methods for efficient large-scale production. Then, the process of getting the drug approved for human use begins. In the United States, drug approval is handled by the Food and Drug Administration (FDA) and involves a series of large-scale experiments using human subjects to make sure the drug is not harmful and effectively treats the condition it aims to treat. This process often takes several years and requires the participation of physicians and scientists, in addition to chemists, to complete testing and gain approval.

An example of a drug that was originally discovered in a living organism is Paclitaxel (Taxol), an anti-cancer drug used to treat breast cancer. This drug was discovered in the bark of the pacific yew tree. Another example is aspirin, originally isolated from willow tree bark. Finding drugs often means testing hundreds of samples of plants, fungi, and other forms of life to see if any biologically active compounds are found within them. Sometimes, traditional medicine can give modern medicine clues to where an active compound can be found. For example, the use of willow bark to make medicine has been known for thousands of years, dating back to ancient Egypt. It was not until the late 1800s, however, that the aspirin molecule, known as acetylsalicylic acid, was purified and marketed for human use.

Occasionally, drugs developed for one use are found to have unforeseen effects that allow these drugs to be used in other, unrelated ways. For example, the drug minoxidil (Rogaine) was originally developed to treat high blood pressure. When tested on humans, it was noticed that individuals taking the drug would grow new hair. Eventually the drug was marketed to men and women with baldness to restore lost hair.

The career of the pharmaceutical chemist may involve detective work, experimentation, and drug development, all with the goal of making human beings healthier.

## SECTION SUMMARY

Matter is anything that occupies space and has mass. It is made up of elements. All of the 92 elements that occur naturally have unique qualities that allow them to combine in various ways to create molecules, which in turn combine to form cells, tissues, organ systems, and organisms. Atoms, which consist of protons, neutrons, and electrons, are the smallest units of an element that retain all of the properties of that element. Electrons can be transferred, shared, or cause charge disparities between atoms to create bonds, including ionic, covalent, and hydrogen bonds, as well as van der Waals interactions.

### Glossary

**anion:** negative ion that is formed by an atom gaining one or more electrons

**atom:** the smallest unit of matter that retains all of the chemical properties of an element

**atomic mass:** calculated mean of the mass number for an element's isotopes

**atomic number:** total number of protons in an atom

**balanced chemical equation:** statement of a chemical reaction with the number of each type of atom equalized for both the products and reactants

**cation:** positive ion that is formed by an atom losing one or more electrons

**chemical bond:** interaction between two or more of the same or different atoms that results in the formation of molecules

**chemical reaction:** process leading to the rearrangement of atoms in molecules

**chemical reactivity:** the ability to combine and to chemically bond with each other

**compound:** substance composed of molecules consisting of atoms of at least two different elements

**covalent bond:** type of strong bond formed between two of the same or different elements; forms when electrons are shared between atoms

**electrolyte:** ion necessary for nerve impulse conduction, muscle contractions and water balance

**electron:** negatively charged subatomic particle that resides outside of the nucleus in the electron orbital; lacks functional mass and has a negative charge of –1 unit

**electron configuration:** arrangement of electrons in an atom's electron shell (for example, $1s^2 2s^2 2p^6$)

**electron orbital:** how electrons are spatially distributed surrounding the nucleus; the area where an electron is most likely to be found

**electron transfer:** movement of electrons from one element to another; important in creation of ionic bonds

**electronegativity:** ability of some elements to attract electrons (often of hydrogen atoms), acquiring partial negative charges in molecules and creating partial positive charges on the hydrogen atoms

**element:** one of 118 unique substances that cannot be broken down into smaller substances; each element has unique properties and a specified number of protons

**equilibrium:** steady state of relative reactant and product concentration in reversible chemical reactions in a closed system

**hydrogen bond:** weak bond between slightly positively charged hydrogen atoms to slightly negatively charged atoms in other molecules

**inert gas:** (also, noble gas) element with filled outer electron shell that is unreactive with other atoms

**ion:** atom or chemical group that does not contain equal numbers of protons and electrons

**ionic bond:** chemical bond that forms between ions with opposite charges (cations and anions)

**irreversible chemical reaction:** chemical reaction where reactants proceed uni-directionally to form products

**isotope:** one or more forms of an element that have different numbers of neutrons

**law of mass action:** chemical law stating that the rate of a reaction is proportional to the concentration of the reacting substances

**mass number:** total number of protons and neutrons in an atom

**matter:** anything that has mass and occupies space

**molecule:** two or more atoms chemically bonded together

**neutron:** uncharged particle that resides in the nucleus of an atom; has a mass of one amu

**noble gas:** see inert gas

**nonpolar covalent bond:** type of covalent bond that forms between atoms when electrons are shared equally between them

**nucleus:** core of an atom; contains protons and neutrons

**octet rule:** rule that atoms are most stable when they hold eight electrons in their outermost shells

**orbital:** region surrounding the nucleus; contains electrons

**periodic table:** organizational chart of elements indicating the atomic number and atomic mass of each element; provides key information about the properties of the elements

**polar covalent bond:** type of covalent bond that forms as a result of unequal sharing of electrons, resulting in the creation of slightly positive and slightly negative charged regions of the molecule

**product:** molecule found on the right side of a chemical equation

**proton:** positively charged particle that resides in the nucleus of an atom; has a mass of one amu and a charge of +1

**radioisotope:** isotope that emits radiation composed of subatomic particles to form more stable elements

**reactant:** molecule found on the left side of a chemical equation

**reversible chemical reaction:** chemical reaction that functions bi-directionally, where products may turn into reactants if their concentration is great enough

**valence shell:** outermost shell of an atom

**van der Waals interaction:** very weak interaction between molecules due to temporary charges attracting atoms that are very close together

# Lewis Dot Formula

## LEWIS SYMBOLS

We use Lewis symbols to describe valence electron configurations of atoms and monatomic ions. A **Lewis symbol** consists of an elemental symbol surrounded by one dot for each of its valence electrons:

•Ca•

Figure 1 shows the Lewis symbols for the elements of the third period of the periodic table.

| Atoms | Electronic Configuration | Lewis Symbol |
|---|---|---|
| sodium | $[Ne]3s^1$ | Na • |
| magnesium | $[Ne]3s^2$ | • Mg • |
| aluminum | $[Ne]3s^2 3p^1$ | • Al • |
| silicon | $[Ne]3s^2 3p^2$ | • Si • |
| phosphorus | $[Ne]3s^2 3p^3$ | • P • |
| sulfur | $[Ne]3s^2 3p^4$ | : S • |
| chlorine | $[Ne]3s^2 3p^5$ | : Cl • |
| argon | $[Ne]3s^2 3p^6$ | : Ar : |

*Figure 1. Lewis symbols illustrating the number of valence electrons for each element in the third period of the periodic table.*

Lewis symbols can also be used to illustrate the formation of cations from atoms, as shown here for sodium and calcium:

Na• ⟶ Na⁺ + e⁻          •Ca• ⟶ Ca²⁺ + 2e⁻
sodium      sodium              calcium       calcium
atom        cation              atom          cation

Likewise, they can be used to show the formation of anions from atoms, as shown below for chlorine and sulfur:

:Cl• + e⁻ ⟶ :Cl:⁻          :S• + 2e⁻ ⟶ :S:²⁻
chlorine     chlorine              sulfur        sulfide
atom         anion                 atom          anion

Figure 2 demonstrates the use of Lewis symbols to show the transfer of electrons during the formation of ionic compounds.

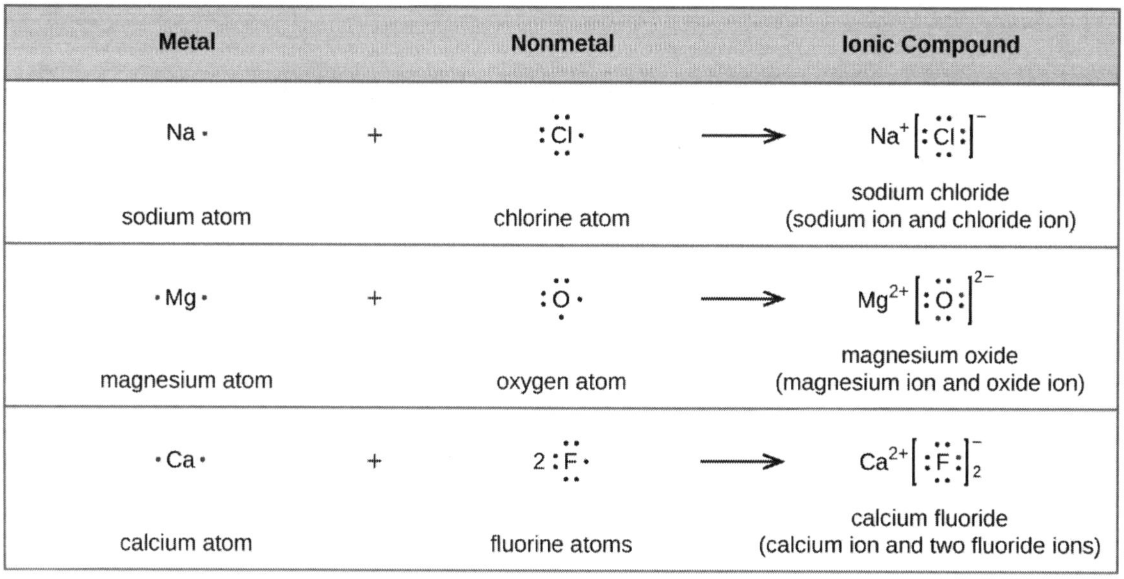

| Metal | Nonmetal | Ionic Compound |
|---|---|---|
| Na· <br> sodium atom | :C̈l· <br> chlorine atom | Na⁺[:C̈l:]⁻ <br> sodium chloride <br> (sodium ion and chloride ion) |
| ·Mg· <br> magnesium atom | :Ö· <br> oxygen atom | Mg²⁺[:Ö:]²⁻ <br> magnesium oxide <br> (magnesium ion and oxide ion) |
| ·Ca· <br> calcium atom | 2 :F̈· <br> fluorine atoms | Ca²⁺[:F̈:]⁻₂ <br> calcium fluoride <br> (calcium ion and two fluoride ions) |

*Figure 2. Cations are formed when atoms lose electrons, represented by fewer Lewis dots, whereas anions are formed by atoms gaining electrons. The total number of electrons does not change.*

## LEWIS STRUCTURES

We also use Lewis symbols to indicate the formation of covalent bonds, which are shown in **Lewis structures**, drawings that describe the bonding in molecules and polyatomic ions. For example, when two chlorine atoms form a chlorine molecule, they share one pair of electrons:

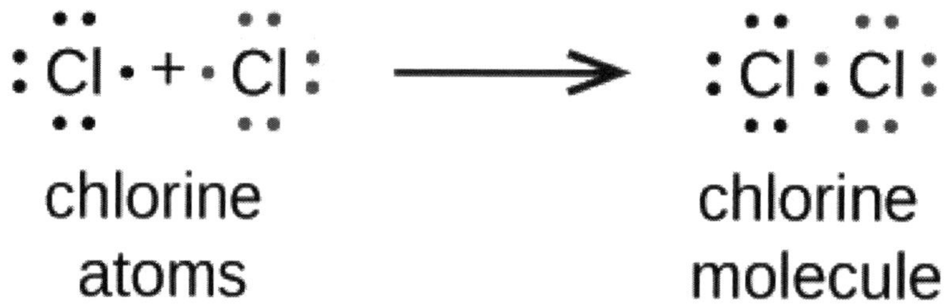

:C̈l· + ·C̈l: ⟶ :C̈l:C̈l:

chlorine atoms         chlorine molecule

The Lewis structure indicates that each Cl atom has three pairs of electrons that are not used in bonding (called **lone pairs**) and one shared pair of electrons (written between the atoms). A dash (or line) is sometimes used to indicate a shared pair of electrons:

H—H          :C̈l—C̈l:

A single shared pair of electrons is called a **single bond**. Each Cl atom interacts with eight valence electrons: the six in the lone pairs and the two in the single bond.

### The Octet Rule

The other halogen molecules ($F_2$, $Br_2$, $I_2$, and $At_2$) form bonds like those in the chlorine molecule: one single bond between atoms and three lone pairs of electrons per atom. This allows each halogen atom to have a noble gas electron configuration. The tendency of main group atoms to form enough bonds to obtain eight valence electrons is known as the **octet rule**.

The number of bonds that an atom can form can often be predicted from the number of electrons needed to reach an octet (eight valence electrons); this is especially true of the nonmetals of the second period of the periodic table (C, N, O, and F). For example, each atom of a group 14 element has four electrons in its outermost shell and therefore requires four more electrons to reach an octet. These four electrons can be gained by forming four covalent bonds, as illustrated here for carbon in $CCl_4$ (carbon tetrachloride) and silicon in $SiH_4$ (silane). Because hydrogen only needs two electrons to fill its valence shell, it is an exception to the octet rule. The transition elements and inner transition elements also do not follow the octet rule:

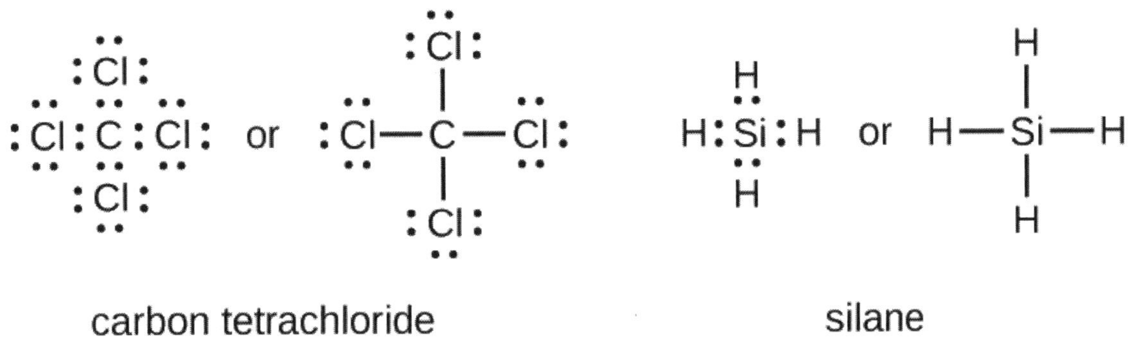

carbon tetrachloride                silane

Group 15 elements such as nitrogen have five valence electrons in the atomic Lewis symbol: one lone pair and three unpaired electrons. To obtain an octet, these atoms form three covalent bonds, as in $NH_3$ (ammonia). Oxygen and other atoms in group 16 obtain an octet by forming two covalent bonds:

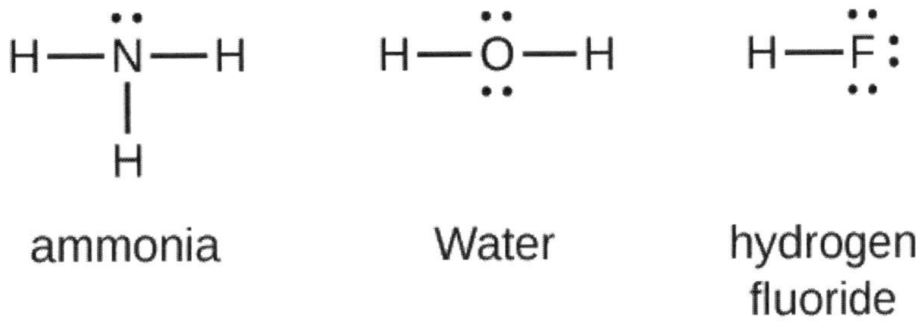

ammonia        Water        hydrogen fluoride

*Double and Triple Bonds*

As previously mentioned, when a pair of atoms shares one pair of electrons, we call this a single bond. However, a pair of atoms may need to share more than one pair of electrons in order to achieve the requisite octet. A **double bond** forms when two pairs of electrons are shared between a pair of atoms, as between the carbon and oxygen atoms in $CH_2O$ (formaldehyde) and between the two carbon atoms in $C_2H_4$ (ethylene):

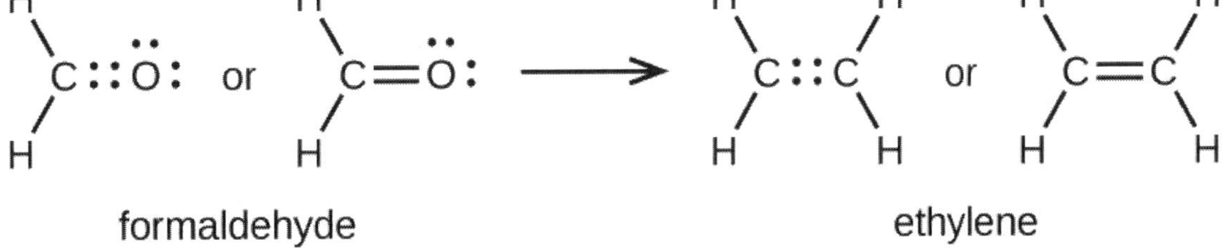

formaldehyde                ethylene

**A triple bond** forms when three electron pairs are shared by a pair of atoms, as in carbon monoxide (CO) and the cyanide ion (CN⁻):

$$:C:::O: \text{ or } :C\equiv O: \longrightarrow :C:::N:^- \text{ or } :C\equiv N:^-$$

carbon monoxide          cyanide ion

## WRITING LEWIS STRUCTURES WITH THE OCTET RULE

For very simple molecules and molecular ions, we can write the Lewis structures by merely pairing up the unpaired electrons on the constituent atoms. See these examples:

$$H\cdot + :\!\overset{..}{\underset{..}{Br}}\!\cdot \longrightarrow H:\!\overset{..}{\underset{..}{Br}}\!:$$

$$2H\cdot + :\!\overset{..}{\underset{.}{S}}\!\cdot \longrightarrow H:\!\overset{..}{\underset{..}{S}}\!: \\ \phantom{2H\cdot + :S\cdot \longrightarrow }H$$

$$\cdot\!\overset{..}{\underset{.}{N}}\!\cdot + \cdot\!\overset{..}{\underset{.}{N}}\!\cdot \longrightarrow :N:::N:$$

# Exercises

## PART 1

1. Why do we use an object's mass, rather than its weight, to indicate the amount of matter it contains?

2. What properties distinguish solids from liquids? Liquids from gases? Solids from gases?

3. How does a heterogeneous mixture differ from a homogeneous mixture? How are they similar?

4. How does a homogeneous mixture differ from a pure substance? How are they similar?

5. How do molecules of elements and molecules of compounds differ? In what ways are they similar

6. How does an atom differ from a molecule? In what ways are they similar?

7. Many of the items you purchase are mixtures of pure compounds. Select three of these commercial products and prepare a list of the ingredients that are pure compounds.

8. Classify each of the following as an element, a compound, or a mixture:

   a. copper

   b. water

   c. nitrogen

   d. sulfur

   e. air

   f. sucrose

   g. a substance composed of molecules each of which contains two iodine atoms

   h. gasoline

9. A sulfur atom and a sulfur molecule are not identical. What is the difference?

10. How are the molecules in oxygen gas, the molecules in hydrogen gas, and water molecules similar? How do they differ?

11. We refer to astronauts in space as weightless, but not without mass. Why?

12. As we drive an automobile, we don't think about the chemicals consumed and produced. Prepare a list of the principal chemicals consumed and produced during the operation of an automobile.

13. Matter is everywhere around us. Make a list by name of fifteen different kinds of matter that you encounter every day. Your list should include (and label at least one example of each) the following: a solid, a liquid, a gas, an element, a compound, a homogenous mixture, a heterogeneous mixture, and a pure substance.

14. When elemental iron corrodes it combines with oxygen in the air to ultimately form red brown iron(III) oxide which we call rust.

   a. If a shiny iron nail with an initial mass of 23.2 g is weighed after being coated in a layer of rust, would you expect the mass to have increased, decreased, or remained the same? Explain.

   b. If the mass of the iron nail increases to 24.1 g, what mass of oxygen combined with the iron?

108  Chemistry of Life

15. As stated in the text, convincing examples that demonstrate the law of conservation of matter outside of the laboratory are few and far between. Indicate whether the mass would increase, decrease, or stay the same for the following scenarios where chemical reactions take place:

  a. Exactly one pound of bread dough is placed in a baking tin. The dough is cooked in an oven at 350 °F releasing a wonderful aroma of freshly baked bread during the cooking process. Is the mass of the baked loaf less than, greater than, or the same as the one pound of original dough? Explain.

  b. When magnesium burns in air a white flaky ash of magnesium oxide is produced. Is the mass of magnesium oxide less than, greater than, or the same as the original piece of magnesium? Explain.

  c. Antoine Lavoisier, the French scientist credited with first stating the law of conservation of matter, heated a mixture of tin and air in a sealed flask to produce tin oxide. Did the mass of the sealed flask and contents decrease, increase, or remain the same after the heating?

## PART 2

1. Classify each of the following changes as physical or chemical:

  a. condensation of steam

  b. burning of gasoline

  c. souring of milk

  d. dissolving of sugar in water

  e. melting of gold

2. Classify each of the following changes as physical or chemical:

  a. coal burning

  b. ice melting

  c. mixing chocolate syrup with milk

  d. explosion of a firecracker

  e. magnetizing of a screwdriver

3. The volume of a sample of oxygen gas changed from 10 mL to 11 mL as the temperature changed. Is this a chemical or physical change?

4. A 2.0-liter volume of hydrogen gas combined with 1.0 liter of oxygen gas to produce 2.0 liters of water vapor. Does oxygen undergo a chemical or physical change?

5. Explain the difference between extensive properties and intensive properties.

6. Identify the following properties as either extensive or intensive.

  a. volume

  b. temperature

  c. humidity

  d. heat

  e. boiling point

7. The density (d) of a substance is an intensive property that is defined as the ratio of its mass (m) to its volume (V). density = $\frac{mass}{volume}$; d = $\frac{m}{V}$. Considering that mass and volume are both extensive properties, explain why their ratio, density, is intensive.

## PART 3

1. Using the periodic table, classify each of the following elements as a metal or a nonmetal, and then further classify each as a main-group (representative) element, transition metal, or inner transition metal:

   a. cobalt

   b. europium

   c. iodine

   d. indium

   e. lithium

   f. oxygen

   g. cadmium

   h. terbium

   i. rhenium

2. Use the periodic table to give the name and symbol for each of the following elements:

   a. the halogen in the same period as the alkali metal with 11 protons

   b. the alkaline earth metal in the same period with the neutral noble gas with 18 electrons

   c. the noble gas in the same row as an isotope with 30 neutrons and 25 protons

   d. the noble gas in the same period as gold

3. Write a symbol for each of the following neutral isotopes. Include the atomic number and mass number for each.

   a. the alkali metal with 11 protons and a mass number of 23

   b. the noble gas element with and 75 neutrons in its nucleus and 54 electrons in the neutral atom

   c. the isotope with 33 protons and 40 neutrons in its nucleus

   d. the alkaline earth metal with 88 electrons and 138 neutrons

4. Write a symbol for each of the following neutral isotopes. Include the atomic number and mass number for each.

   a. the chalcogen with a mass number of 125

   b. the halogen whose longest-lived isotope is radioactive

   c. the noble gas, used in lighting, with 10 electrons and 10 neutrons

   d. the lightest alkali metal with three neutrons

## 110 Chemistry of Life

5. Give the number of protons, electrons, and neutrons in neutral atoms of each of the following isotopes:

   a. $^{7}_{3}Li$

   b. $^{125}_{52}Te$

   c. $^{109}_{47}Ag$

   d. $^{15}_{7}N$

   e. $^{31}_{15}P$

6. Using the periodic table, predict whether the following compounds are ionic or covalent:

   a. $SO_2$

   b. $CaF_2$

   c. $N_2H_4$

   d. $Al_2(SO_4)_3$

# Check Your Knowledge: Self-Test

1. Classify each of the following as an element, a compound, or a mixture:

   a. iron

   b. oxygen

   c. mercury oxide

   d. pancake syrup

   e. carbon dioxide

   f. a substance composed of molecules each of which contains one hydrogen atom and one chlorine atom

   g. baking soda

   h. baking powder

2. Yeast converts glucose to ethanol and carbon dioxide during anaerobic fermentation as depicted in the simple chemical equation here: glucose → ethanol + carbon dioxide

   a. If 200.0 g of glucose is fully converted, what will be the total mass of ethanol and carbon dioxide produced?

   b. If the fermentation is carried out in an open container, would you expect the mass of the container and contents after fermentation to be less than, greater than, or the same as the mass of the container and contents before fermentation? Explain.

   c. If 97.7 g of carbon dioxide is produced, what mass of ethanol is produced?

3. Classify the six underlined properties in the following paragraph as chemical or physical: Fluorine is a pale yellow gas that reacts with most substances. The free element melts at −220 °C and boils at −188 °C. Finely divided metals burn in fluorine with a bright flame. Nineteen grams of fluorine will react with 1.0 gram of hydrogen.

4. How does an element differ from a compound? How are they similar?

5. Compare and contrast ionic, covalent, and hydrogen bonds.

6. What are the subatomic particles of an atom?

7. How are polar bonds different from nonpolar bonds?

8. What is the difference between groups and periods?

9. Draw the Bohr's model and the Lewis dot formula for the following atoms, elements, and compounds:

   a. Sodium

   b. Chlorine

   c. Sodium Chloride

   d. Oxygen

   e. $O_2$

   f. $N_2$

   g. $H_2O$

10. Using the periodic table, classify each of the following elements as a metal or a nonmetal, and then further classify each as a main-group (representative) element, transition metal, or inner transition metal:

    a. uranium

    b. bromine

    c. strontium

    d. neon

    e. gold

    f. americium

    g. rhodium

    h. sulfur

    i. carbon

    j. potassium

11. Using the periodic table, identify the lightest member of each of the following groups:

    a. noble gases

    b. alkaline earth metals

    c. alkali metals

    d. chalcogens

    1. Using the periodic table, identify the heaviest member of each of the following groups:

       a. alkali metals

       b. chalcogens

       c. noble gases

       d. alkaline earth metals

12. Use the periodic table to give the name and symbol for each of the following elements:

    a. the noble gas in the same period as germanium

    b. the alkaline earth metal in the same period as selenium

    c. the halogen in the same period as lithium

    d. the chalcogen in the same period as cadmium

13. Give the number of protons, electrons, and neutrons in neutral atoms of each of the following isotopes:

    a. $^{10}_{5}B$

    b. $^{199}_{80}Hg$

    c. $^{63}_{29}Cu$

    d. $^{13}_{6}C$

    e. $^{77}_{34}Se$

14. Predict whether the following compounds are ionic or molecular:

    1. KI, the compound used as a source of iodine in table salt
    2. $H_2O_2$, the bleach and disinfectant hydrogen peroxide
    3. $CHCl_3$, the anesthetic chloroform
    4. $Li_2CO_3$, a source of lithium in antidepressants

# Chapter 4: Stoichiometry and The Mole

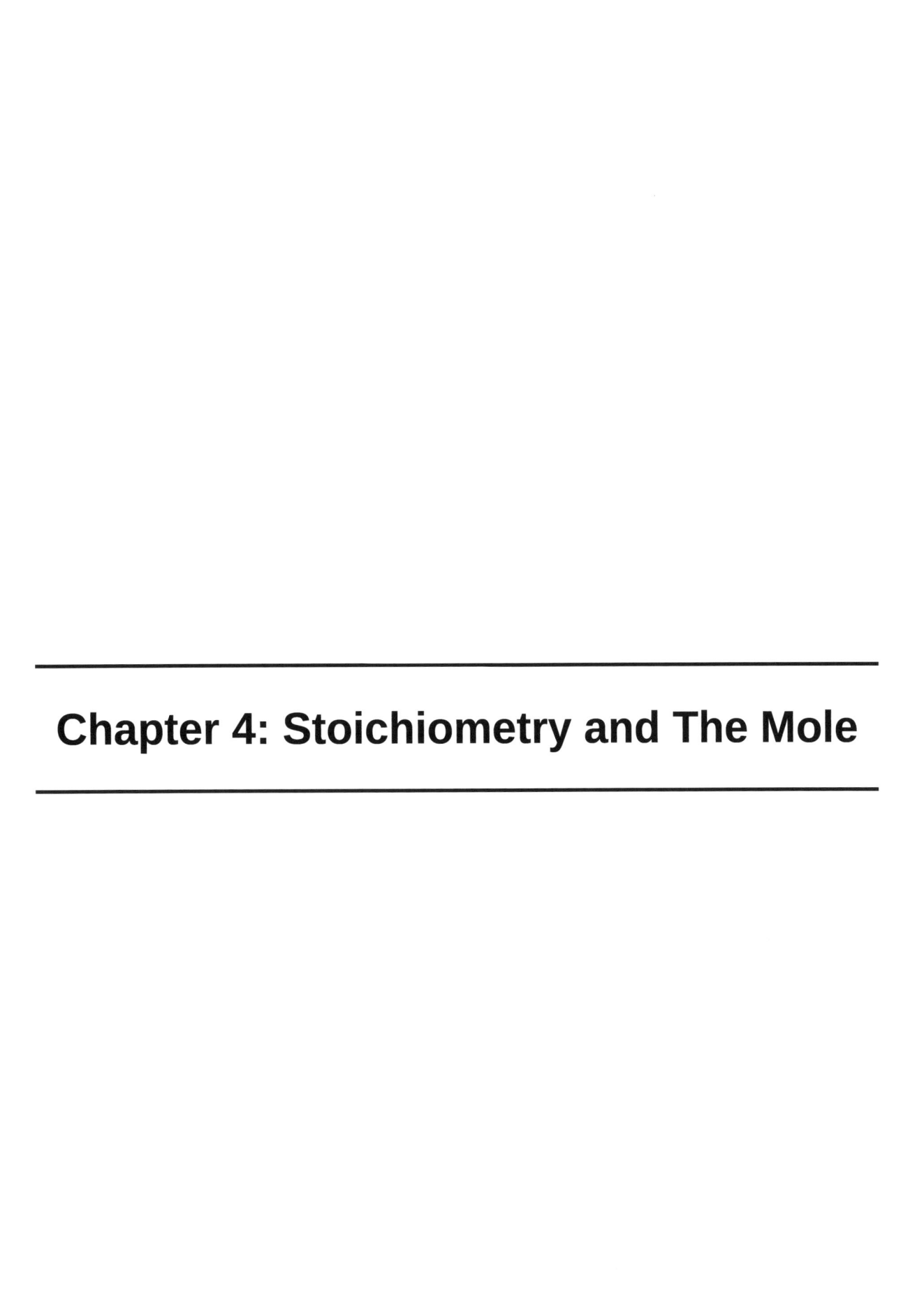

# Chemical Equations

The preceding chapter introduced the use of element symbols to represent individual atoms. When atoms gain or lose electrons to yield ions, or combine with other atoms to form molecules, their symbols are modified or combined to generate chemical formulas that appropriately represent these species. Extending this symbolism to represent both the identities and the relative quantities of substances undergoing a chemical (or physical) change involves writing and balancing a **chemical equation**. Consider as an example the reaction between one methane molecule ($CH_4$) and two diatomic oxygen molecules ($O_2$) to produce one carbon dioxide molecule ($CO_2$) and two water molecules ($H_2O$). The chemical equation representing this process is provided in the upper half of Figure, with space-filling molecular models shown in the lower half of the figure.

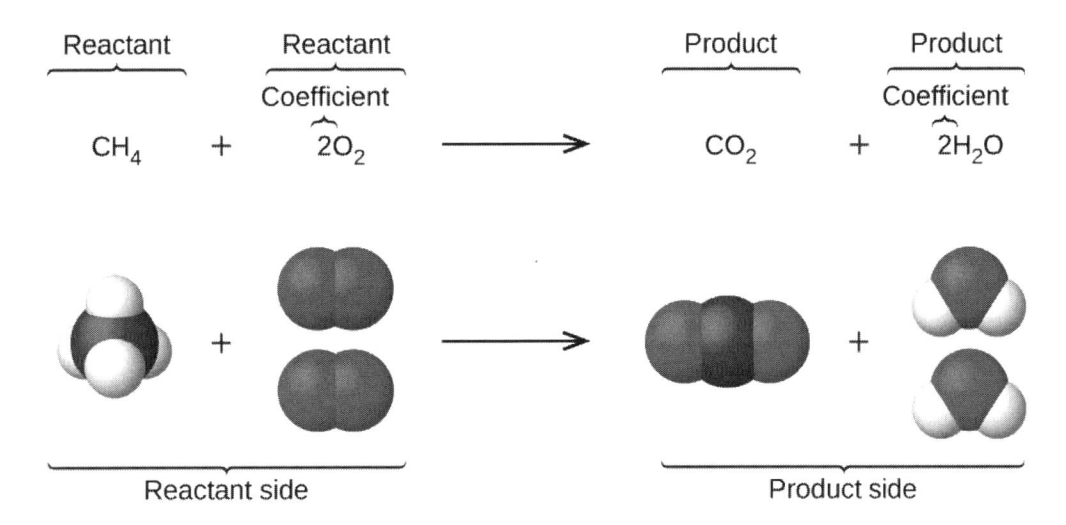

*Figure 1. The reaction between methane and oxygen to yield carbon dioxide in water (shown at bottom) may be represented by a chemical equation using formulas (top).*

This example illustrates the fundamental aspects of any chemical equation:

1. The substances undergoing reaction are called **reactants**, and their formulas are placed on the left side of the equation.

2. The substances generated by the reaction are called **products**, and their formulas are placed on the right side of the equation.

3. Plus signs (+) separate individual reactant and product formulas, and an arrow (→) separates the reactant and product (left and right) sides of the equation.

4. The relative numbers of reactant and product species are represented by **coefficients** (numbers placed immediately to the left of each formula). A coefficient of 1 is typically omitted.

It is common practice to use the smallest possible whole-number coefficients in a chemical equation, as is done in this example. Realize, however, that these coefficients represent the *relative* numbers of

118   Stoichiometry and The Mole

reactants and products, and, therefore, they may be correctly interpreted as ratios. Methane and oxygen react to yield carbon dioxide and water in a 1:2:1:2 ratio. This ratio is satisfied if the numbers of these molecules are, respectively, 1-2-1-2, or 2-4-2-4, or 3-6-3-6, and so on. Likewise, these coefficients may be interpreted with regard to any amount (number) unit, and so this equation may be correctly read in many ways, including:

- *One* methane molecule and *two* oxygen molecules react to yield *one* carbon dioxide molecule and *two* water molecules.

- *One dozen* methane molecules and *two dozen* oxygen molecules react to yield *one dozen* carbon dioxide molecules and *two dozen* water molecules.

- *One mole* of methane molecules and *2 moles* of oxygen molecules react to yield *1 mole* of carbon dioxide molecules and *2 moles* of water molecules.

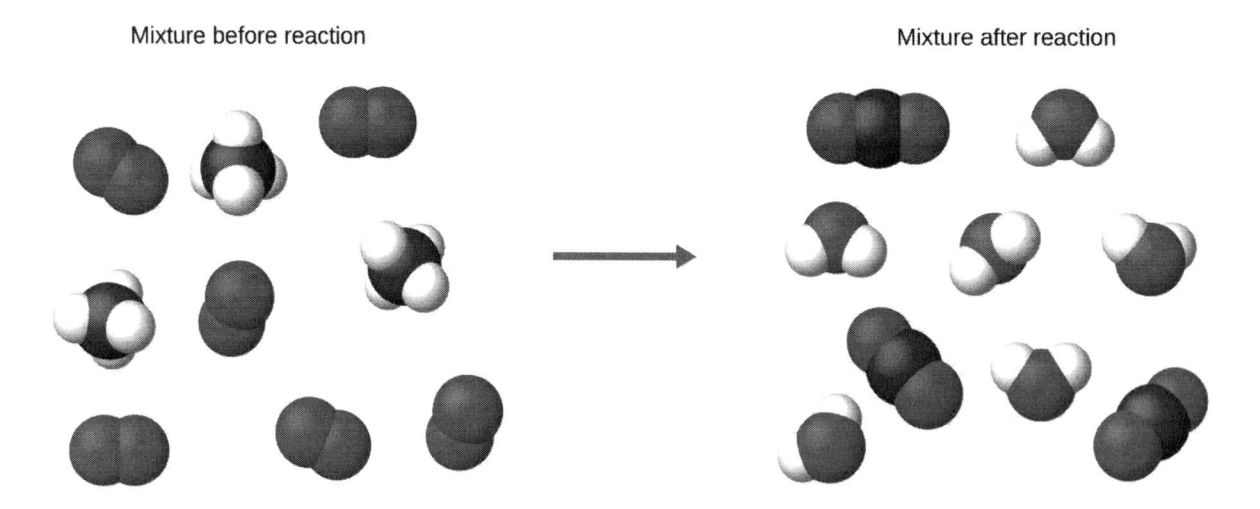

*Figure 2. Regardless of the absolute number of molecules involved, the ratios between numbers of molecules are the same as that given in the chemical equation.*

## BALANCING EQUATIONS

A **balanced chemical is equation** has equal numbers of atoms for each element involved in the reaction are represented on the reactant and product sides. This is a requirement the equation must satisfy to be consistent with the law of conservation of matter. It may be confirmed by simply summing the numbers of atoms on either side of the arrow and comparing these sums to ensure they are equal. Note that the number of atoms for a given element is calculated by multiplying the coefficient of any formula containing that element by the element's subscript in the formula. If an element appears in more than one formula on a given side of the equation, the number of atoms represented in each must be computed and then added together. For example, both product species in the example reaction, $CO_2$ and $H_2O$, contain the element oxygen, and so the number of oxygen atoms on the product side of the equation is

$$\left(1CO_2 \text{ molecule} \times \frac{2 \text{ O atoms}}{CO_2 \text{ molecule}}\right) + \left(2H_2O \text{ molecule} \times \frac{1 \text{ O atom}}{H_2O \text{ molecule}}\right) = 4 \text{ O atoms}$$

The equation for the reaction between methane and oxygen to yield carbon dioxide and water is confirmed to be balanced per this approach, as shown here:

$$CH_4 + 2O_2 \rightarrow CO_2 + 2H_2O$$

| Element | Reactants | Products | Balanced? |
|---------|-----------|----------|-----------|
| C | 1 × 1 = 1 | 1 × 1 = 1 | 1 = 1, yes |
| H | 4 × 1 = 4 | 2 × 2 = 4 | 4 = 4, yes |
| O | 2 × 2 = 4 | (1 × 2) + (2 × 1) = 4 | 4 = 4, yes |

A balanced chemical equation often may be derived from a qualitative description of some chemical reaction by a fairly simple approach known as balancing by inspection. Consider as an example the decomposition of water to yield molecular hydrogen and oxygen. This process is represented qualitatively by an *unbalanced* chemical equation:

$$H_2O \rightarrow H_2 + O_2 \text{ (unbalanced)}$$

Comparing the number of H and O atoms on either side of this equation confirms its imbalance:

| Element | Reactants | Products | Balanced? |
|---------|-----------|----------|-----------|
| H | 1 × 2 = 2 | 1 × 2 = 2 | 2 = 2, yes |
| O | 1 × 1 = 1 | 1 × 2 = 2 | 1 ≠ 2, no |

The numbers of H atoms on the reactant and product sides of the equation are equal, but the numbers of O atoms are not. To achieve balance, the *coefficients* of the equation may be changed as needed. Keep in mind, of course, that the *formula subscripts* define, in part, the identity of the substance, and so these cannot be changed without altering the qualitative meaning of the equation. For example, changing the reactant formula from $H_2O$ to $H_2O_2$ would yield balance in the number of atoms, but doing so also changes the reactant's identity (it's now hydrogen peroxide and not water). The O atom balance may be achieved by changing the coefficient for $H_2O$ to 2.

$$2H_2O \rightarrow H_2 + O_2 \text{ (unbalanced)}$$

| Element | Reactants | Products | Balanced? |
|---------|-----------|----------|-----------|
| H | 2 × 2 = 4 | 1 × 2 = 2 | 4 ≠ 2, no |
| O | 2 × 1 = 2 | 1 × 2 = 2 | 2 = 2, yes |

The H atom balance was upset by this change, but it is easily reestablished by changing the coefficient for the $H_2$ product to 2.

$$2H_2O \rightarrow 2H_2 + O_2 \text{ (balanced)}$$

| Element | Reactants | Products | Balanced? |
|---------|-----------|----------|-----------|
| H | 2 × 2 = 4 | 2 × 2 = 2 | 4 = 4, yes |
| O | 2 × 1 = 2 | 1 × 2 = 2 | 2 = 2, yes |

These coefficients yield equal numbers of both H and O atoms on the reactant and product sides, and the balanced equation is, therefore:

$$2H_2O \rightarrow 2H_2 + O_2$$

### Example 1: Balancing Chemical Equations

Write a balanced equation for the reaction of molecular nitrogen ($N_2$) and oxygen ($O_2$) to form dinitrogen pentoxide.

# Formula Mass and the Mole Concept

> **Learning Objectives**
>
> By the end of this section, you will be able to:
> - Calculate formula masses for covalent and ionic compounds
> - Define the amount unit mole and the related quantity Avogadro's number
> - Explain the relation between mass, moles, and numbers of atoms or molecules, and perform calculations deriving these quantities from one another

We can argue that modern chemical science began when scientists started exploring the quantitative as well as the qualitative aspects of chemistry. For example, Dalton's atomic theory was an attempt to explain the results of measurements that allowed him to calculate the relative masses of elements combined in various compounds. Understanding the relationship between the masses of atoms and the chemical formulas of compounds allows us to quantitatively describe the composition of substances.

## FORMULA MASS

In an earlier chapter, we described the development of the atomic mass unit, the concept of average atomic masses, and the use of chemical formulas to represent the elemental makeup of substances. These ideas can be extended to calculate the **formula mass** of a substance by summing the average atomic masses of all the atoms represented in the substance's formula.

### Formula Mass for Covalent Substances

For covalent substances, the formula represents the numbers and types of atoms composing a single molecule of the substance; therefore, the formula mass may be correctly referred to as a molecular mass. Consider chloroform ($CHCl_3$), a covalent compound once used as a surgical anesthetic and now primarily used in the production of the "anti-stick" polymer, Teflon. The molecular formula of chloroform indicates that a single molecule contains one carbon atom, one hydrogen atom, and three

| Element | Quantity | | Average atomic mass (amu) | | Subtotal (amu) |
|---|---|---|---|---|---|
| C | 1 | × | 12.01 | = | 12.01 |
| H | 1 | × | 1.008 | = | 1.008 |
| Cl | 3 | × | 35.45 | = | 106.35 |
| | | | | Molecular mass | 119.37 |

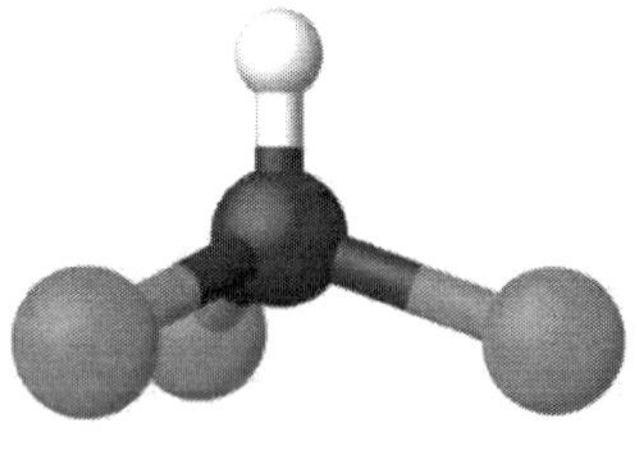

*Figure 1. The average mass of a chloroform molecule, $CHCl_3$, is 119.37 amu, which is the sum of the average atomic masses of each of its constituent atoms. The model shows the molecular structure of chloroform.*

Formula Mass and the Mole Concept   121

chlorine atoms. The average molecular mass of a chloroform molecule is therefore equal to the sum of the average atomic masses of these atoms. Figure 1 outlines the calculations used to derive the molecular mass of chloroform, which is 119.37 amu.

Likewise, the molecular mass of an aspirin molecule, $C_9H_8O_4$, is the sum of the atomic masses of nine carbon atoms, eight hydrogen atoms, and four oxygen atoms, which amounts to 180.15 amu (Figure 2).

| Element | Quantity |   | Average atomic mass (amu) |   | Subtotal (amu) |
|---|---|---|---|---|---|
| C | 9 | × | 12.01 | = | 108.09 |
| H | 8 | × | 1.008 | = | 8.064 |
| O | 4 | × | 16.00 | = | 64.00 |
|   |   |   | Molecular mass |   | 180.15 |

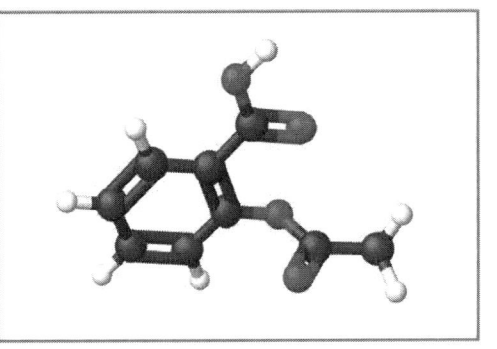

*Figure 2. The average mass of an aspirin molecule is 180.15 amu. The model shows the molecular structure of aspirin, $C_9H_8O_4$.*

### Example 1: Computing Molecular Mass for a Covalent Compound

Ibuprofen, $C_{13}H_{18}O_2$, is a covalent compound and the active ingredient in several popular nonprescription pain medications, such as Advil and Motrin. What is the molecular mass (amu) for this compound?

## Formula Mass for Ionic Compounds

Ionic compounds are composed of discrete cations and anions combined in ratios to yield electrically neutral bulk matter. The formula mass for an ionic compound is calculated in the same way as the formula mass for covalent compounds: by summing the average atomic masses of all the atoms in the compound's formula. Keep in mind, however, that the formula for an ionic compound does not represent the composition of a discrete molecule, so it may not correctly be referred to as the "molecular mass."

As an example, consider sodium chloride, NaCl, the chemical name for common table salt. Sodium chloride is an ionic compound composed of sodium cations, $Na^+$, and chloride anions, $Cl^-$, combined in a 1:1 ratio. The formula mass for this compound is computed as 58.44 amu (see Figure 3).

| Element | Quantity |   | Average atomic mass (amu) |   | Subtotal |
|---|---|---|---|---|---|
| Na | 1 | × | 22.99 | = | 22.99 |
| Cl | 1 | × | 35.45 | = | 35.45 |
|   |   |   | Formula mass |   | 58.44 |

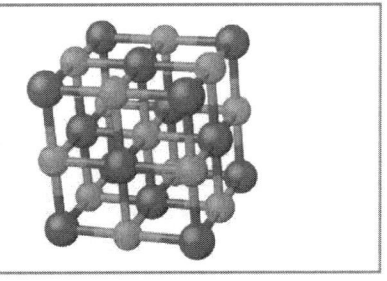

*Figure 3. Table salt, NaCl, contains an array of sodium and chloride ions combined in a 1:1 ratio. Its formula mass is 58.44 amu.*

Note that the average masses of neutral sodium and chlorine atoms were used in this computation, rather than the masses for sodium cations and chlorine anions. This approach is perfectly acceptable when computing the formula mass of an ionic compound. Even though a sodium cation has a slightly smaller mass than a sodium atom (since it is missing an electron), this difference will be offset by the fact that a chloride anion is slightly more massive than a chloride atom (due to the extra electron). Moreover, the mass of an electron is negligibly small with respect to the mass of a typical atom. Even when calculating the mass of an isolated ion, the missing or additional electrons can generally be ignored, since their contribution to the overall mass is negligible, reflected only in the nonsignificant digits that will be lost when the computed mass is properly rounded. The few exceptions to this guideline are very light ions derived from elements with precisely known atomic masses.

### Example 2: Computing Formula Mass for an Ionic Compound

Aluminum sulfate, $Al_2(SO_4)_3$, is an ionic compound that is used in the manufacture of paper and in various water purification processes. What is the formula mass (amu) of this compound?

## THE MOLE

The identity of a substance is defined not only by the types of atoms or ions it contains, but by the quantity of each type of atom or ion. For example, water, $H_2O$, and hydrogen peroxide, $H_2O_2$, are alike in that their respective molecules are composed of hydrogen and oxygen atoms. However, because a hydrogen peroxide molecule contains two oxygen atoms, as opposed to the water molecule, which has only one, the two substances exhibit very different properties. Today, we possess sophisticated instruments that allow the direct measurement of these defining microscopic traits; however, the same traits were originally derived from the measurement of macroscopic properties (the masses and volumes of bulk quantities of matter) using relatively simple tools (balances and volumetric glassware). This experimental approach required the introduction of a new unit for amount of substances, the *mole*, which remains indispensable in modern chemical science.

The mole is an amount unit similar to familiar units like pair, dozen, gross, etc. It provides a specific measure of *the number* of atoms or molecules in a bulk sample of matter. A **mole** is defined as *the amount of substance containing the same number of discrete entities (atoms, molecules, ions, etc.) as the number of atoms in a sample of pure $^{12}C$ weighing exactly 12 g.* One Latin connotation for the word "mole" is "large mass" or "bulk," which is consistent with its use as the name for this unit. The mole provides a link between an easily measured macroscopic property, bulk mass, and an extremely important fundamental property, number of atoms, molecules, and so forth.

The number of entities composing a mole has been experimentally determined to be $6.02214179 \times 10^{23}$, a fundamental constant named **Avogadro's number ($N_A$)** or the Avogadro constant in honor of Italian scientist Amedeo Avogadro. This constant is properly reported with an explicit unit of "per mole," a conveniently rounded version being $6.022 \times 10^{23}/mol$.

Consistent with its definition as an amount unit, 1 mole of any element contains the same number of atoms as 1 mole of any other element. The masses of 1 mole of different elements, however, are different, since the masses of the individual atoms are drastically different. The **molar mass** of an element (or compound) is the mass in grams of 1 mole of that substance, a property expressed in units of grams per mole (g/mol) (see Figure 4).

Formula Mass and the Mole Concept 123

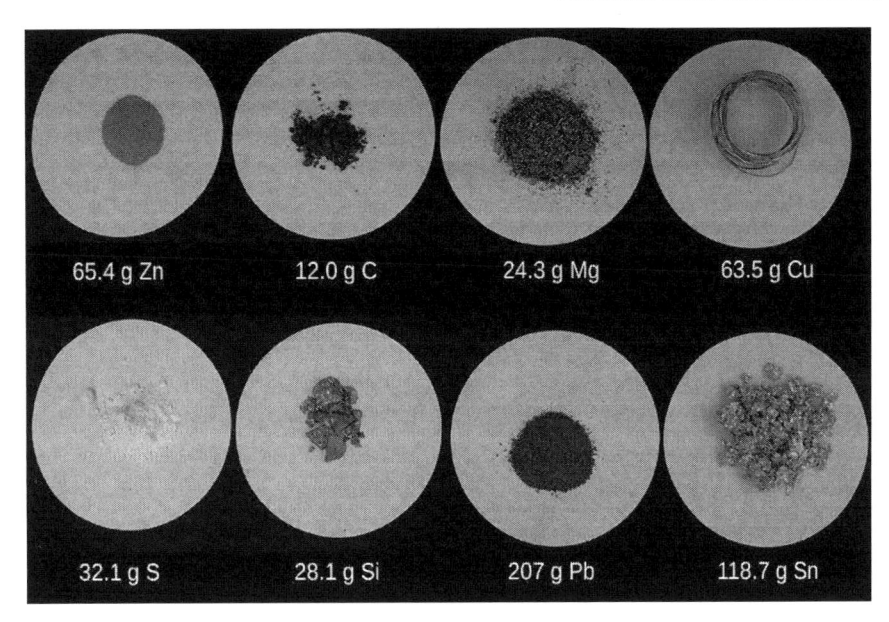

*Figure 4. Each sample contains 6.022 × 10²³ atoms—1.00 mol of atoms. From left to right (top row): 65.4g zinc, 12.0g carbon, 24.3g magnesium, and 63.5g copper. From left to right (bottom row): 32.1g sulfur, 28.1g silicon, 207g lead, and 118.7g tin. (credit: modification of work by Mark Ott)*

Because the definitions of both the mole and the atomic mass unit are based on the same reference substance, $^{12}C$, *the molar mass of any substance is numerically equivalent to its atomic or formula weight in amu.* Per the amu definition, a single $^{12}C$ atom weighs 12 amu (its atomic mass is 12 amu). According to the definition of the mole, 12 g of $^{12}C$ contains 1 mole of $^{12}C$ atoms (its molar mass is 12 g/mol). This relationship holds for all elements, since their atomic masses are measured relative to that of the amu-reference substance, $^{12}C$. Extending this principle, the molar mass of a compound in grams is likewise numerically equivalent to its formula mass in amu (Figure 5).

*Figure 5. Each sample contains 6.02 × 10²³ molecules or formula units—1.00 mol of the compound or element. Clock-wise from the upper left: 130.2g of $C_8H_{17}OH$ (1-octanol, formula mass 130.2 amu), 454.9g of $HgI_2$ (mercury(II) iodide, formula mass 459.9 amu), 32.0g of $CH_3OH$ (methanol, formula mass 32.0 amu) and 256.5g of $S_8$ (sulfur, formula mass 256.6 amu). (credit: Sahar Atwa)*

## 124 Stoichiometry and The Mole

| Element | Average Atomic Mass (amu) | Molar Mass (g/mol) | Atoms/Mole |
|---|---|---|---|
| C | 12.01 | 12.01 | $6.022 \times 10^{23}$ |
| H | 1.008 | 1.008 | $6.022 \times 10^{23}$ |
| O | 16.00 | 16.00 | $6.022 \times 10^{23}$ |
| Na | 22.99 | 22.99 | $6.022 \times 10^{23}$ |
| Cl | 33.45 | 33.45 | $6.022 \times 10^{23}$ |

While atomic mass and molar mass are numerically equivalent, keep in mind that they are vastly different in terms of scale, as represented by the vast difference in the magnitudes of their respective units (amu versus g). To appreciate the enormity of the mole, consider a small drop of water weighing about 0.03 g (see Figure 6). The number of molecules in a single droplet of water is roughly 100 billion times greater than the number of people on earth.

Although this represents just a tiny fraction of 1 mole of water (~18 g), it contains more water molecules than can be clearly imagined. If the molecules were distributed equally among the roughly seven billion people on earth, each person would receive more than 100 billion molecules.

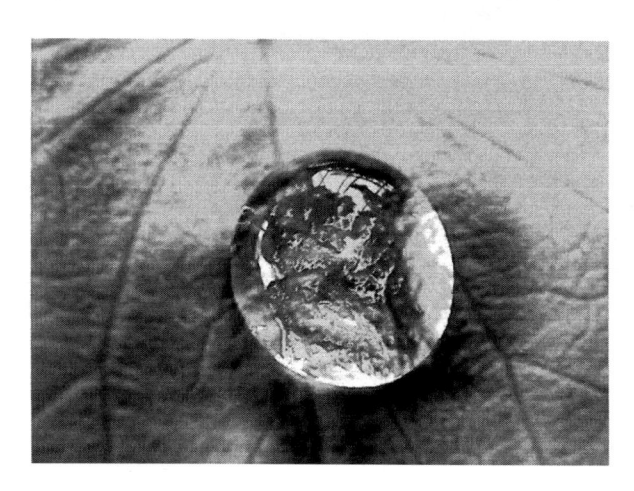

*Figure 6. A single drop of water.*

The relationships between formula mass, the mole, and Avogadro's number can be applied to compute various quantities that describe the composition of substances and compounds. For example, if we

---

The mole is used in chemistry to represent $6.022 \times 10^{23}$ of something, but it can be difficult to conceptualize such a large number. Watch this video to learn more.

https://youtu.be/TEl4jeETVmg

know the mass and chemical composition of a substance, we can determine the number of moles and calculate number of atoms or molecules in the sample. Likewise, if we know the number of moles of a substance, we can derive the number of atoms or molecules and calculate the substance's mass.

---

**Example 3: Deriving Moles From Grams for an Element**

According to nutritional guidelines from the US Department of Agriculture, the estimated average requirement for dietary potassium is 4.7 g. What is the estimated average requirement of potassium in moles?

---

**Example 4: Deriving Grams From Moles for an Element**

A liter of air contains $9.2 \times 10^{-4}$ mol argon. What is the mass of Ar in a liter of air?

---

**Example 5: Deriving Number of Atoms From Mass for an Element**

Copper is commonly used to fabricate electrical wire (Figure 7). How many copper atoms are in 5.00 g of copper wire?

*Figure 7. Copper wire is composed of many, many atoms of Cu. (credit: Emilian Robert Vicol)*

### Example 6: Deriving Moles from Grams for a Compound

Our bodies synthesize protein from amino acids. One of these amino acids is glycine, which has the molecular formula $C_2H_5O_2N$. How many moles of glycine molecules are contained in 28.35 g of glycine?

### Example 7: Deriving Grams from Moles for a Compound

Vitamin C is a covalent compound with the molecular formula $C_6H_8O_6$. The recommended daily dietary allowance of vitamin C for children aged 4–8 years is $1.42 \times 10^{-4}$ mol. What is the mass of this allowance in grams?

### Example 8: Deriving The Number of Atoms and Molecules from The Mass of a Compound

A packet of an artificial sweetener contains 40.0 mg of saccharin ($C_7H_5NO_3S$), which has the structural formula:

Given that saccharin has a molar mass of 183.18 g/mol, how many saccharin molecules are in a 40.0-mg (0.0400-g) sample of saccharin? How many carbon atoms are in the same sample?

## Counting Neurotransmitter Molecules in The Brain

The brain is the control center of the central nervous system (Figure 8). It sends and receives signals to and from muscles and other internal organs to monitor and control their functions; it processes stimuli detected by sensory organs to guide interactions with the external world; and it houses the complex physiological processes that give rise to our intellect and emotions. The broad field of neuroscience spans all aspects of the structure and function of the central nervous system, including research on the anatomy and physiology of the brain. Great progress has been made in brain research over the past few decades, and the BRAIN Initiative, a federal initiative announced in 2013, aims to accelerate and capitalize on these advances through the concerted efforts of various industrial, academic, and government agencies (more details available at the White House's website).

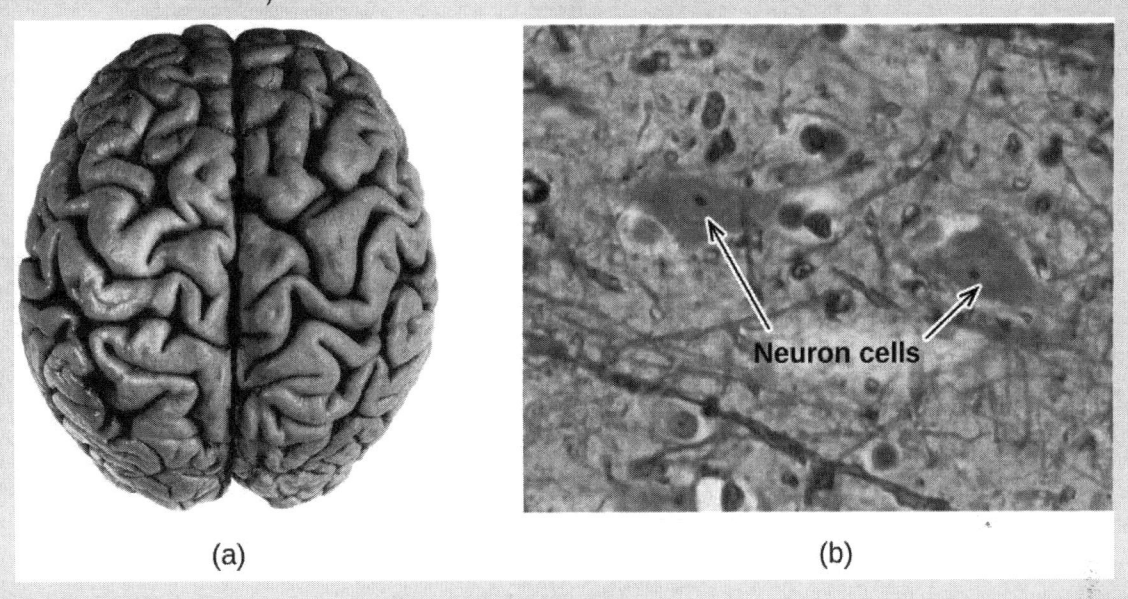

*Figure 8. (a) A typical human brain weighs about 1.5 kg and occupies a volume of roughly 1.1 L. (b) Information is transmitted in brain tissue and throughout the central nervous system by specialized cells called neurons (micrograph shows cells at 1600× magnification).*

Specialized cells called neurons transmit information between different parts of the central nervous system by way of electrical and chemical signals. Chemical signaling occurs at the interface between different neurons when one of the cells releases molecules (called neurotransmitters) that diffuse across the small gap between the cells (called the synapse) and bind to the surface of the other cell. These neurotransmitter molecules are stored in small intracellular structures called vesicles that fuse to the cell wall and then break open to release their contents when the neuron is appropriately stimulated. This process is called exocytosis (see Figure 9). One neurotransmitter that has been very extensively studied is dopamine, $C_8H_{11}NO_2$. Dopamine is involved in various neurological processes that impact a wide variety of human behaviors. Dysfunctions in the dopamine systems of the brain underlie serious neurological diseases such as Parkinson's and schizophrenia.

*(Continued)*

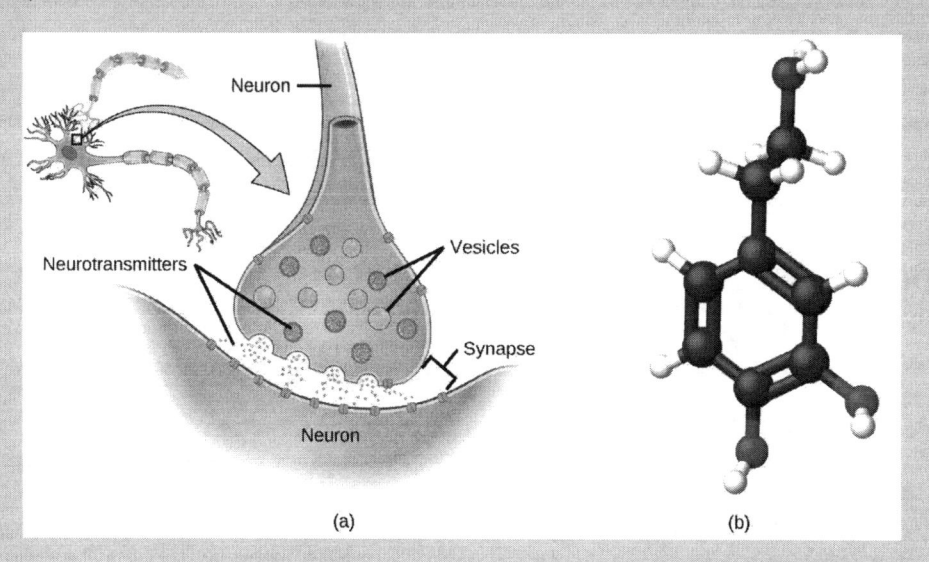

*Figure 9. (a) Chemical signals are transmitted from neurons to other cells by the release of neurotransmitter molecules into the small gaps (synapses) between the cells. (b) Dopamine, $C_8H_{11}NO_2$, is a neurotransmitter involved in a number of neurological processes.*

One important aspect of the complex processes related to dopamine signaling is the number of neurotransmitter molecules released during exocytosis. Since this number is a central factor in determining neurological response (and subsequent human thought and action), it is important to know how this number changes with certain controlled stimulations, such as the administration of drugs. It is also important to understand the mechanism responsible for any changes in the number of neurotransmitter molecules released—for example, some dysfunction in exocytosis, a change in the number of vesicles in the neuron, or a change in the number of neurotransmitter molecules in each vesicle.

Significant progress has been made recently in directly measuring the number of dopamine molecules stored in individual vesicles and the amount actually released when the vesicle undergoes exocytosis. Using miniaturized probes that can selectively detect dopamine molecules in very small amounts, scientists have determined that the vesicles of a certain type of mouse brain neuron contain an average of 30,000 dopamine molecules per vesicle (about $5 \times 10^{-20}$ mol or 50 zmol). Analysis of these neurons from mice subjected to various drug therapies shows significant changes in the average number of dopamine molecules contained in individual vesicles, increasing or decreasing by up to three-fold, depending on the specific drug used. These studies also indicate that not all of the dopamine in a given vesicle is released during exocytosis, suggesting that it may be possible to regulate the fraction released using pharmaceutical therapies.[1]

---

1. Omiatek, Donna M., Amanda J. Bressler, Ann-Sofie Cans, Anne M. Andrews, Michael L. Heien, and Andrew G. Ewing. "The Real Catecholamine Content of Secretory Vesicles in the CNS Revealed by Electrochemical Cytometry." *Scientific Report* 3 (2013): 1447, accessed January 14, 2015, doi:10.1038/srep01447.

## Key Concepts and Summary

The formula mass of a substance is the sum of the average atomic masses of each atom represented in the chemical formula and is expressed in atomic mass units. The formula mass of a covalent compound is also called the molecular mass. A convenient amount unit for expressing very large numbers of atoms or molecules is the mole. Experimental measurements have determined the number of entities composing 1 mole of substance to be $6.022 \times 10^{23}$, a quantity called Avogadro's number. The mass in grams of 1 mole of substance is its molar mass. Due to the use of the same reference substance in defining the atomic mass unit and the mole, the formula mass (amu) and molar mass (g/mol) for any substance are numerically equivalent (for example, one $H_2O$ molecule weighs approximately 18 amu and 1 mole of $H_2O$ molecules weighs approximately 18 g).

## Glossary

**Avogadro's number ($N_A$):** experimentally determined value of the number of entities comprising 1 mole of substance, equal to $6.022 \times 10^{23}$ mol$^{-1}$

**formula mass:** sum of the average masses for all atoms represented in a chemical formula; for covalent compounds, this is also the molecular mass

**molar mass:** mass in grams of 1 mole of a substance

**mole:** amount of substance containing the same number of atoms, molecules, ions, or other entities as the number of atoms in exactly 12 grams of $^{12}C$

# Stoichiometry

## Learning Objectives

By the end of this section, you will be able to:

- Explain the concept of stoichiometry as it pertains to chemical reactions
- Use balanced chemical equations to derive stoichiometric factors relating amounts of reactants and products
- Perform stoichiometric calculations involving mass, moles, and solution molarity

A balanced chemical equation provides a great deal of information in a very succinct format. Chemical formulas provide the identities of the reactants and products involved in the chemical change, allowing classification of the reaction. Coefficients provide the relative numbers of these chemical species, allowing a quantitative assessment of the relationships between the amounts of substances consumed and produced by the reaction. These quantitative relationships are known as the reaction's **stoichiometry**, a term derived from the Greek words *stoicheion* (meaning "element") and *metron* (meaning "measure"). In this module, the use of balanced chemical equations for various stoichiometric applications is explored.

The general approach to using stoichiometric relationships is similar in concept to the way people go about many common activities. Food preparation, for example, offers an appropriate comparison. A recipe for making eight pancakes calls for 1 cup pancake mix, $\frac{3}{4}$ cup milk, and one egg. The "equation" representing the preparation of pancakes per this recipe is

$$1 \text{ cup mix} + \frac{3}{4} \text{ cup milk} + 1 \text{ egg} \rightarrow 8 \text{ pancakes}$$

If two dozen pancakes are needed for a big family breakfast, the ingredient amounts must be increased proportionally according to the amounts given in the recipe. For example, the number of eggs required to make 24 pancakes is

$$24 \text{ pancakes} \times \frac{1 \text{ egg}}{8 \text{ pancakes}} = 3 \text{ eggs}$$

Balanced chemical equations are used in much the same fashion to determine the amount of one reactant required to react with a given amount of another reactant, or to yield a given amount of product, and so forth. The coefficients in the balanced equation are used to derive **stoichiometric factors** that permit computation of the desired quantity. To illustrate this idea, consider the production of ammonia by reaction of hydrogen and nitrogen:

$$N_2(g) + 3H_2(g) \rightarrow 2NH_3(g)$$

This equation shows ammonia molecules are produced from hydrogen molecules in a 2:3 ratio, and stoichiometric factors may be derived using any amount (number) unit:

$$\frac{2\text{NH}_3 \text{ molecules}}{3\text{H}_2 \text{ molecules}} \quad \text{or} \quad \frac{2 \text{ doz NH}_3 \text{ molecules}}{3 \text{ doz H}_2 \text{ molecules}} \quad \text{or} \quad \frac{2 \text{ mol NH}_3 \text{ molecules}}{3 \text{ mol H}_2 \text{ molecules}}$$

These stoichiometric factors can be used to compute the number of ammonia molecules produced from a given number of hydrogen molecules, or the number of hydrogen molecules required to produce a given number of ammonia molecules. Similar factors may be derived for any pair of substances in any chemical equation.

---

### Example 1: Moles of Reactant Required In a Reaction

How many moles of $I_2$ are required to react with 0.429 mol of Al according to the following equation (see Figure 1)?

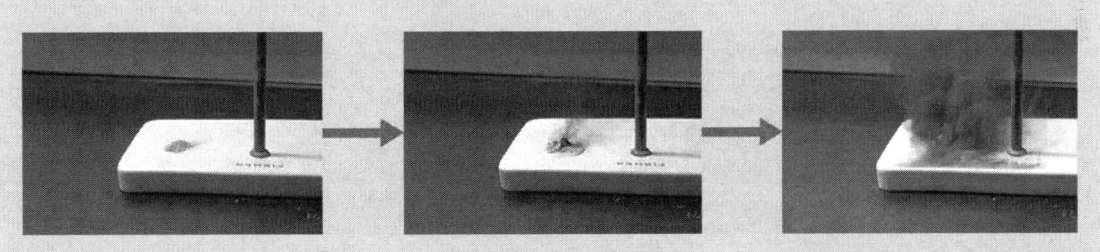

$$2\text{Al} + 3\text{I}_2 \rightarrow 2\text{AlI}_3$$

*Figure 1. Aluminum and iodine react to produce aluminum iodide. The heat of the reaction vaporizes some of the solid iodine as a purple vapor. (credit: modification of work by Mark Ott)*

---

### Example 2: Number of Product Molecules Generated By a Reaction

How many carbon dioxide molecules are produced when 0.75 mol of propane is combusted according to this equation?

$$\text{C}_3\text{H}_8 + 5\text{O}_2 \rightarrow 3\text{CO}_2 + 4\text{H}_2\text{O}$$

---

These examples illustrate the ease with which the amounts of substances involved in a chemical reaction of known stoichiometry may be related. Directly measuring numbers of atoms and molecules is, however, not an easy task, and the practical application of stoichiometry requires that we use the more readily measured property of mass.

132 Stoichiometry and The Mole

---

### Example 3: Relating Masses of Reactants and Products

What mass of sodium hydroxide, NaOH, would be required to produce 16 g of the antacid milk of magnesia [magnesium hydroxide, Mg(OH)²] by the following reaction?

$$MgCl_2(aq) + 2NaOH(aq) \rightarrow Mg(OH)_2(s) + 2NaCl(aq)$$

---

### Example 4: Relating Masses of Reactants

What mass of oxygen gas, $O_2$, from the air is consumed in the combustion of 702 g of octane, $C_8H_{18}$, one of the principal components of gasoline?

$$2C_8H_{18} + 25O_2 \rightarrow 16CO_2 + 18H_2O$$

---

These examples illustrate just a few instances of reaction stoichiometry calculations. Numerous variations on the beginning and ending computational steps are possible depending upon what particular quantities are provided and sought (volumes, solution concentrations, and so forth). Regardless of the details, all these calculations share a common essential component: the use of stoichiometric factors derived from balanced chemical equations. Figure 2 provides a general outline of the various computational steps associated with many reaction stoichiometry calculations.

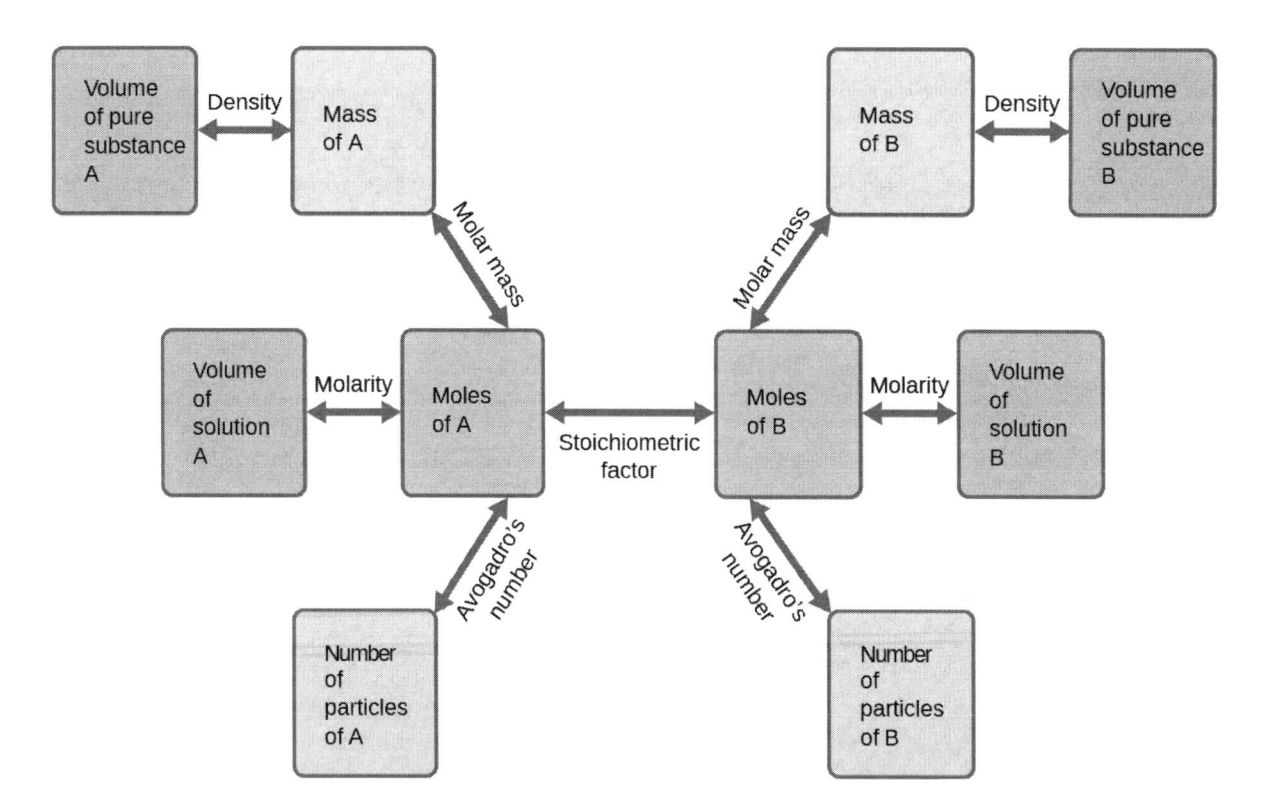

*Figure 2. The flow chart depicts the various computational steps involved in most reaction stoichiometry calculations.*

## Airbags

Airbags (Figure 3) are a safety feature provided in most automobiles since the 1990s. The effective operation of an airbag requires that it be rapidly inflated with an appropriate amount (volume) of gas when the vehicle is involved in a collision. This requirement is satisfied in many automotive airbag systems through use of explosive chemical reactions, one common choice being the decomposition of sodium azide, $NaN_3$. When sensors in the vehicle detect a collision, an electrical current is passed through a carefully measured amount of $NaN_3$ to initiate its decomposition:

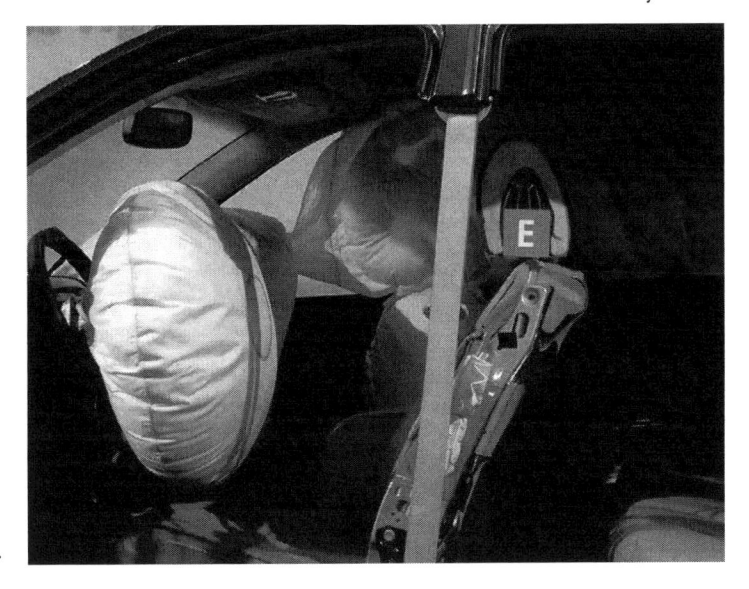

*Figure 3. Airbags deploy upon impact to minimize serious injuries to passengers. (credit: Jon Seidman)*

$$2NaN_3(s) \rightarrow 3N_2(g) + 2Na(s)$$

This reaction is very rapid, generating gaseous nitrogen that can deploy and fully inflate a typical airbag in a fraction of a second (~0.03–0.1 s). Among many engineering considerations, the amount of sodium azide used must be appropriate for generating enough nitrogen gas to fully inflate the air bag and ensure its proper function. For example, a small mass (~100 g) of $NaN_3$ will generate approximately 50 L of $N_2$.

For more information about the chemistry and physics behind airbags and for helpful diagrams on how airbags work, go to How Stuff Works: https://auto.howstuffworks.com/car-driving-safety/safety-regulatory-devices/airbag.htm.

### Key Concepts and Summary

A balanced chemical equation may be used to describe a reaction's stoichiometry (the relationships between amounts of reactants and products). Coefficients from the equation are used to derive stoichiometric factors that subsequently may be used for computations relating reactant and product masses, molar amounts, and other quantitative properties.

### Glossary

**stoichiometric factor:** ratio of coefficients in a balanced chemical equation, used in computations relating amounts of reactants and products

**stoichiometry:** relationships between the amounts of reactants and products of a chemical reaction

# Determining Empirical and Molecular Formulas

> **Learning Objectives**
>
> By the end of this section, you will be able to:
> - Compute the percent composition of a compound
> - Determine the empirical formula of a compound
> - Determine the molecular formula of a compound

In the previous section, we discussed the relationship between the bulk mass of a substance and the number of atoms or molecules it contains (moles). Given the chemical formula of the substance, we were able to determine the amount of the substance (moles) from its mass, and vice versa. But what if the chemical formula of a substance is unknown? In this section, we will explore how to apply these very same principles in order to derive the chemical formulas of unknown substances from experimental mass measurements.

## PERCENT COMPOSITION

The elemental makeup of a compound defines its chemical identity, and chemical formulas are the most succinct way of representing this elemental makeup. When a compound's formula is unknown, measuring the mass of each of its constituent elements is often the first step in the process of determining the formula experimentally. The results of these measurements permit the calculation of the compound's **percent composition**, defined as the percentage by mass of each element in the compound. For example, consider a gaseous compound composed solely of carbon and hydrogen. The percent composition of this compound could be represented as follows:

$$\%H = \frac{\text{mass H}}{\text{mass compound}} \times 100\%$$

$$\%C = \frac{\text{mass C}}{\text{mass compound}} \times 100\%$$

If analysis of a 10.0-g sample of this gas showed it to contain 2.5 g H and 7.5 g C, the percent composition would be calculated to be 25% H and 75% C:

$$\%H = \frac{2.5\text{g H}}{10.0\text{g compound}} \times 100\% = 25\%$$

$$\%C = \frac{7.5\text{g C}}{10.0\text{g compound}} \times 100\% = 75\%$$

Determining Empirical and Molecular Formulas    135

> ### Example 1: Calculation of Percent Composition
>
> Analysis of a 12.04-g sample of a liquid compound composed of carbon, hydrogen, and nitrogen showed it to contain 7.34 g C, 1.85 g H, and 2.85 g N. What is the percent composition of this compound?

### Determining Percent Composition from Formula Mass

Percent composition is also useful for evaluating the relative abundance of a given element in different compounds of known formulas. As one example, consider the common nitrogen-containing fertilizers ammonia ($NH_3$), ammonium nitrate ($NH_4NO_3$), and urea ($CH_4N_2O$). The element nitrogen is the active ingredient for agricultural purposes, so the mass percentage of nitrogen in the compound is a practical and economic concern for consumers choosing among these fertilizers. For these sorts of applications, the percent composition of a compound is easily derived from its formula mass and the atomic masses of its constituent elements. A molecule of $NH_3$ contains one N atom weighing 14.01 amu and three H atoms weighing a total of (3 × 1.008 amu) = 3.024 amu The formula mass of ammonia is therefore (14.01 amu + 3.024 amu) = 17.03 amu, and its percent composition is:

$$\%N = \frac{14.01 \text{amu N}}{17.03 \text{amu NH}_3} \times 100\% = 82.27\%$$

$$\%H = \frac{3.024 \text{amu N}}{17.03 \text{amu NH}_3} \times 100\% = 17.76\%$$

This same approach may be taken considering a pair of molecules, a dozen molecules, or a mole of molecules, etc. The latter amount is most convenient and would simply involve the use of molar masses instead of atomic and formula masses, as demonstrated in the example problem below. As long as we know the chemical formula of the substance in question, we can easily derive percent composition from the formula mass or molar mass.

> ### Example 2: Determining Percent Composition from a Molecular Formula
>
> Aspirin is a compound with the molecular formula $C_9H_8O_4$. What is its percent composition?

## DETERMINATION OF EMPIRICAL FORMULAS

As previously mentioned, the most common approach to determining a compound's chemical formula is to first measure the masses of its constituent elements. However, we must keep in mind that chemical formulas represent the relative *numbers*, not masses, of atoms in the substance. Therefore, any experimentally derived data involving mass must be used to derive the corresponding numbers of atoms in the compound. To accomplish this, we can use molar masses to convert the mass of each element to a number of moles. We then consider the moles of each element relative to each other, converting these numbers into a whole-number ratio that can be used to derive the empirical formula of the substance. Consider a sample of compound determined to contain 1.71 g C and 0.287 g H. The corresponding numbers of atoms (in moles) are:

$$1.17\text{g C} \times \frac{1\text{mol C}}{12.01\text{g C}} = 0.142\text{mol C}$$

$$0.287\text{g H} \times \frac{1\text{mol H}}{1.008\text{g H}} = 0.284\text{mol H}$$

Thus, we can accurately represent this compound with the formula $C_{0.142}H_{0.248}$. Of course, per accepted convention, formulas contain whole-number subscripts, which can be achieved by dividing each subscript by the smaller subscript:

$$C_{\frac{0.142}{0.142}}H_{\frac{0.248}{0.142}} \text{ or } CH_2$$

(Recall that subscripts of "1" are not written but rather assumed if no other number is present.)

The empirical formula for this compound is thus $CH_2$. This may or not be the compound's *molecular formula* as well; however, we would need additional information to make that determination (as discussed later in this section).

Consider as another example a sample of compound determined to contain 5.31 g Cl and 8.40 g O. Following the same approach yields a tentative empirical formula of:

$$Cl_{0.150}O_{0.525} = Cl_{\frac{0.150}{0.150}}O_{\frac{0.525}{0.150}} = ClO_{3.5}$$

In this case, dividing by the smallest subscript still leaves us with a decimal subscript in the empirical formula. To convert this into a whole number, we must multiply each of the subscripts by two, retaining the same atom ratio and yielding $Cl_2O_7$ as the final empirical formula.

In summary, empirical formulas are derived from experimentally measured element masses by:

1. Deriving the number of moles of each element from its mass

2. Dividing each element's molar amount by the smallest molar amount to yield subscripts for a tentative empirical formula

3. Multiplying all coefficients by an integer, if necessary, to ensure that the smallest whole-number ratio of subscripts is obtained

Figure 1 outlines this procedure in flow chart fashion for a substance containing elements A and X.

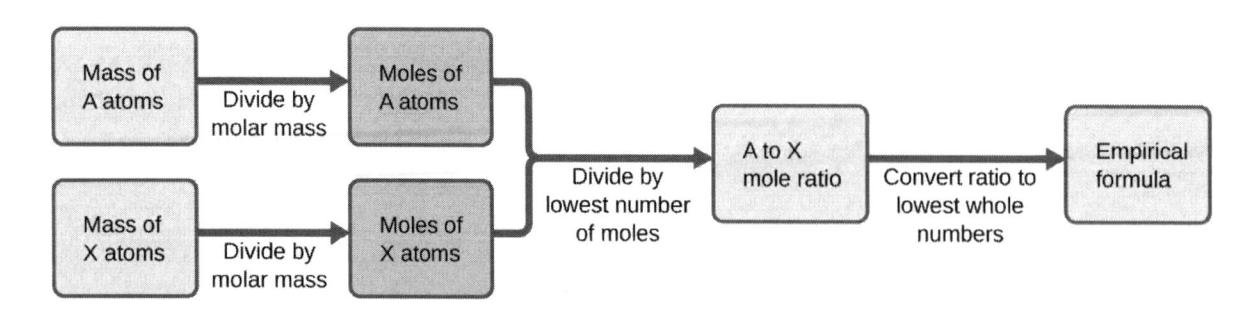

*Figure 1. The empirical formula of a compound can be derived from the masses of all elements in the sample.*

### Example 3: Determining A Compound's Empirical Formula from The Masses of Its Elements

A sample of the black mineral hematite (Figure 2), an oxide of iron found in many iron ores, contains 34.97 g of iron and 15.03 g of oxygen. What is the empirical formula of hematite?

Figure 2. Hematite is an iron oxide that is used in jewelry. (credit: Mauro Cateb)

For additional worked examples illustrating the derivation of empirical formulas, watch the brief video clip below.

https://youtu.be/mdNYDMoQ6As

### Deriving Empirical Formulas from Percent Composition

Finally, with regard to deriving empirical formulas, consider instances in which a compound's percent composition is available rather than the absolute masses of the compound's constituent elements. In such cases, the percent composition can be used to calculate the masses of elements present in any convenient mass of compound; these masses can then be used to derive the empirical formula in the usual fashion.

### Example 4: Determining An Empirical Formula from Percent Composition

The bacterial fermentation of grain to produce ethanol forms a gas with a percent composition of 27.29% C and 72.71% O (Figure 3). What is the empirical formula for this gas?

*Figure 3. An oxide of carbon is removed from these fermentation tanks through the large copper pipes at the top. (credit: "Dual Freq"/Wikipedia)*

## DERIVATION OF MOLECULAR FORMULAS

Recall that empirical formulas are symbols representing the relative numbers of a compound's elements. Determining the absolute numbers of atoms that compose a single molecule of a covalent compound requires knowledge of both its empirical formula and its molecular mass or molar mass. These quantities may be determined experimentally by various measurement techniques. Molecular mass, for example, is often derived from the mass spectrum of the compound (see discussion of this technique in the previous chapter on atoms and molecules). Molar mass can be measured by a number of experimental methods, many of which will be introduced in later chapters of this text.

Molecular formulas are derived by comparing the compound's molecular or molar mass to its **empirical formula mass**. As the name suggests, an empirical formula mass is the sum of the average atomic masses of all the atoms represented in an empirical formula. If we know the molecular (or molar) mass of the substance, we can divide this by the empirical formula mass in order to identify the number of empirical formula units per molecule, which we designate as $n$:

$$\frac{\text{molecular or molar mass}\left(\text{amu or }\frac{g}{mol}\right)}{\text{empirical formula mass}\left(\text{amu or }\frac{g}{mol}\right)} = n \text{ formula units/molecule}$$

The molecular formula is then obtained by multiplying each subscript in the empirical formula by $n$, as shown below for the generic empirical formula $A_xB_y$:

$$(A_xB_y)_n = A_{nx}B_{nx}$$

For example, consider a covalent compound whose empirical formula is determined to be $CH_2O$. The empirical formula mass for this compound is approximately 30 amu (the sum of 12 amu for one C atom, 2 amu for two H atoms, and 16 amu for one O atom). If the compound's molecular mass is determined to be 180 amu, this indicates that molecules of this compound contain six times the number of atoms represented in the empirical formula:

$$\frac{180 \text{ amu/molecule}}{30 \frac{\text{amu}}{\text{formula unit}}} = 6 \text{ formula units/molecule}$$

Molecules of this compound are then represented by molecular formulas whose subscripts are six times greater than those in the empirical formula:

$$(CH_2O)_6 = C_6H_{12}O_6$$

Note that this same approach may be used when the molar mass (g/mol) instead of the molecular mass (amu) is used. In this case, we are merely considering one mole of empirical formula units and molecules, as opposed to single units and molecules.

### Example 5: Determination of The Molecular Formula for Nicotine

Nicotine, an alkaloid in the nightshade family of plants that is mainly responsible for the addictive nature of cigarettes, contains 74.02% C, 8.710% H, and 17.27% N. If 40.57 g of nicotine contains 0.2500 mol nicotine, what is the molecular formula?

### Key Concepts and Summary

The chemical identity of a substance is defined by the types and relative numbers of atoms composing its fundamental entities (molecules in the case of covalent compounds, ions in the case of ionic compounds). A compound's percent composition provides the mass percentage of each element in the compound, and it is often experimentally determined and used to derive the compound's empirical formula. The empirical formula mass of a covalent compound may be compared to the compound's molecular or molar mass to derive a molecular formula.

**Key Equations**

- $\% X = \frac{\text{mass X}}{\text{mass compound}} \times 100\%$

- $\frac{\text{molecular or molar mass} \left( \text{amu or } \frac{g}{\text{mol}} \right)}{\text{empirical formula mass} \left( \text{amu or } \frac{g}{\text{mol}} \right)} = n \text{ formula units/molecule}$

- $(A_xB_y)_n = A_{nx}B_{ny}$

## Glossary

**empirical formula mass:** sum of average atomic masses for all atoms represented in an empirical formula

**percent composition:** percentage by mass of the various elements in a compound

# Mole-Mass and Mole-Mole Conversions

## Learning Objectives

By the end of this section, you will be able to:

- From a given number of moles of a substance, calculate the mass of another substance involved using the balanced chemical equation.

- From a given mass of a substance, calculate the moles of another substance involved using the balanced chemical equation.

- From a given mass of a substance, calculate the mass of another substance involved using the balanced chemical equation.

Mole-mole calculations are not the only type of calculations that can be performed using balanced chemical equations. Recall that the molar mass can be determined from a chemical formula and used as a conversion factor. We can add that conversion factor as another step in a calculation to make a mole-mass calculation, where we start with a given number of moles of a substance and calculate the mass of another substance involved in the chemical equation, or vice versa.

For example, suppose we have the balanced chemical equation

$$2Al + 3Cl_2 \rightarrow 2AlCl_3$$

Suppose we know we have 123.2 g of $Cl_2$. How can we determine how many moles of $AlCl_3$ we will get when the reaction is complete? First and foremost, *chemical equations are not balanced in terms of grams; they are balanced in terms of moles.* So to use the balanced chemical equation to relate an amount of $Cl_2$ to an amount of $AlCl_3$, we need to convert the given amount of $Cl_2$ into moles. We know how to do this by simply using the molar mass of $Cl_2$ as a conversion factor. The molar mass of $Cl_2$ (which we get from the atomic mass of Cl from the periodic table) is 70.90 g/mol. We must invert this fraction so that the units cancel properly:

$$123.2 \; g\,Cl_2 \times \frac{1 \; mol \; Cl_2}{70.90 \; g\,Cl_2} = 1.738 \; mol \; Cl_2$$

Now that we have the quantity in moles, we can use the balanced chemical equation to construct a conversion factor that relates the number of moles of Cl2 to the number of moles of AlCl3. The numbers in the conversion factor come from the coefficients in the balanced chemical equation:

$$\frac{2 \; mol \; AlCl_3}{3 \; mol \, Cl_2}$$

Using this conversion factor with the molar quantity we calculated above, we get

142  Stoichiometry and The Mole

$$1.738 \; \cancel{mol \, Cl_2} \times \frac{2 \; mol \; AlCl_3}{3 \; \cancel{mol \, Cl_2}} = 1.159 \; mol \; AlCl_3$$

So, we will get 1.159 mol of $AlCl_3$ if we react 123.2 g of $Cl_2$.

In this last example, we did the calculation in two steps. However, it is mathematically equivalent to perform the two calculations sequentially on one line:

$$123.2 \; \cancel{g \, Cl_2} \times \frac{1 \; mol \; Cl_2}{70.90 \; \cancel{g \, Cl_2}} \times \frac{2 \; mol \; AlCl_3}{3 \; \cancel{mol \, Cl_2}} = 1.159 \; mol \; AlCl_3$$

The units still cancel appropriately, and we get the same numerical answer in the end. Sometimes the answer may be slightly different from doing it one step at a time because of rounding of the intermediate answers, but the final answers should be effectively the same.

### Example 1

How many moles of HCl will be produced when 249 g of $AlCl_3$ are reacted according to this chemical equation?

$$2AlCl_3 + 3H_2O(\ell) \rightarrow Al_2O_3 + 6HCl(g)$$

**Solution**

We will do this in two steps: convert the mass of $AlCl_3$ to moles and then use the balanced chemical equation to find the number of moles of HCl formed. The molar mass of $AlCl_3$ is 133.33 g/mol, which we have to invert to get the appropriate conversion factor:

$$249 \; \cancel{g \, AlCl_3} \times \frac{1 \; mol \; AlCl_3}{133.33 \; \cancel{g \, AlCl_3}} = 1.87 \; mol \; AlCl_3$$

Now we can use this quantity to determine the number of moles of HCl that will form. From the balanced chemical equation, we construct a conversion factor between the number of moles of $AlCl_3$ and the number of moles of HCl:

$$\frac{6 \, mol \, HCl}{2 \, mol \, AlCl_3}$$

Applying this conversion factor to the quantity of $AlCl_3$, we get

$$1.87 \; \cancel{mol \, AlCl_3} \times \frac{6 \; mol \; HCl}{2 \; \cancel{mol \, AlCl_3}} = 5.61 \; mol \; HCl$$

Alternatively, we could have done this in one line:

$$249 \; \cancel{g \, AlCl_3} \times \frac{1 \; mol \; AlCl_3}{133.33 \; \cancel{g \, AlCl_3}} \times \frac{6 \; mol \; HCl}{2 \; \cancel{mol \, AlCl_3}} = 5.60 \; mol \; HCl$$

The last digit in our final answer is slightly different because of rounding differences, but the answer is essentially the same.

A variation of the mole-mass calculation is to start with an amount in moles and then determine an amount of another substance in grams. The steps are the same but are performed in reverse order.

---

### Example 2

How many grams of $NH_3$ will be produced when 33.9 mol of $H_2$ are reacted according to this chemical equation?

$$N_2(g) + 3H_2(g) \rightarrow 2NH_3(g)$$

**Solution**

The conversions are the same, but they are applied in a different order. Start by using the balanced chemical equation to convert to moles of another substance and then use its molar mass to determine the mass of the final substance. In two steps, we have

$$33.9 \text{ mol } H_2 \times \frac{2 \text{ mol } NH_3}{3 \text{ mol } H_2} = 22.6 \text{ mol } NH_3$$

Now, using the molar mass of $NH_3$, which is 17.03 g/mol, we get

$$22.6 \text{ mol } NH_3 \times \frac{17.03 \text{ g } NH_3}{1 \text{ mol } NH_3} = 385 \text{ g } NH_3$$

---

It should be a trivial task now to extend the calculations to mass-mass calculations, in which we start with a mass of some substance and end with the mass of another substance in the chemical reaction. For this type of calculation, the molar masses of two different substances must be used—be sure to keep track of which is which. Again, however, it is important to emphasize that before the balanced chemical reaction is used, the mass quantity must first be converted to moles. Then the coefficients of the balanced chemical reaction can be used to convert to moles of another substance, which can then be converted to a mass.

For example, let us determine the number of grams of $SO_3$ that can be produced by the reaction of 45.3 g of $SO_2$ and $O_2$:

$$2SO_2(g) + O_2(g) \rightarrow 2SO_3(g)$$

First, we convert the given amount, 45.3 g of $SO_2$, to moles of $SO_2$ using its molar mass (64.06 g/mol):

$$45.3 \text{ g } SO_2 \times \frac{1 \text{ mol } SO_2}{64.06 \text{ g } SO_2} = 0.707 \text{ mol } SO_2$$

Second, we use the balanced chemical reaction to convert from moles of $SO_2$ to moles of $SO_3$:

## 144  Stoichiometry and The Mole

$$0.707 \; mol \, SO_2 \times \frac{2 \; mol \; SO_3}{2 \; mol \; SO_2} = 0.707 \; mol \; SO_3$$

Finally, we use the molar mass of $SO_3$ (80.06 g/mol) to convert to the mass of $SO_3$:

$$0.707 \; mol \, SO_3 \times \frac{80.06 \; g \; SO_3}{1 \; mol \; SO_3} = 56.6 \; g \; SO_3$$

We can also perform all three steps sequentially, writing them on one line as

$$45.3 \; g \, SO_2 \times \frac{1 \; mol \; SO_2}{64.06 \; g \; SO_2} \times \frac{2 \; mol \; SO_3}{2 \; mol \; SO_2} \times \frac{80.06 \; g \, SO_3}{1 \; mol \; SO_3} = 56.6 \; g \; SO_3$$

We get the same answer. Note how the initial and all the intermediate units cancel, leaving grams of $SO_3$, which is what we are looking for, as our final answer.

### Example 3

What mass of Mg will be produced when 86.4 g of K are reacted?

$$MgCl_2(s) + 2K(s) \rightarrow Mg(s) + 2KCl(s)$$

**Solution**

We will simply follow the steps

$massK \rightarrow molK \rightarrow molMg \rightarrow massMg$

In addition to the balanced chemical equation, we need the molar masses of K (39.09 g/mol) and Mg (24.31 g/mol). In one line,

$$86.4 \; g \, K \times \frac{1 \; mol \; K}{39.09 \; g \, K} \times \frac{1 \; mol \; Mg}{2 \; mol \; K} \times \frac{24.31 \; g \; Mg}{1 \; mol \; Mg} = 26.87 \; g \; Mg$$

### Key Takeaways

- Mole quantities of one substance can be related to mass quantities using a balanced chemical equation.
- Mass quantities of one substance can be related to mass quantities using a balanced chemical equation.
- In all cases, quantities of a substance must be converted to moles before the balanced chemical equation can be used to convert to moles of another substance

# Exercises

## PART 1

1. What is the total mass (amu) of carbon in each of the following molecules?

    a. CH$_4$

    b. CHCl$_3$

    c. C$_{12}$H$_{10}$O$_6$

    d. CH$_3$CH$_2$CH$_2$CH$_2$CH$_3$

2. What is the total mass of hydrogen in each of the molecules?

    a. CH$_4$

    b. CHCl$_3$

    c. C$_{12}$H$_{10}$O$_6$

    d. CH$_3$CH$_2$CH$_2$CH$_2$CH$_3$

3. Calculate the molecular or formula mass of each of the following:

    a. P$_4$

    b. H$_2$O

    c. Ca(NO$_3$)$_2$

    d. CH$_3$CO$_2$H (acetic acid)

    e. C$_{12}$H$_{22}$O$_{11}$ (sucrose, cane sugar).

4. Determine the molecular mass of the following compounds:

    a.

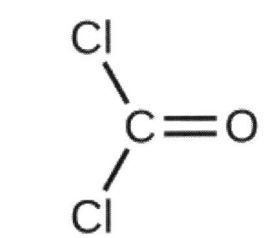

    ![Cl$_2$C=O structure]

    b.

    H—C≡C—H

    c.

    ![(Br)(H)C=C(Br)(H) structure]

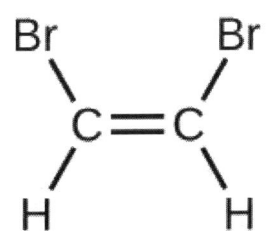

## 146 Stoichiometry and The Mole

d.

$$O=S(=O)(-O-H)(-O-H)$$

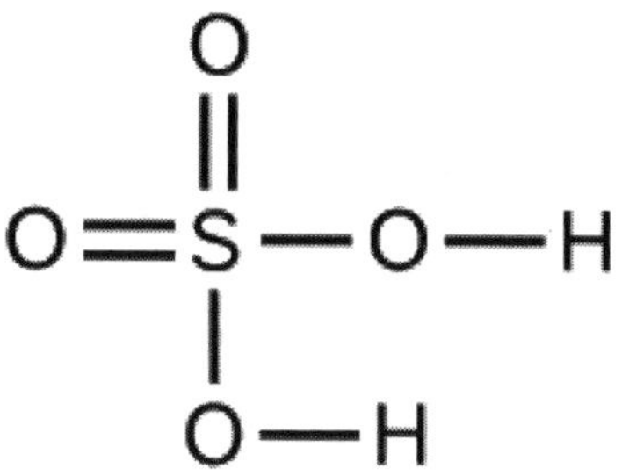

5. Determine the molecular mass of the following compounds:

a.

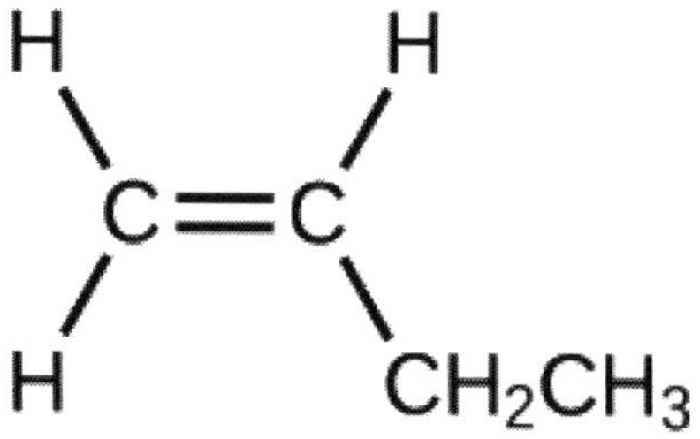

$$\begin{array}{c} H \\ \diagdown \\ C=C \\ \diagup \quad \diagdown \\ H \quad\quad CH_2CH_3 \end{array} \begin{array}{c} H \\ \diagup \end{array}$$

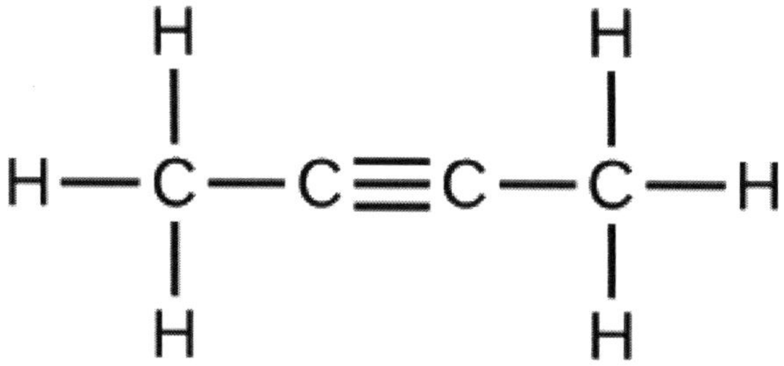

b.

$$H-\underset{\underset{H}{|}}{\overset{\overset{H}{|}}{C}}-C\equiv C-\underset{\underset{H}{|}}{\overset{\overset{H}{|}}{C}}-H$$

c.

$$Cl-\underset{\underset{H}{|}}{\overset{\overset{Cl}{|}}{Si}}-\underset{\underset{H}{|}}{\overset{\overset{Cl}{|}}{Si}}-Cl$$

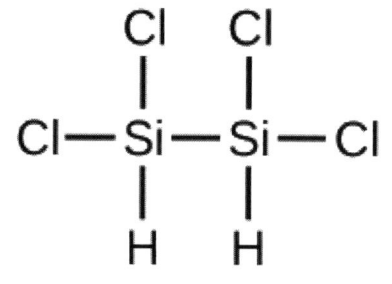

d.

$$O=\underset{\underset{O-H}{|}}{\overset{\overset{O-H}{|}}{P}}-O-H$$

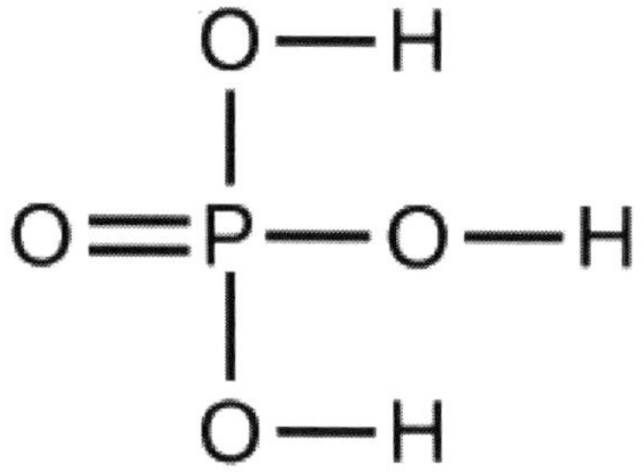

6. Which molecule has a molecular mass of 28.05 amu?

a. $H-C\equiv C-H$

b.

$$\begin{array}{c} H \quad\quad\quad H \\ \diagdown \quad\quad \diagup \\ C=C \\ \diagup \quad\quad \diagdown \\ H \quad\quad\quad H \end{array}$$

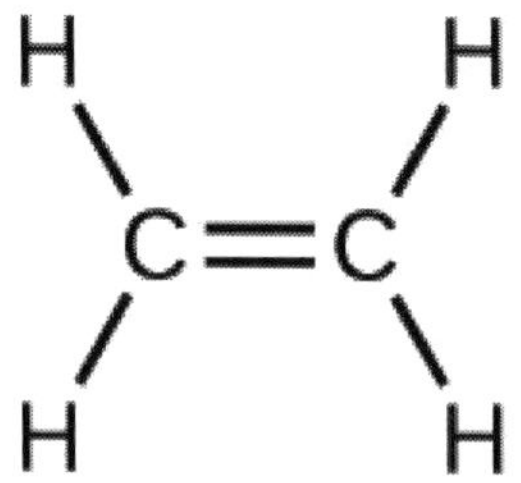

c.

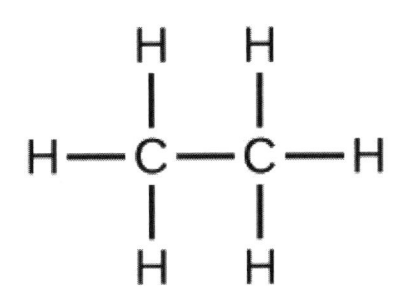

```
      H   H
      |   |
  H — C — C — H
      |   |
      H   H
```

7. Write a sentence that describes how to determine the number of moles of a compound in a known mass of the compound if we know its molecular formula.

8. Compare 1 mole of $H_2$, 1 mole of $O_2$, and 1 mole of $F_2$.

   a. Which has the largest number of molecules? Explain why.

   b. Which has the greatest mass? Explain why.

9. Which contains the greatest mass of oxygen: 0.75 mol of ethanol ($C_2H_5OH$), 0.60 mol of formic acid ($HCO_2H$), or 1.0 mol of water ($H_2O$)? Explain why.

10. Which contains the greatest number of moles of oxygen atoms: 1 mol of ethanol ($C_2H_5OH$), 1 mol of formic acid ($HCO_2H$), or 1 mol of water ($H_2O$)? Explain why.

11. How are the molecular mass and the molar mass of a compound similar and how are they different?

12. Calculate the molar mass of each of the following compounds:

    a. hydrogen fluoride, HF

    b. ammonia, $NH_3$

    c. nitric acid, $HNO_3$

    d. silver sulfate, $Ag_2SO_4$

    e. boric acid, $B(OH)_3$

13. Calculate the molar mass of each of the following:

    a. $S_8$

    b. $C_5H_{12}$

    c. $Sc_2(SO_4)_3$

    d. $CH_3COCH_3$ (acetone)

    e. $C_6H_{12}O_6$ (glucose)

14. Calculate the empirical or molecular formula mass and the molar mass of each of the following minerals:

    a. limestone, $CaCO_3$

    b. halite, NaCl

    c. beryl, $Be_3Al_2Si_6O_{18}$

    d. malachite, $Cu_2(OH)_2CO_3$

    e. turquoise, $CuAl_6(PO_4)_4(OH)_8(H_2O)_4$

### 148  Stoichiometry and The Mole

15. Calculate the molar mass of each of the following:

    a. the anesthetic halothane, $C_2HBrClF_3$

    b. the herbicide paraquat, $C_{12}H_{14}N_2Cl_2$

    c. caffeine, $C_8H_{10}N_4O_2$

    d. urea, $CO(NH_2)_2$

    e. a typical soap, $C_{17}H_{35}CO_2Na$

16. Determine the number of moles of compound and the number of moles of each type of atom in each of the following:

    a. 25.0 g of propylene, $C_3H_6$

    b. $3.06 \times 10^{-3}$ g of the amino acid glycine, $C_2H_5NO_2$

    c. 25 lb of the herbicide Treflan, $C_{13}H_{16}N_2O_4F$ (1 lb = 454 g)

    d. 0.125 kg of the insecticide Paris Green, $Cu_4(AsO_3)_2(CH_3CO_2)_2$

    e. 325 mg of aspirin, $C_6H_4(CO_2H)(CO_2CH_3)$

17. Determine the mass of each of the following:

    a. 0.0146 mol KOH

    b. 10.2 mol ethane, $C_2H_6$

    c. $1.6 \times 10^{-3}$ mol $Na_2SO_4$

    d. $6.854 \times 10^3$ mol glucose, $C_6H_{12}O_6$

    e. 2.86 mol $Co(NH_3)_6Cl_3$

18. Determine the number of moles of the compound and determine the number of moles of each type of atom in each of the following:

    a. 2.12 g of potassium bromide, KBr

    b. 0.1488 g of phosphoric acid, $H_3PO_4$

    c. 23 kg of calcium carbonate, $CaCO_3$

    d. 78.452 g of aluminum sulfate, $Al_2(SO_4)_3$

    e. 0.1250 mg of caffeine, $C_8H_{10}N_4O_2$

19. Determine the mass of each of the following:

    a. 2.345 mol LiCl

    b. 0.0872 mol acetylene, $C_2H_2$

    c. $3.3 \times 10^{-2}$ mol $Na_2CO_3$

    d. $1.23 \times 10^3$ mol fructose, $C_6H_{12}O_6$

    e. 0.5758 mol $FeSO_4(H_2O)_7$

20. The approximate minimum daily dietary requirement of the amino acid leucine, $C_6H_{13}NO_2$, is 1.1 g. What is this requirement in moles?

21. Determine the mass in grams of each of the following:

    a. 0.600 mol of oxygen atoms

    b. 0.600 mol of oxygen molecules, $O_2$

    c. 0.600 mol of ozone molecules, $O_3$

22. A 55-kg woman has $7.5 \times 10^{-3}$ mol of hemoglobin (molar mass = 64,456 g/mol) in her blood. How many hemoglobin molecules is this? What is this quantity in grams?

23. Determine the number of atoms and the mass of zirconium, silicon, and oxygen found in 0.3384 mol of zircon, $ZrSiO_4$, a semiprecious stone.

24. Determine which of the following contains the greatest mass of hydrogen: 1 mol of $CH_4$, 0.6 mol of $C_6H_6$, or 0.4 mol of $C_3H_8$.

25. Determine which of the following contains the greatest mass of aluminum: 122 g of $AlPO_4$, 266 g of $Al_2Cl_6$, or 225 g of $Al_2S_3$.

26. Diamond is one form of elemental carbon. An engagement ring contains a diamond weighing 1.25 carats (1 carat = 200 mg). How many atoms are present in the diamond?

27. The Cullinan diamond was the largest natural diamond ever found (January 25, 1905). It weighed 3104 carats (1 carat = 200 mg). How many carbon atoms were present in the stone

28. One 55-gram serving of a particular cereal supplies 270 mg of sodium, 11% of the recommended daily allowance. How many moles and atoms of sodium are in the recommended daily allowance?

29. A certain nut crunch cereal contains 11.0 grams of sugar (sucrose, $C_{12}H_{22}O_{11}$) per serving size of 60.0 grams. How many servings of this cereal must be eaten to consume 0.0278 moles of sugar?

30. A tube of toothpaste contains 0.76 g of sodium monofluorophosphate ($Na_2PO_3F$) in 100 mL

    a. What mass of fluorine atoms in mg was present?

    b. How many fluorine atoms were present?

31. Which of the following represents the least number of molecules?

    a. 20.0 g of $H_2O$ (18.02 g/mol)

    b. 77.0 g of $CH_4$ (16.06 g/mol)

    c. 68.0 g of $CaH_2$ (42.09 g/mol)

    d. 100.0 g of $N_2O$ (44.02 g/mol)

    e. 84.0 g of HF (20.01 g/mol)

## PART 2

1. Write the balanced equation, then outline the steps necessary to determine the information requested in each of the following:

    a. The number of moles and the mass of chlorine, $Cl_2$, required to react with 10.0 g of sodium metal, Na, to produce sodium chloride, NaCl.

    b. The number of moles and the mass of oxygen formed by the decomposition of 1.252 g of mercury(II) oxide.

c. The number of moles and the mass of sodium nitrate, $NaNO_3$, required to produce 128 g of oxygen. ($NaNO_2$ is the other product.)

d. The number of moles and the mass of carbon dioxide formed by the combustion of 20.0 kg of carbon in an excess of oxygen.

e. The number of moles and the mass of copper(II) carbonate needed to produce 1.500 kg of copper(II) oxide. ($CO_2$ is the other product.)

f. The number of moles and the mass of $BrCH_2CH_2Br$ formed by the reaction of 12.85 g of $CH_2=CH_2$ with an excess of $Br_2$.

2. Determine the number of moles and the mass requested for each reaction in Exercise 1.

3. Write the balanced equation, then outline the steps necessary to determine the information requested in each of the following:

   a. The number of moles and the mass of Mg required to react with 5.00 g of HCl and produce $MgCl_2$ and $H_2$.

   b. The number of moles and the mass of oxygen formed by the decomposition of 1.252 g of silver(I) oxide.

   c. The number of moles and the mass of magnesium carbonate, $MgCO_3$, required to produce 283 g of carbon dioxide. (MgO is the other product.)

   d. The number of moles and the mass of water formed by the combustion of 20.0 kg of acetylene, $C_2H_2$, in an excess of oxygen.

   e. The number of moles and the mass of barium peroxide, $BaO_2$, needed to produce 2.500 kg of barium oxide, BaO ($O_2$ is the other product.)

   f. The number of moles and the mass of $CH_2=CH_2$ required to react with $H_2O$ to produce 9.55 g of $CH_3CH_2OH$.

4. Determine the number of moles and the mass requested for each reaction in Exercise 3.

5. $H_2$ is produced by the reaction of 118.5 mL of a 0.8775-M solution of $H_3PO_4$ according to the following equation:

   $$2Cr + 2H_3PO_4 \rightarrow 3H_2 + 2CrPO_4.$$

   a. Outline the steps necessary to determine the number of moles and mass of $H_2$.

   b. Perform the calculations outlined.

6. Gallium chloride is formed by the reaction of 2.6 L of a 1.44 M solution of HCl according to the following equation:

   $$2Ga + 6HCl \rightarrow 2GaCl_3 + 3H_2.$$

   a. Outline the steps necessary to determine the number of moles and mass of gallium chloride.

   b. Perform the calculations outlined.

7. $I_2$ is produced by the reaction of 0.4235 mol of $CuCl_2$ according to the following equation:

   $$2CuCl_2 + 4KI \rightarrow 2CuI + 4KCl + I_2.$$

a. How many molecules of I₂ are produced?

b. What mass of I₂ is produced?

8. Silver is often extracted from ores as K[Ag(CN)₂] and then recovered by the reaction

   $$2K[Ag(CN)_2](aq) + Zn(s) \rightarrow 2Ag(s) + Zn(CN)_2(aq) + 2KCN(aq)$$

   a. How many molecules of Zn(CN)₂ are produced by the reaction of 35.27 g of K[Ag(CN)₂]?

   b. What mass of Zn(CN)₂ is produced?

9. What mass of silver oxide, Ag₂O, is required to produce 25.0 g of silver sulfadiazine, AgC₁₀H₉N₄SO₂, from the reaction of silver oxide and sulfadiazine?

   $$2C_{10}H_{10}N_4SO_2 + Ag_2O \rightarrow 2AgC_{10}H_9N_4SO_2 + H_2O$$

10. Carborundum is silicon carbide, SiC, a very hard material used as an abrasive on sandpaper and in other applications. It is prepared by the reaction of pure sand, SiO₂, with carbon at high temperature. Carbon monoxide, CO, is the other product of this reaction. Write the balanced equation for the reaction, and calculate how much SiO₂ is required to produce 3.00 kg of SiC.

11. Automotive air bags inflate when a sample of sodium azide, NaN₃, is very rapidly decomposed. $2NaN_3(s) \rightarrow 2Na(s) + 3N_2(g)$ What mass of sodium azide is required to produce 2.6 ft³ (73.6 L) of nitrogen gas with a density of 1.25 g/L?

12. Urea, CO(NH₂)₂, is manufactured on a large scale for use in producing urea-formaldehyde plastics and as a fertilizer. What is the maximum mass of urea that can be manufactured from the CO₂ produced by combustion of 1.00×10³kg1.00×103kg of carbon followed by the reaction? $CO_2(g) + 2NH_3(g) \rightarrow CO(NH_2)_2(s) + H_2O(l)$

13. In an accident, a solution containing 2.5 kg of nitric acid was spilled. Two kilograms of Na₂CO₃ was quickly spread on the area and CO₂ was released by the reaction. Was sufficient Na₂CO₃ used to neutralize all of the acid?

14. A compact car gets 37.5 miles per gallon on the highway. If gasoline contains 84.2% carbon by mass and has a density of 0.8205 g/mL, determine the mass of carbon dioxide produced during a 500-mile trip (3.785 liters per gallon).

15. What volume of a 0.750 M solution of hydrochloric acid, a solution of HCl, can be prepared from the HCl produced by the reaction of 25.0 g of NaCl with an excess of sulfuric acid? $NaCl(s) + H_2SO_4(l) \rightarrow HCl(g) + NaHSO_4(s)$

16. What volume of a 0.2089 M KI solution contains enough KI to react exactly with the Cu(NO₃)₂ in 43.88 mL of a 0.3842 M solution of Cu(NO₃)₂? $2Cu(NO_3)_2 + 4KI \rightarrow 2CuI + I_2 + 4KNO_3$

17. A mordant is a substance that combines with a dye to produce a stable fixed color in a dyed fabric. Calcium acetate is used as a mordant. It is prepared by the reaction of acetic acid with calcium hydroxide. $2CH_3CO_2H + Ca(OH)_2 \rightarrow Ca(CH_3CO_2)_2 + 2H_2O$ What mass of Ca(OH)₂ is required to react with the acetic acid in 25.0 mL of a solution having a density of 1.065 g/mL and containing 58.0% acetic acid by mass?

18. The toxic pigment called white lead, Pb₃(OH)₂(CO₃)₂, has been replaced in white paints by rutile, TiO₂. How much rutile (g) can be prepared from 379 g of an ore that contains 88.3% ilmenite (FeTiO₃) by mass? $2FeTiO_3 + 4HCl + Cl_2 \rightarrow 2FeCl_3 + 2TiO_2 + 2H_2O$

## PART 3

1. Calculate the following to four significant figures:

   a. the percent composition of ammonia, $NH_3$

   b. the percent composition of photographic "hypo," $Na_2S_2O_3$

   c. the percent of calcium ion in $Ca_3(PO_4)_2$

2. Determine the following to four significant figures:

   a. the percent composition of hydrazoic acid, $HN_3$

   b. the percent composition of TNT, $C_6H_2(CH_3)(NO_2)_3$

   c. the percent of $SO_4^{2-}$ in $Al_2(SO_4)_3$

3. Determine the percent ammonia, $NH_3$, in $Co(NH_3)_6Cl_3$, to three significant figures.

4. Determine the percent water in $CuSO_4 \cdot 5H_2O$ to three significant figures.

## PART 4

1. What information do we need to determine the molecular formula of a compound from the empirical formula?

2. Determine the empirical formulas for compounds with the following percent compositions:

   a. 15.8% carbon and 84.2% sulfur

   b. 40.0% carbon, 6.7% hydrogen, and 53.3% oxygen

3. Determine the empirical formulas for compounds with the following percent compositions:

   a. 43.6% phosphorus and 56.4% oxygen

   b. 28.7% K, 1.5% H, 22.8% P, and 47.0% O

4. Polymers are large molecules composed of simple units repeated many times. Thus, they often have relatively simple empirical formulas. Calculate the empirical formulas of the following polymers:

   a. Lucite (Plexiglas); 59.9% C, 8.06% H, 32.0% O

   b. Saran; 24.8% C, 2.0% H, 73.1% Cl

   c. polyethylene; 86% C, 14% H

   d. polystyrene; 92.3% C, 7.7% H

   e. Orlon; 67.9% C, 5.70% H, 26.4% N

5. A compound of carbon and hydrogen contains 92.3% C and has a molar mass of 78.1 g/mol. What is its molecular formula?

6. Dichloroethane, a compound that is often used for dry cleaning, contains carbon, hydrogen, and chlorine. It has a molar mass of 99 g/mol. Analysis of a sample shows that it contains 24.3% carbon and 4.1% hydrogen. What is its molecular formula?

7. Determine the empirical and molecular formula for chrysotile asbestos. Chrysotile has the following percent composition: 28.03% Mg, 21.60% Si, 1.16% H, and 49.21% O. The molar mass for chrysotile is 520.8 g/mol.

8. A major textile dye manufacturer developed a new yellow dye. The dye has a percent composition of 75.95% C, 17.72% N, and 6.33% H by mass with a molar mass of about 240 g/mol. Determine the molecular formula of the dye.

## PART 5

1. What mass of $CO_2$ is produced by the combustion of 1.00 mol of $CH_4$?

    $CH_4(g) + 2O_2(g) \rightarrow CO_2(g) + 2H_2O(l)$

2. What mass of $H_2O$ is produced by the combustion of 1.00 mol of $CH_4$?

    $CH_4(g) + 2O_2(g) \rightarrow CO_2(g) + 2H_2O(l)$

3. What mass of HgO is required to produce 0.692 mol of $O_2$?

    $2HgO(s) \rightarrow 2Hg(l) + O_2(g)$

4. What mass of $NaHCO_3$ is needed to produce 2.659 mol of $CO_2$?

    $2NaHCO_3(s) \rightarrow Na_2CO_3(s) + H_2O(l) + CO_2(g)$

5. How many moles of Al can be produced from 10.87 g of Ag?

    $Al(NO_3)_3(s) + 3Ag \rightarrow Al + 3AgNO_3$

6. How many moles of HCl can be produced from 0.226 g of $SOCl_2$?

    $SOCl_2(l) + H_2O(l) \rightarrow SO_2(g) + 2HCl(g)$

7. How many moles of $O_2$ are needed to prepare 1.00 g of $Ca(NO_3)_2$?

    $Ca(s) + N_2(g) + 3O_2(g) \rightarrow Ca(NO_3)_2(s)$

8. How many moles of $C_2H_5OH$ are needed to generate 106.7 g of $H_2O$?

    $C_2H_5OH(l) + 3O_2(g) \rightarrow 2CO_2(g) + 3H_2O(l)$

9. What mass of $O_2$ can be generated by the decomposition of 100.0 g of $NaClO_3$?

    $2NaClO_3 \rightarrow 2NaCl(s) + 3O_2(g)$

10. What mass of $Li_2O$ is needed to react with 1,060 g of $CO_2$?

    $Li_2O(aq) + CO_2(g) \rightarrow Li_2CO_3(aq)$

11. What mass of $Fe_2O_3$ must be reacted to generate 324 g of $Al_2O_3$?

    $Fe_2O_3(s) + 2Al(s) \rightarrow 2Fe(s) + Al_2O_3(s)$

12. What mass of Fe is generated when 100.0 g of Al are reacted?

    $Fe_2O_3(s) + 2Al(s) \rightarrow 2Fe(s) + Al_2O_3(s)$

13. What mass of $MnO_2$ is produced when 445 g of $H_2O$ are reacted?

    $H_2O(l) + 2MnO_4^-(aq) + Br^-(aq) \rightarrow BrO_3^-(aq) + 2MnO_2(s) + 2OH^-(aq)$

14. What mass of $PbSO_4$ is produced when 29.6 g of $H_2SO_4$ are reacted?

    $Pb(s) + PbO_2(s) + 2H_2SO_4(aq) \rightarrow 2PbSO_4(s) + 2H_2O(l)$

15. If 83.9 g of ZnO are formed, what mass of $Mn_2O_3$ is formed with it?

    $Zn(s) + 2MnO_2(s) \rightarrow ZnO(s) + Mn_2O_3(s)$

## Stoichiometry and The Mole

16. If 14.7 g of $NO_2$ are reacted, what mass of $H_2O$ is reacted with it?

    $3NO_2(g) + H_2O(l) \rightarrow 2HNO_3(aq) + NO(g)$

17. If 88.4 g of $CH_2S$ are reacted, what mass of HF is produced?

    $CH_2S + 6F_2 \rightarrow CF_4 + 2HF + SF_6$

18. If 100.0 g of $Cl_2$ are needed, what mass of NaOCl must be reacted?

    $NaOCl + HCl \rightarrow NaOH + Cl_2$

# Check Your Knowledge: Self-Test

1. Write a balanced equation for the decomposition of ammonium nitrate to form molecular nitrogen, molecular oxygen, and water. (Hint: Balance oxygen last, since it is present in more than one molecule on the right side of the equation.)

2. Acetaminophen, $C_8H_9NO_2$, is a covalent compound and the active ingredient in several popular nonprescription pain medications, such as Tylenol. What is the molecular mass (amu) for this compound?

3. Calcium phosphate, $Ca_3(PO_4)_2$, is an ionic compound and a common anti-caking agent added to food products. What is the formula mass (amu) of calcium phosphate?

4. Beryllium is a light metal used to fabricate transparent X-ray windows for medical imaging instruments. How many moles of Be are in a thin-foil window weighing 3.24 g?

5. What is the mass of 2.561 mol of gold?

6. A prospector panning for gold in a river collects 15.00 g of pure gold. How many Au atoms are in this quantity of gold?

7. How many moles of sucrose, $C_{12}H_{22}O_{11}$, are in a 25-g sample of sucrose?

8. What is the mass of 0.443 mol of hydrazine, $N_2H_4$?

9. How many $C_4H_{10}$ molecules are contained in 9.213 g of this compound? How many hydrogen atoms?

10. How many moles of $Ca(OH)_2$ are required to react with 1.36 mol of $H_3PO_4$ to produce $Ca_3(PO_4)_2$ according to the equation $3Ca(OH)_2 + 2H_3PO_4 \rightarrow Ca_3(PO_4)_2 + 6H_2O$

11. many $NH_3$ molecules are produced by the reaction of 4.0 mol of $Ca(OH)_2$ according to the following equation:
    $(NH_4)_2SO_4 + Ca(OH)_2 \rightarrow 2NH_3 + CaSO_4 + 2H_2O.$

12. What mass of gallium oxide, $Ga_2O_3$, can be prepared from 29.0 g of gallium metal? The equation for the reaction is $4Ga + 3O_2 \rightarrow 2Ga_2O_3$.

13. What mass of CO is required to react with 25.13 g of $Fe_2O_3$ according to the equation $Fe_2O_3 + 3CO \rightarrow 2Fe + 3CO_2$?

14. A 24.81-g sample of a gaseous compound containing only carbon, oxygen, and chlorine is determined to contain 3.01 g C, 4.00 g O, and 17.81 g Cl. What is this compound's percent composition?

15. To three significant digits, what is the mass percentage of iron in the compound $Fe_2O_3$?

16. What is the empirical formula of a compound if a sample contains 0.130 g of nitrogen and 0.370 g of oxygen?

17. What is the empirical formula of a compound containing 40.0% C, 6.71% H, and 53.28% O?

18. What is the molecular formula of a compound with a percent composition of 49.47% C, 5.201% H, 28.84% N, and 16.48% O, and a molecular mass of 194.2 amu?

156  Stoichiometry and The Mole

19. How many moles of $Al_2O_3$ will be produced when 23.9 g of $H_2O$ are reacted according to this chemical equation?

$$2AlCl_3 + 3H_2O(\ell) \rightarrow Al_2O_3 + 6HCl(g)$$

20. How many grams of $N_2$ are needed to produce 2.17 mol of $NH_3$ when reacted according to this chemical equation?

$$N_2(g) + 3H_2(g) \rightarrow 2NH_3(g)$$

30.4 g (Note: here we go from a product to a reactant, showing that mole-mass problems can begin and end with any substance in the chemical equation.)

21. What mass of $H_2$ will be produced when 122 g of Zn are reacted?

$$Zn(s) + 2HCl(aq) \rightarrow ZnCl_2(aq) + H_2(g)$$

# Chapter 5: Water

# Water

### Learning Objectives

By the end of this section, you will be able to:
- Describe the properties of water that are critical to maintaining life
- Explain why water is an excellent solvent
- Provide examples of water's cohesive and adhesive properties

Why do scientists spend time looking for water on other planets? Why is water so important? It is because water is essential to life as we know it. Water is one of the more abundant molecules and the one most critical to life on Earth. Approximately 60–70 percent of the human body is made up of water. Without it, life as we know it simply would not exist.

The polarity of the water molecule and its resulting hydrogen bonding make water a unique substance with special properties that are intimately tied to the processes of life. Life originally evolved in a watery environment, and most of an organism's cellular chemistry and metabolism occur inside the watery contents of the cell's cytoplasm. Special properties of water are its high heat capacity and heat of vaporization, its ability to dissolve polar molecules, its cohesive and adhesive properties, and its dissociation into ions that leads to the generation of pH. Understanding these characteristics of water helps to elucidate its importance in maintaining life.

## WATER'S POLARITY

One of water's important properties is that it is composed of polar molecules: the hydrogen and oxygen within water molecules ($H_2O$) form polar covalent bonds. While there is no net charge to a water molecule, the polarity of water creates a slightly positive charge on hydrogen and a slightly negative charge on oxygen, contributing to water's properties of attraction. Water's charges are generated because oxygen is more electronegative than hydrogen, making it more likely that a shared electron would be found near the oxygen nucleus than the hydrogen nucleus, thus generating the partial negative charge near the oxygen.

As a result of water's polarity, each water molecule attracts other water molecules because of the opposite charges between water molecules, forming hydrogen bonds. Water also attracts or is attracted

*Figure 1: Oil and water do not mix. As this macro image of oil and water shows, oil does not dissolve in water but forms droplets instead. This is due to it being a nonpolar compound. (credit: Gautam Dogra).*

to other polar molecules and ions. A polar substance that interacts readily with or dissolves in water is referred to as hydrophilic (hydro- = "water"; -philic = "loving"). In contrast, non-polar molecules such as oils and fats do not interact well with water, as shown in [Figure 1] and separate from it rather than dissolve in it, as we see in salad dressings containing oil and vinegar (an acidic water solution). These nonpolar compounds are called hydrophobic (hydro- = "water"; -phobic = "fearing").

## WATER'S STATES: GAS, LIQUID, AND SOLID

The formation of hydrogen bonds is an important quality of the liquid water that is crucial to life as we know it. As water molecules make hydrogen bonds with each other, water takes on some unique chemical characteristics compared to other liquids and, since living things have a high water content, understanding these chemical features is key to understanding life. In liquid water, hydrogen bonds are constantly formed and broken as the water molecules slide past each other. The breaking of these bonds is caused by the motion (kinetic energy) of the water molecules due to the heat contained in the system. When the heat is raised as water is boiled, the higher kinetic energy of the water molecules causes the hydrogen bonds to break completely and allows water molecules to escape into the air as gas (steam or water vapor). On the other hand, when the temperature of water is reduced and water freezes, the water molecules form a crystalline structure maintained by hydrogen bonding (there is not enough energy to break the hydrogen bonds) that makes ice less dense than liquid water, a phenomenon not seen in the solidification of other liquids.

Water's lower density in its solid form is due to the way hydrogen bonds are oriented as it freezes: the water molecules are pushed farther apart compared to liquid water. With most other liquids, solidification when the temperature drops includes the lowering of kinetic energy between molecules, allowing them to pack even more tightly than in liquid form and giving the solid a greater density than the liquid.

The lower density of ice, illustrated and pictured in [Figure 2], an anomaly, causes it to float at the surface of liquid water, such as in an iceberg or in the ice cubes in a glass of ice water. In lakes and ponds, ice will form on the surface of the water creating an insulating barrier that protects the animals and plant life in the pond from freezing. Without this layer of insulating ice, plants and animals living

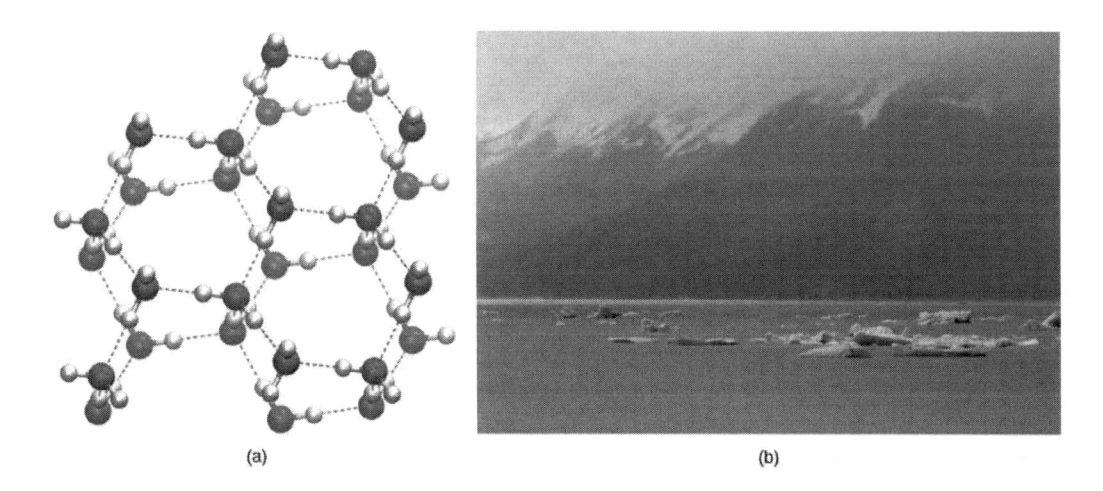

*Figure 2: Hydrogen bonding makes ice less dense than liquid water. The (a) lattice structure of ice makes it less dense than the freely flowing molecules of liquid water, enabling it to (b) float on water. (credit a: modification of work by Jane Whitney, image created using Visual Molecular Dynamics (VMD) software1; credit b: modification of work by Carlos Ponte) Figure 2: Hydrogen bonding makes ice less dense than liquid water. The (a) lattice structure of ice makes it less dense than the freely flowing molecules of liquid water, enabling it to (b) float on water. (credit a: modification of work by Jane Whitney, image created using Visual Molecular Dynamics (VMD) software1; credit b: modification of work by Carlos Ponte)*

in the pond would freeze in the solid block of ice and could not survive. The detrimental effect of freezing on living organisms is caused by the expansion of ice relative to liquid water. The ice crystals that form upon freezing rupture the delicate membranes essential for the function of living cells, irreversibly damaging them. Cells can only survive freezing if the water in them is temporarily replaced by another liquid like glycerol.

> Click here to see a 3-D animation of the structure of an ice lattice: http://janewhitney.com/ice_movie_resources.

## WATER'S HIGH HEAT CAPACITY

Water's high heat capacity is a property caused by hydrogen bonding among water molecules. Water has the highest specific heat capacity of any liquids. Specific heat is defined as the amount of heat one gram of a substance must absorb or lose to change its temperature by one degree Celsius. For water, this amount is one calorie. It therefore takes water a long time to heat and long time to cool. In fact, the specific heat capacity of water is about five times more than that of sand. This explains why the land cools faster than the sea. Due to its high heat capacity, water is used by warm blooded animals to more evenly disperse heat in their bodies: it acts in a similar manner to a car's cooling system, transporting heat from warm places to cool places, causing the body to maintain a more even temperature.

## WATER'S HEAT OF VAPORIZATION

Water also has a high heat of vaporization, the amount of energy required to change one gram of a liquid substance to a gas. A considerable amount of heat energy (586 cal) is required to accomplish this change in water. This process occurs on the surface of water. As liquid water heats up, hydrogen bonding makes it difficult to separate the liquid water molecules from each other, which is required for it to enter its gaseous phase (steam). As a result, water acts as a heat sink or heat reservoir and requires much more heat to boil than does a liquid such as ethanol (grain alcohol), whose hydrogen bonding with other ethanol molecules is weaker than water's hydrogen bonding. Eventually, as water reaches its boiling point of 100° Celsius (212° Fahrenheit), the heat is able to break the hydrogen bonds between the water molecules, and the kinetic energy (motion) between the water molecules allows them to escape from the liquid as a gas. Even when below its boiling point, water's individual molecules acquire enough energy from other water molecules such that some surface water molecules can escape and vaporize: this process is known as evaporation.

The fact that hydrogen bonds need to be broken for water to evaporate means that a substantial amount of energy is used in the process. As the water evaporates, energy is taken up by the process, cooling the environment where the evaporation is taking place. In many living organisms, including in humans, the evaporation of sweat, which is 90 percent water, allows the organism to cool so that homeostasis of body temperature can be maintained.

## WATER'S SOLVENT PROPERTIES

Since water is a polar molecule with slightly positive and slightly negative charges, ions and polar molecules can readily dissolve in it. Therefore, water is referred to as a solvent, a substance capable of dissolving other polar molecules and ionic compounds. The charges associated with these molecules will form hydrogen bonds with water, surrounding the particle with water molecules. This is referred to as a sphere of hydration, or a hydration shell, as illustrated in [Figure 3] and serves to keep the particles separated or dispersed in the water.

162  Water

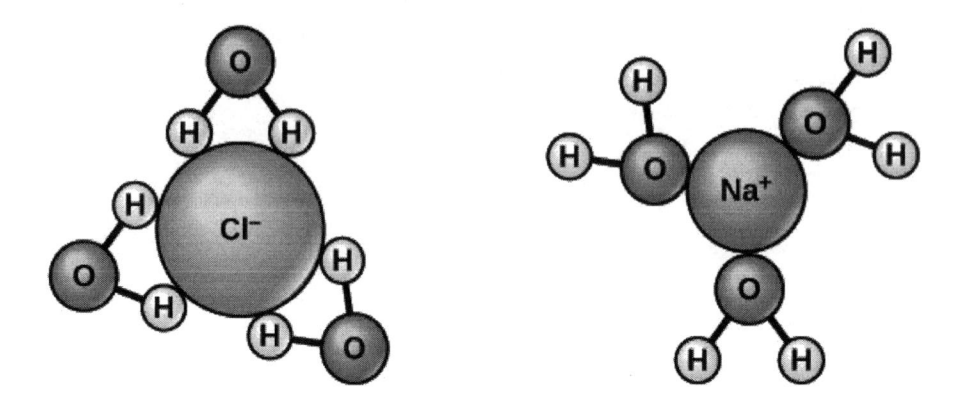

*Figure 3: When table salt (NaCl) is mixed in water, spheres of hydration are formed around the ions.*

When ionic compounds are added to water, the individual ions react with the polar regions of the water molecules and their ionic bonds are disrupted in the process of dissociation. Dissociation occurs when atoms or groups of atoms break off from molecules and form ions. Consider table salt (NaCl, or sodium chloride): when NaCl crystals are added to water, the molecules of NaCl dissociate into Na$^+$ and Cl$^-$ ions, and spheres of hydration form around the ions, illustrated in [Figure 3]. The positively charged sodium ion is surrounded by the partially negative charge of the water molecule's oxygen. The negatively charged chloride ion is surrounded by the partially positive charge of the hydrogen on the water molecule.

## CONCENTRATIONS OF SOLUTES

Various mixtures of solutes and water are described in chemistry. The concentration of a given solute is the number of particles of that solute in a given space (oxygen makes up about 21 percent of atmospheric air). In the bloodstream of humans, glucose concentration is usually measured in milligram (mg) per deciliter (dL), and in a healthy adult averages about 100 mg/dL. Another method of measuring the concentration of a solute is by its molarilty—which is moles (M) of the molecules per liter (L). The mole of an element is its atomic weight, while a mole of a compound is the sum of the atomic weights of its components, called the molecular weight. An often-used example is calculating a mole of glucose, with the chemical formula $C_6H_{12}O_6$. Using the periodic table, the atomic weight of carbon (C) is 12.011 grams (g), and there are six carbons in glucose, for a total atomic weight of 72.066 g. Doing the same calculations for hydrogen (H) and oxygen (O), the molecular weight equals 180.156g (the "gram molecular weight" of glucose). When water is added to make one liter of solution, you have one mole (1M) of glucose. This is particularly useful in chemistry because of the relationship of moles to "Avogadro's number." A mole of any solution has the same number of particles in it: $6.02 \times 10^{23}$. Many substances in the bloodstream and other tissue of the body are measured in thousandths of a mole, or millimoles (mM).

**A colloid** is a mixture that is somewhat like a heavy solution. The solute particles consist of tiny clumps of molecules large enough to make the liquid mixture opaque (because the particles are large enough to scatter light). Familiar examples of colloids are milk and cream. In the thyroid glands, the thyroid hormone is stored as a thick protein mixture also called a colloid.

**A suspension** is a liquid mixture in which a heavier substance is suspended temporarily in a liquid, but over time, settles out. This separation of particles from a suspension is called sedimentation. An example of sedimentation occurs in the blood test that establishes sedimentation rate, or sed rate. The test measures how quickly red blood cells in a test tube settle out of the watery portion of blood (known as plasma) over a set period of time. Rapid sedimentation of blood cells does not normally happen in the healthy body, but aspects of certain diseases can cause blood cells to clump together, and these heavy clumps of blood cells settle to the bottom of the test tube more quickly than do normal blood cells.

## WATER'S COHESIVE AND ADHESIVE PROPERTIES

Have you ever filled a glass of water to the very top and then slowly added a few more drops? Before it overflows, the water forms a dome-like shape above the rim of the glass. This water can stay above the glass because of the property of cohesion. In cohesion, water molecules are attracted to each other (because of hydrogen bonding), keeping the molecules together at the liquid-gas (water-air) interface, although there is no more room in the glass.

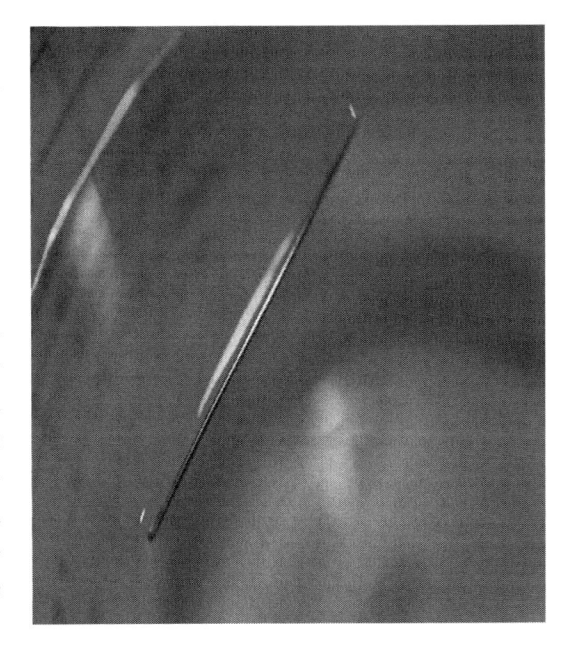

*Figure 4: The weight of the needle is pulling the surface downward; at the same time, the surface tension is pulling it up, suspending it on the surface of the water and keeping it from sinking. Notice the indentation in the water around the needle. (credit: Cory Zanker)*

Cohesion allows for the development of surface tension, the capacity of a substance to withstand being ruptured when placed under tension or stress. This is also why water forms droplets when placed on a dry surface rather than being flattened out by gravity. When a small scrap of paper is placed onto the droplet of water, the paper floats on top of the water droplet even though paper is denser (heavier) than the water. Cohesion and surface tension keep the hydrogen bonds of water molecules intact and support the item floating on the top. It's even possible to "float" a needle on top of a glass of water if it is placed gently without breaking the surface tension, as shown in [Figure 4].

These cohesive forces are related to water's property of adhesion, or the attraction between water molecules and other molecules. This attraction is sometimes stronger than water's cohesive forces, especially when the water is exposed to charged surfaces such as those found on the inside of thin glass tubes known as capillary tubes. Adhesion is observed when water "climbs" up the tube placed in a glass of water: notice that the water appears to be higher on the sides of the tube than in the middle. This is because the water molecules are attracted to the charged glass walls of the capillary more than they are to each other and therefore adhere to it. This type of adhesion is called capillary action, and is illustrated in [Figure 5].

Why are cohesive and adhesive forces important for life? Cohesive and adhesive forces are important for the transport of water from the roots to the leaves in plants. These forces create a "pull" on the water column. This pull results from the tendency of water molecules being evaporated on the surface of the plant to stay connected to water molecules below them, and so they are pulled along. Plants use this natural phenomenon to help transport water from their roots to their leaves. Without these properties of water, plants would be unable to receive the water and the dissolved minerals they require. In another example, insects such as the water strider, shown in [Figure 6], use the surface tension of water to stay afloat on the surface layer of water and even mate there.

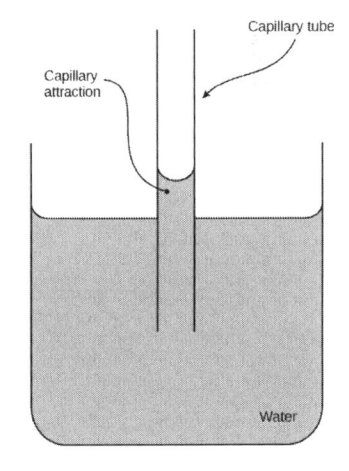

*Figure 5: Capillary action in a glass tube is caused by the adhesive forces exerted by the internal surface of the glass exceeding the cohesive forces between the water molecules themselves. (credit: modification of work by Pearson-Scott Foresman, donated to the Wikimedia Foundation)*

## THE ROLE OF WATER IN CHEMICAL REACTIONS

Two types of chemical reactions involve the creation or the consumption of water: dehydration synthesis and hydrolysis.

- In **dehydration synthesis**, one reactant gives up an atom of hydrogen and another reactant gives up a hydroxyl group (OH) in the synthesis of a new product. In the formation of their covalent bond, a molecule of water is released as a byproduct. This is also sometimes referred to as a condensation reaction.

- In **hydrolysis**, a molecule of water disrupts a compound, breaking its bonds. The water is itself split into H and OH. One portion of the severed compound then bonds with the hydrogen atom, and the other portion bonds with the hydroxyl group.

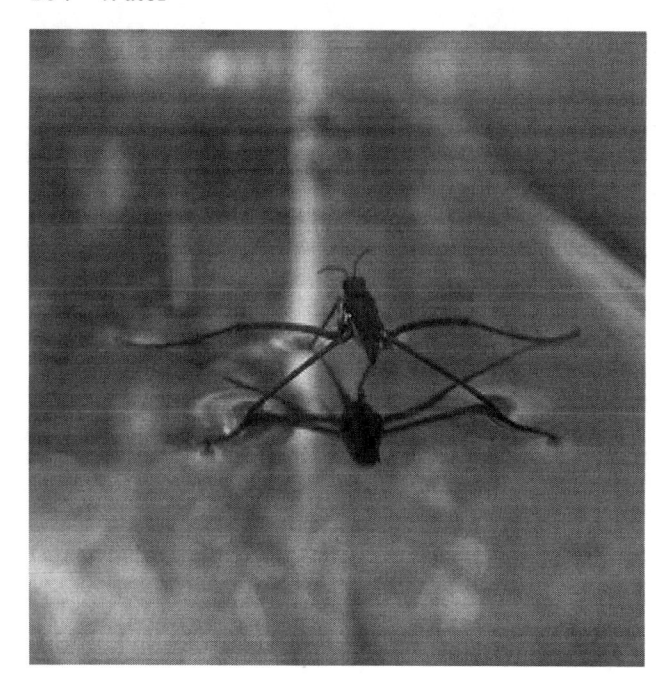

*Figure 6: Water's cohesive and adhesive properties allow this water strider (Gerris sp.) to stay afloat. (credit: Tim Vickers)*

These reactions are reversible, and play an important role in the chemistry of organic compounds (which will be discussed shortly).

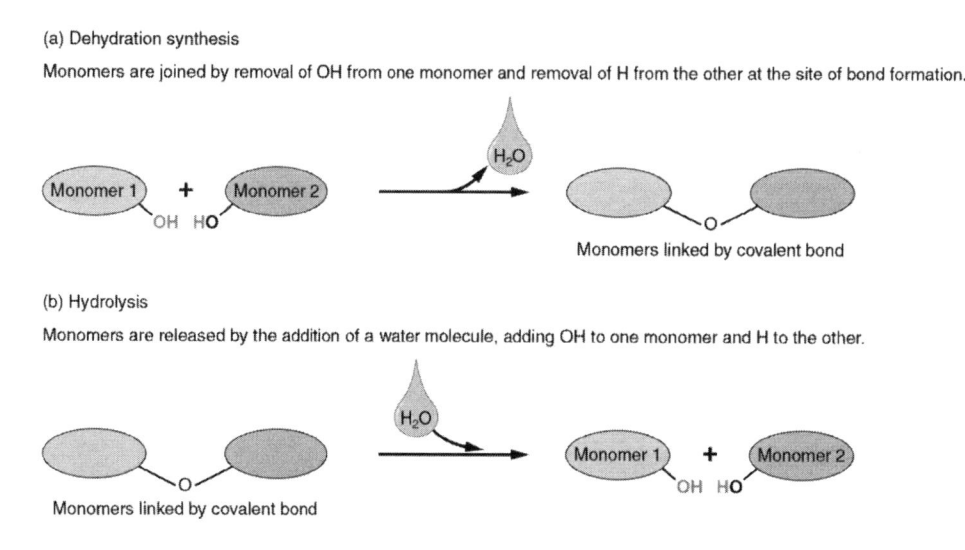

*Figure 1. Dehydration Synthesis and Hydrolysis. Monomers, the basic units for building larger molecules, form polymers (two or more chemically-bonded monomers). (a) In dehydration synthesis, two monomers are covalently bonded in a reaction in which one gives up a hydroxyl group and the other a hydrogen atom. A molecule of water is released as a byproduct during dehydration reactions. (b) In hydrolysis, the covalent bond between two monomers is split by the addition of a hydrogen atom to one and a hydroxyl group to the other, which requires the contribution of one molecule of water.*

### Salts

Recall that salts are formed when ions form ionic bonds. In these reactions, one atom gives up one or more electrons, and thus becomes positively charged, whereas the other accepts one or more electrons and becomes negatively charged. You can now define a salt as a substance that, when dissolved in water, dissociates into ions other than H$^+$ or OH$^-$. This fact is important in distinguishing salts from acids and bases, discussed next.

A typical salt, NaCl, dissociates completely in water (Figure 2). The positive and negative regions on the water molecule (the hydrogen and oxygen ends respectively) attract the negative chloride and positive sodium ions, pulling them away from each other. Again, whereas nonpolar and polar covalently bonded compounds break apart into molecules in solution, salts dissociate into ions. These ions are electrolytes; they are capable of conducting an electrical current in solution. This property is critical to the function of ions in transmitting nerve impulses and prompting muscle contraction.

Many other salts are important in the body. For example, bile salts produced by the liver help break apart dietary fats, and calcium phosphate salts form the mineral portion of teeth and bones.

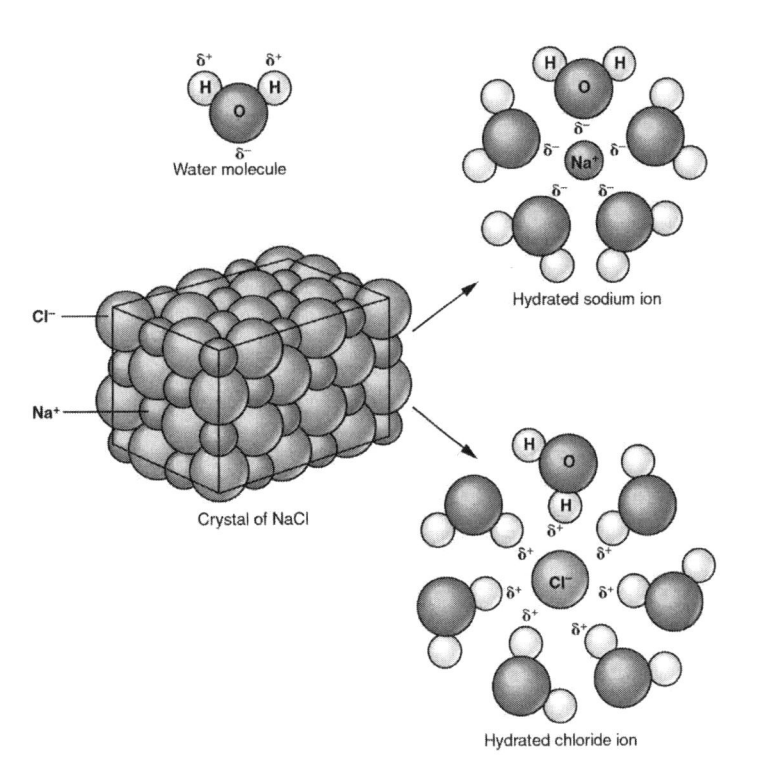

*Figure 2. Dissociation of Sodium Chloride in Water. Notice that the crystals of sodium chloride dissociate not into molecules of NaCl, but into Na⁺ cations and Cl⁻ anions, each completely surrounded by water molecules.*

### Glossary

**adhesion:** attraction between water molecules and other molecules

**calorie:** amount of heat required to change the temperature of one gram of water by one degree Celsius

**capillary action:** occurs because water molecules are attracted to charges on the inner surfaces of narrow tubular structures such as glass tubes, drawing the water molecules to the sides of the tubes

**cohesion:** intermolecular forces between water molecules caused by the polar nature of water; responsible for surface tension

**dissociation:** release of an ion from a molecule such that the original molecule now consists of an ion and the charged remains of the original, such as when water dissociates into $H^+$ and $OH^-$

**evaporation:** separation of individual molecules from the surface of a body of water, leaves of a plant, or the skin of an organism

**heat of vaporization of water:** high amount of energy required for liquid water to turn into water vapor

**hydrophilic:** describes ions or polar molecules that interact well with other polar molecules such as water

**hydrophobic:** describes uncharged non-polar molecules that do not interact well with polar molecules such as water

**litmus paper:** (also, pH paper) filter paper that has been treated with a natural water-soluble dye that changes its color as the pH of the environment changes so it can be used as a pH indicator

**pH paper:** see litmus paper

**pH scale:** scale ranging from zero to 14 that is inversely proportional to the concentration of hydrogen ions in a solution

**solvent:** substance capable of dissolving another substance

**specific heat capacity:** the amount of heat one gram of a substance must absorb or lose to change its temperature by one degree Celsius

**sphere of hydration:** when a polar water molecule surrounds charged or polar molecules thus keeping them dissolved and in solution

**surface tension:** tension at the surface of a body of liquid that prevents the molecules from separating; created by the attractive cohesive forces between the molecules of the liquid

# Molarity

### Learning Objectives

By the end of this section, you will be able to:
- Describe the fundamental properties of solutions
- Calculate solution concentrations using molarity
- Perform dilution calculations using the dilution equation

In preceding sections, we focused on the composition of substances: samples of matter that contain only one type of element or compound. However, mixtures—samples of matter containing two or more substances physically combined—are more commonly encountered in nature than are pure substances. Similar to a pure substance, the relative composition of a mixture plays an important role in determining its properties. The relative amount of oxygen in a planet's atmosphere determines its ability to sustain aerobic life. The relative amounts of iron, carbon, nickel, and other elements in steel (a mixture known as an "alloy") determine its physical strength and resistance to corrosion. The relative amount of the active ingredient in a medicine determines its effectiveness in achieving the desired pharmacological effect. The relative amount of sugar in a beverage determines its sweetness (see Figure 1). In this section, we will describe one of the most common ways in which the relative compositions of mixtures may be quantified.

## SOLUTIONS

We have previously defined solutions as homogeneous mixtures, meaning that the composition of the mixture (and therefore its properties) is uniform throughout its entire volume. Solutions occur frequently in nature and have also been implemented in many forms of manmade technology. We will explore a more thorough treatment of solution properties in the chapter on solutions and colloids, but here we will introduce some of the basic properties of solutions.

The relative amount of a given solution component is known as its **concentration**. Often, though not always, a solution contains one component with a concentration that is significantly greater than that of all other components. This component is called the **solvent** and may be viewed as the medium in

*Figure 1. Sugar is one of many components in the complex mixture known as coffee. The amount of sugar in a given amount of coffee is an important determinant of the beverage's sweetness. (credit: Jane Whitney)*

168  Water

which the other components are dispersed, or **dissolved**. Solutions in which water is the solvent are, of course, very common on our planet. A solution in which water is the solvent is called an **aqueous solution**.

A **solute** is a component of a solution that is typically present at a much lower concentration than the solvent. Solute concentrations are often described with qualitative terms such as **dilute** (of relatively low concentration) and **concentrated** (of relatively high concentration).

Concentrations may be quantitatively assessed using a wide variety of measurement units, each convenient for particular applications. **Molarity (M)** is a useful concentration unit for many applications in chemistry. Molarity is defined as the number of moles of solute in exactly 1 liter (1 L) of the solution:

$$M = \frac{\text{mol solute}}{\text{L solution}}$$

### Example 1: Calculating Molar Concentrations

A 355-mL soft drink sample contains 0.133 mol of sucrose (table sugar). What is the molar concentration of sucrose in the beverage?

**Check Your Learning**

A teaspoon of table sugar contains about 0.01 mol sucrose. What is the molarity of sucrose if a teaspoon of sugar has been dissolved in a cup of tea with a volume of 200 mL?

### Example 2: Deriving Moles and Volumes from Molar Concentrations

How much sugar (mol) is contained in a modest sip (~10 mL) of the soft drink from Example 1?

**Check Your Learning**

What volume (mL) of the sweetened tea described in Example 1 contains the same amount of sugar (mol) as 10 mL of the soft drink in this example?

When performing calculations stepwise, as in Example 4, it is important to refrain from rounding any intermediate calculation results, which can lead to rounding errors in the final result. In Example 4, the molar amount of NaCl computed in the first step, 1.325 mol, would be properly rounded to 1.32 mol if it were to be reported; however, although the last digit (5) is not significant, it must be retained as a guard digit in the intermediate calculation. If we had not retained this guard digit, the final calculation for the mass of NaCl would have been 77.1 g, a difference of 0.3 g.

In addition to retaining a guard digit for intermediate calculations, we can also avoid rounding errors by performing computations in a single step (see Example 5). This eliminates intermediate steps so that only the final result is rounded.

## Example 3: Calculating Molar Concentrations from The Mass of Solute

Distilled white vinegar (Figure 2) is a solution of acetic acid, $CH_3CO_2H$, in water. A 0.500-L vinegar solution contains 25.2 g of acetic acid. What is the concentration of the acetic acid solution in units of molarity?

*Figure 2. Distilled white vinegar is a solution of acetic acid in water.*

**Check Your Learning**

Calculate the molarity of 6.52 g of $CoCl_2$ (128.9 g/mol) dissolved in an aqueous solution with a total volume of 75.0 mL.

## Example 4: Determining The Mass of Solute in A Given Volume of Solution

How many grams of NaCl are contained in 0.250 L of a 5.30-$M$ solution?

**Check Your Learning**

How many grams of $CaCl_2$ (110.98 g/mol) are contained in 250.0 mL of a 0.200-$M$ solution of calcium chloride?

## Example 5: Determining The Volume of Solution Containing A Given Mass of Solute

In Example 3, we found the typical concentration of vinegar to be 0.839 $M$. What volume of vinegar contains 75.6 g of acetic acid?

**Check Your Learning**

What volume of a 1.50-$M$ KBr solution contains 66.0 g KBr?

## DILUTION OF SOLUTIONS

**Dilution** is the process whereby the concentration of a solution is lessened by the addition of solvent. For example, we might say that a glass of iced tea becomes increasingly diluted as the ice melts. The water from the melting ice increases the volume of the solvent (water) and the overall volume of the solution (iced tea), thereby reducing the relative concentrations of the solutes that give the beverage its taste (Figure 3).

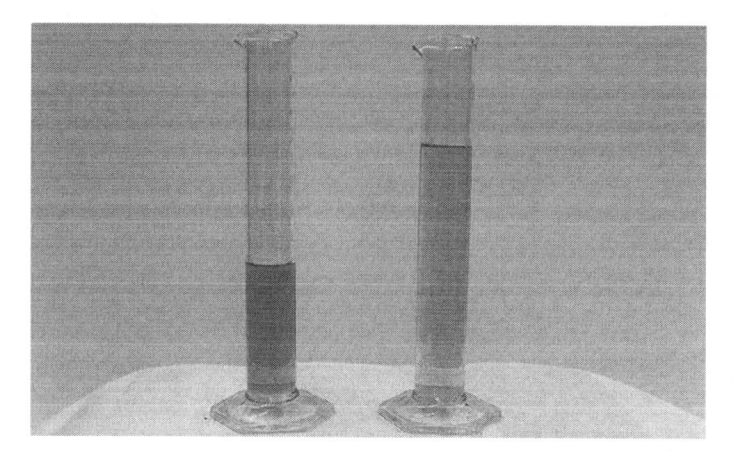

*Figure 3. Both solutions contain the same mass of copper nitrate. The solution on the right is more dilute because the copper nitrate is dissolved in more solvent. (credit: Mark Ott)*

Dilution is also a common means of preparing solutions of a desired concentration. By adding solvent to a measured portion of a more concentrated *stock solution*, we can achieve a particular concentration. For example, commercial pesticides are typically sold as solutions in which the active ingredients are far more concentrated than is appropriate for their application. Before they can be used on crops, the pesticides must be diluted. This is also a very common practice for the preparation of a number of common laboratory reagents (Figure 4).

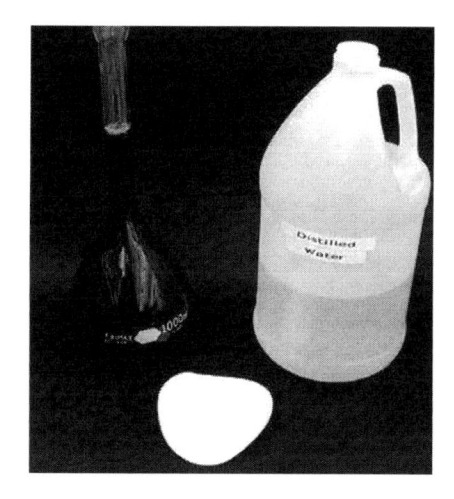

*Figure 4. A solution of KMnO4 is prepared by mixing water with 4.74 g of KMnO4 in a flask. (credit: modification of work by Mark Ott)*

A simple mathematical relationship can be used to relate the volumes and concentrations of a solution before and after the dilution process. According to the definition of molarity, the molar amount of solute in a solution is equal to the product of the solution's molarity and its volume in liters: $n = ML$.

Expressions like these may be written for a solution before and after it is diluted:

$$n_1 = M_1 L_1$$
$$n_2 = M_2 L_2$$

where the subscripts "1" and "2" refer to the solution before and after the dilution, respectively. Since the dilution process *does not change the amount of solute in the solution*, $n_1 = n_2$. Thus, these two equations may be set equal to one another:

$$M_1L_1 = M_2L_2$$

This relation is commonly referred to as the dilution equation. Although we derived this equation using molarity as the unit of concentration and liters as the unit of volume, other units of concentration and volume may be used, so long as the units properly cancel per the factor-label method. Reflecting this versatility, the dilution equation is often written in the more general form:

$$C_1V_1 = C_2V_2$$

where $C$ and $V$ are concentration and volume, respectively.

> Use the **PhET simulation for Concentration** to explore the relations between solute amount, solution volume, and concentration and to confirm the dilution equation.

### Example 6: Determining The Concentration of A Diluted Solution

If 0.850 L of a 5.00-$M$ solution of copper nitrate, $Cu(NO_3)_2$, is diluted to a volume of 1.80 L by the addition of water, what is the molarity of the diluted solution?

**Check Your Learning**

What is the concentration of the solution that results from diluting 25.0 mL of a 2.04-$M$ solution of $CH_3OH$ to 500.0 mL?

### Example 7: Volume of A Diluted Solution

What volume of 0.12 $M$ HBr can be prepared from 11 mL (0.011 L) of 0.45 $M$ HBr?

**Check Your Learning**

A laboratory experiment calls for 0.125 $M$ $HNO_3$. What volume of 0.125 $M$ $HNO_3$ can be prepared from 0.250 L of 1.88 $M$ $HNO_3$?

### Example 8: Volume of A Concentrated Solution Needed for Dilution

What volume of 1.59 $M$ KOH is required to prepare 5.00 L of 0.100 $M$ KOH?

**Check Your Learning**

What volume of a 0.575-$M$ solution of glucose, $C_6H_{12}O_6$, can be prepared from 50.00 mL of a 3.00-$M$ glucose solution?

## Key Concepts and Summary

Solutions are homogeneous mixtures. Many solutions contain one component, called the solvent, in which other components, called solutes, are dissolved. An aqueous solution is one for which the solvent is water. The concentration of a solution is a measure of the relative amount of solute in a given amount of solution. Concentrations may be measured using various units, with one very useful unit being molarity, defined as the number of moles of solute per liter of solution. The solute concentration of a solution may be decreased by adding solvent, a process referred to as dilution. The dilution equation is a simple relation between concentrations and volumes of a solution before and after dilution.

**Key Equations**

- $M = \dfrac{\text{mol solute}}{\text{L solution}}$
- $C_1 V_1 = C_2 V_2$

## Glossary

**aqueous solution:** solution for which water is the solvent

**concentrated:** qualitative term for a solution containing solute at a relatively high concentration

**concentration:** quantitative measure of the relative amounts of solute and solvent present in a solution

**dilute:** qualitative term for a solution containing solute at a relatively low concentration

**dilution:** process of adding solvent to a solution in order to lower the concentration of solutes

**dissolved:** describes the process by which solute components are dispersed in a solvent

**molarity (M):** unit of concentration, defined as the number of moles of solute dissolved in 1 liter of solution

**solute:** solution component present in a concentration less than that of the solvent

**solvent:** solution component present in a concentration that is higher relative to other components

# Other Units for Solution Concentrations

### Learning Objectives

By the end of this section, you will be able to:

- Define the concentration units of mass percentage, volume percentage, mass-volume percentage, parts-per-million (ppm), and parts-per-billion (ppb)
- Perform computations relating a solution's concentration and its components' volumes and/or masses using these units

In the previous section, we introduced molarity, a very useful measurement unit for evaluating the concentration of solutions. However, molarity is only one measure of concentration. In this section, we will introduce some other units of concentration that are commonly used in various applications, either for convenience or by convention.

## MASS PERCENTAGE

Earlier in this chapter, we introduced percent composition as a measure of the relative amount of a given element in a compound. Percentages are also commonly used to express the composition of mixtures, including solutions. The **mass percentage** of a solution component is defined as the ratio of the component's mass to the solution's mass, expressed as a percentage:

$$\text{mass percentage} = \frac{\text{mass of component}}{\text{mass of solution}} \times 100\%$$

We are generally most interested in the mass percentages of solutes, but it is also possible to compute the mass percentage of solvent.

Mass percentage is also referred to by similar names such as *percent mass, percent weight, weight/weight percent*, and other variations on this theme. The most common symbol for mass percentage is simply the percent sign, %, although more detailed symbols are often used including %mass, %weight, and (w/w)%. Use of these more detailed symbols can prevent confusion of mass percentages with other types of percentages, such as volume percentages (to be discussed later in this section).

Mass percentages are popular concentration units for consumer products. The label of a typical liquid bleach bottle (Figure 1) cites the concentration of its active ingredient, sodium hypochlorite (NaOCl), as being 7.4%. A 100.0-g sample of bleach would therefore contain 7.4 g of NaOCl.

*Figure 1. Liquid bleach is an aqueous solution of sodium hypochlorite (NaOCl). This brand has a concentration of 7.4% NaOCl by mass.*

---

### Example 1: Calculation of Percent By Mass

A 5.0-g sample of spinal fluid contains 3.75 mg (0.00375 g) of glucose. What is the percent by mass of glucose in spinal fluid?

**Check Your Learning**

A bottle of a tile cleanser contains 135 g of HCl and 775 g of water. What is the percent by mass of HCl in this cleanser?

---

### Example 2: Calculations Using Mass Percentage

"Concentrated" hydrochloric acid is an aqueous solution of 37.2% HCl that is commonly used as a laboratory reagent. The density of this solution is 1.19 g/mL. What mass of HCl is contained in 0.500 L of this solution?

**Check Your Learning**

What volume of concentrated HCl solution contains 125 g of HCl?

---

## VOLUME PERCENTAGE

Liquid volumes over a wide range of magnitudes are conveniently measured using common and relatively inexpensive laboratory equipment. The concentration of a solution formed by dissolving a liquid solute in a liquid solvent is therefore often expressed as a **volume percentage**, %vol or (v/v)%:

$$\text{volume percentage} = \frac{\text{volume solute}}{\text{volume solution}} \times 100\%$$

Other Units for Solution Concentrations 175

> ### Example 3: Calculations Using Volume Percentage
>
> Rubbing alcohol (isopropanol) is usually sold as a 70%vol aqueous solution. If the density of isopropyl alcohol is 0.785 g/mL, how many grams of isopropyl alcohol are present in a 355 mL bottle of rubbing alcohol?
>
> **Check Your Learning**
>
> Wine is approximately 12% ethanol ($CH_3CH_2OH$) by volume. Ethanol has a molar mass of 46.06 g/mol and a density 0.789 g/mL. How many moles of ethanol are present in a 750-mL bottle of wine?

## MASS-VOLUME PERCENTAGE

"Mixed" percentage units, derived from the mass of solute and the volume of solution, are popular for certain biochemical and medical applications. A **mass–volume percent** is a ratio of a solute's mass to the solution's volume expressed as a percentage. The specific units used for solute mass and solution volume may vary, depending on the solution. For example, physiological saline solution, used to prepare intravenous fluids, has a concentration of 0.9% mass/volume (m/v), indicating that the composition is 0.9 g of solute per 100 mL of solution. The concentration of glucose in blood (commonly referred to as "blood sugar") is also typically expressed in terms of a mass-volume ratio. Though not expressed explicitly as a percentage, its concentration is usually given in milligrams of glucose per deciliter (100 mL) of blood (Figure 2).

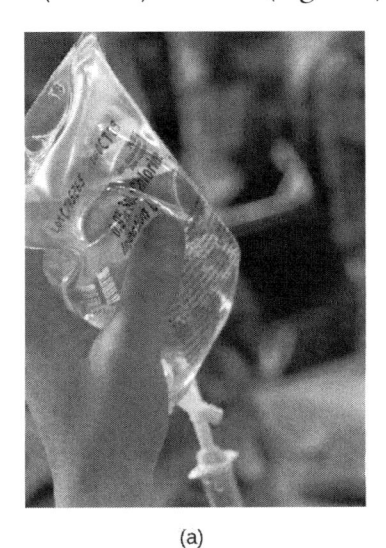

(a)   (b)

*Figure 2. "Mixed" mass-volume units are commonly encountered in medical settings. (a) The NaCl concentration of physiological saline is 0.9% (m/v). (b) This device measures glucose levels in a sample of blood. The normal range for glucose concentration in blood (fasting) is around 70–100 mg/dL. (credit a: modification of work by "The National Guard"/Flickr; credit b: modification of work by Biswarup Ganguly)*

## PARTS PER MILLION AND PARTS PER BILLION

Very low solute concentrations are often expressed using appropriately small units such as **parts per million (ppm)** or **parts per billion (ppb)**. Like percentage ("part per hundred") units, ppm and ppb may be defined in terms of masses, volumes, or mixed mass-volume units. There are also ppm and ppb units defined with respect to numbers of atoms and molecules.

176  Water

The mass-based definitions of ppm and ppb are given below:

$$\text{ppm} = \frac{\text{mass solute}}{\text{mass solution}} \times 10^6 \, \text{ppm}$$

$$\text{ppb} = \frac{\text{mass solute}}{\text{mass solution}} \times 10^9 \, \text{ppb}$$

Both ppm and ppb are convenient units for reporting the concentrations of pollutants and other trace contaminants in water. Concentrations of these contaminants are typically very low in treated and natural waters, and their levels cannot exceed relatively low concentration thresholds without causing adverse effects on health and wildlife. For example, the EPA has identified the maximum safe level of fluoride ion in tap water to be 4 ppm. Inline water filters are designed to reduce the concentration of fluoride and several other trace-level contaminants in tap water (Figure 3).

(a)                                    (b)

*Figure 3. (a) In some areas, trace-level concentrations of contaminants can render unfiltered tap water unsafe for drinking and cooking. (b) Inline water filters reduce the concentration of solutes in tap water. (credit a: modification of work by Jenn Durfey; credit b: modification of work by "vastateparkstaff"/Wikimedia commons)*

### Example 4: Calculation of Parts Per Million and Parts Per Billion Concentrations

According to the EPA, when the concentration of lead in tap water reaches 15 ppb, certain remedial actions must be taken. What is this concentration in ppm? At this concentration, what mass of lead (μg) would be contained in a typical glass of water (300 mL)?

**Check Your Learning**

A 50.0-g sample of industrial wastewater was determined to contain 0.48 mg of mercury. Express the mercury concentration of the wastewater in ppm and ppb units.

## Key Concepts and Summary

In addition to molarity, a number of other solution concentration units are used in various applications. Percentage concentrations based on the solution components' masses, volumes, or both are useful for expressing relatively high concentrations, whereas lower concentrations are conveniently expressed using ppm or ppb units. These units are popular in environmental, medical, and other fields where mole-based units such as molarity are not as commonly used.

**Key Equations**

- $\text{Percent by mass} = \dfrac{\text{mass of solute}}{\text{mass of solution}} \times 100$

- $\text{ppm} = \dfrac{\text{mass solute}}{\text{mass solution}} \times 10^6 \text{ ppm}$

- $\text{ppb} = \dfrac{\text{mass solute}}{\text{mass solution}} \times 10^9 \text{ ppb}$

## Glossary

**mass percentage:** ratio of solute-to-solution mass expressed as a percentage

**mass-volume percent:** ratio of solute mass to solution volume, expressed as a percentage

**parts per billion (ppb):** ratio of solute-to-solution mass multiplied by $10^9$

**parts per million (ppm):** ratio of solute-to-solution mass multiplied by $10^6$

**volume percentage:** ratio of solute-to-solution volume expressed as a percentage

# Electrolytes

### Learning Objectives

By the end of this section, you will be able to:

- Define and give examples of electrolytes
- Distinguish between the physical and chemical changes that accompany dissolution of ionic and covalent electrolytes
- Relate electrolyte strength to solute-solvent attractive forces

When some substances are dissolved in water, they undergo either a physical or a chemical change that yields ions in solution. These substances constitute an important class of compounds called **electrolytes**. Substances that do not yield ions when dissolved are called **nonelectrolytes**. If the physical or chemical process that generates the ions is essentially 100% efficient (all of the dissolved compound yields ions), then the substance is known as a **strong electrolyte**. If only a relatively small fraction of the dissolved substance undergoes the ion-producing process, it is called a **weak electrolyte**.

Substances may be identified as strong, weak, or nonelectrolytes by measuring the electrical conductance of an aqueous solution containing the substance. To conduct electricity, a substance must contain freely mobile, charged species. Most familiar is the conduction of electricity through metallic wires, in which case the mobile, charged entities are electrons. Solutions may also conduct electricity if they contain dissolved ions, with conductivity increasing as ion concentration increases. Applying a voltage to electrodes immersed in a solution permits assessment of the relative concentration of dissolved ions, either quantitatively, by measuring the electrical current flow, or qualitatively, by observing the brightness of a light bulb included in the circuit (Figure 1).

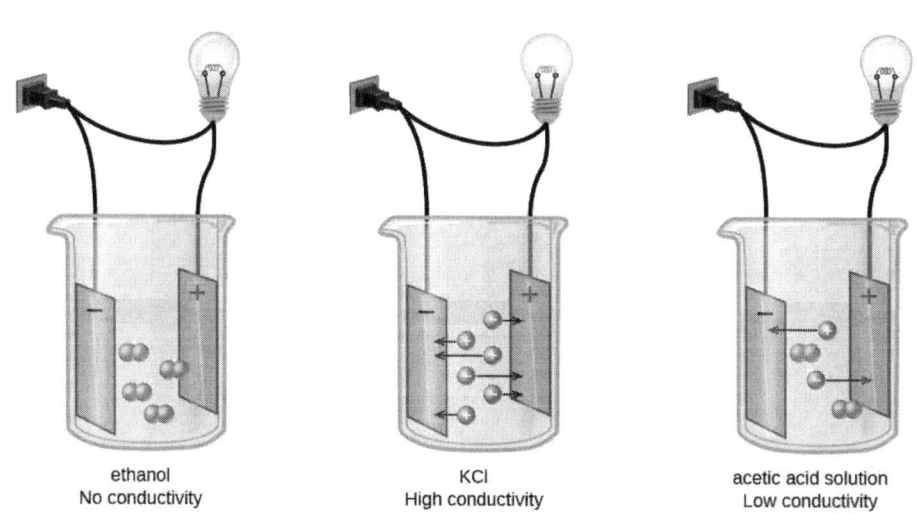

*Figure 1. Solutions of nonelectrolytes such as ethanol do not contain dissolved ions and cannot conduct electricity. Solutions of electrolytes contain ions that permit the passage of electricity. The conductivity of an electrolyte solution is related to the strength of the electrolyte.*

## IONIC ELECTROLYTES

Water and other polar molecules are attracted to ions, as shown in Figure 2. The electrostatic attraction between an ion and a molecule with a dipole is called an **ion–dipole attraction**. These attractions play an important role in the dissolution of ionic compounds in water.

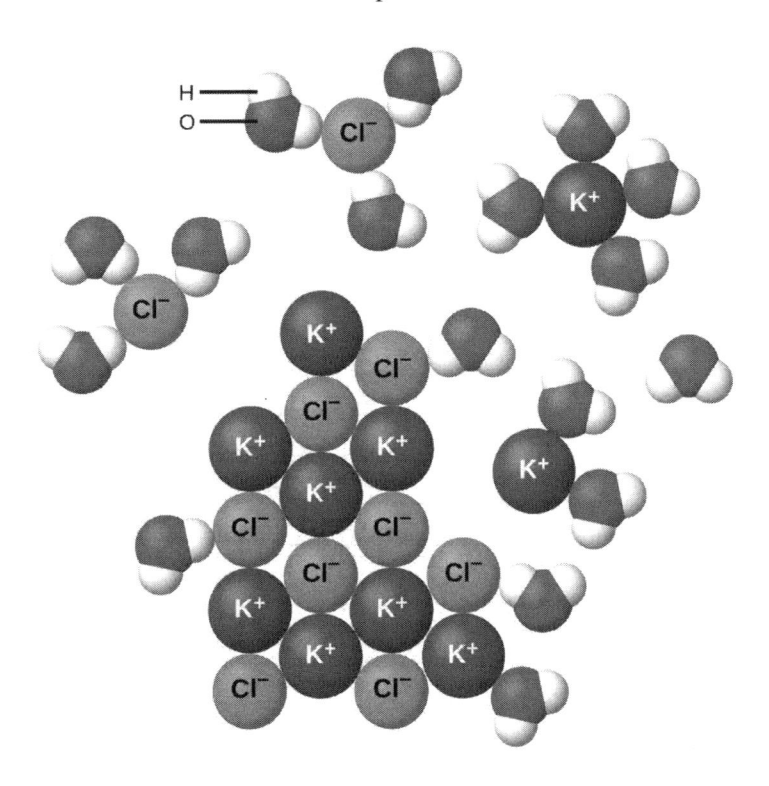

*Figure 2. As potassium chloride (KCl) dissolves in water, the ions are hydrated. The polar water molecules are attracted by the charges on the $K^+$ and $Cl^-$ ions. Water molecules in front of and behind the ions are not shown.*

When ionic compounds dissolve in water, the ions in the solid separate and disperse uniformly throughout the solution because water molecules surround and solvate the ions, reducing the strong electrostatic forces between them. This process represents a physical change known as **dissociation**. Under most conditions, ionic compounds will dissociate nearly completely when dissolved, and so they are classified as strong electrolytes.

Let us consider what happens at the microscopic level when we add solid KCl to water. Ion-dipole forces attract the positive (hydrogen) end of the polar water molecules to the negative chloride ions at the surface of the solid, and they attract the negative (oxygen) ends to the positive potassium ions. The water molecules penetrate between individual $K^+$ and $Cl^-$ ions and surround them, reducing the strong interionic forces that bind the ions together and letting them move off into solution as solvated ions, as Figure 2 shows. The reduction of the electrostatic attraction permits the independent motion of each hydrated ion in a dilute solution, resulting in an increase in the disorder of the system as the ions change from their fixed and ordered positions in the crystal to mobile and much more disordered states in solution. This increased disorder is responsible for the dissolution of many ionic compounds, including KCl, which dissolve with absorption of heat.

In other cases, the electrostatic attractions between the ions in a crystal are so large, or the ion–dipole attractive forces between the ions and water molecules are so weak, that the increase in disorder cannot compensate for the energy required to separate the ions, and the crystal is insoluble. Such is the case for compounds such as calcium carbonate (limestone), calcium phosphate (the inorganic component of bone), and iron oxide (rust).

## COVALENT ELECTROLYTES

Pure water is an extremely poor conductor of electricity because it is only very slightly ionized—only about two out of every 1 billion molecules ionize at 25 °C. Water ionizes when one molecule of water gives up a proton to another molecule of water, yielding hydronium and hydroxide ions.

$$H_2O(l) + H_2O(l) \rightleftharpoons H_3O^+(aq) + OH^-(aq)$$

In some cases, we find that solutions prepared from covalent compounds conduct electricity because the solute molecules react chemically with the solvent to produce ions. For example, pure hydrogen chloride is a gas consisting of covalent HCl molecules. This gas contains no ions. However, when we dissolve hydrogen chloride in water, we find that the solution is a very good conductor. The water molecules play an essential part in forming ions: Solutions of hydrogen chloride in many other solvents, such as benzene, do not conduct electricity and do not contain ions.

Hydrogen chloride is an *acid*, and so its molecules react with water, transferring H⁺ ions to form hydronium ions ($H_3O^+$) and chloride ions (Cl⁻):

This reaction is essentially 100% complete for HCl (i.e., it is a *strong acid* and, consequently, a strong electrolyte). Likewise, weak acids and bases that only react partially generate relatively low concentrations of ions when dissolved in water and are classified as weak electrolytes. The reader may wish to review the discussion of strong and weak acids provided in the earlier chapter of this text on reaction classes and stoichiometry.

### Key Concepts and Summary

Substances that dissolve in water to yield ions are called electrolytes. Electrolytes may be covalent compounds that chemically react with water to produce ions (for example, acids and bases), or they may be ionic compounds that dissociate to yield their constituent cations and anions, when dissolved. Dissolution of an ionic compound is facilitated by ion-dipole attractions between the ions of the compound and the polar water molecules. Soluble ionic substances and strong acids ionize completely and are strong electrolytes, while weak acids and bases ionize to only a small extent and are weak electrolytes. Nonelectrolytes are substances that do not produce ions when dissolved in water.

> ## Glossary
>
> **dissociation:** physical process accompanying the dissolution of an ionic compound in which the compound's constituent ions are solvated and dispersed throughout the solution
>
> **electrolyte:** substance that produces ions when dissolved in water
>
> **ion–dipole attraction:** electrostatic attraction between an ion and a polar molecule
>
> **nonelectrolyte:** substance that does not produce ions when dissolved in water
>
> **strong electrolyte:** substance that dissociates or ionizes completely when dissolved in water
>
> **weak electrolyte:** substance that ionizes only partially when dissolved in water

# pH, Acids, Bases and Buffers

**Learning Objectives**

By the end of this section, you will be able to:

- Discuss the role of acids, bases, and buffers in homeostasis

## PH, ACIDS, BASED AND BUFFERS

Acids and bases, like salts, dissociate in water into electrolytes. Acids and bases can very much change the properties of the solutions in which they are dissolved.

### Acids

An **acid** is a substance that releases hydrogen ions ($H^+$) in solution (Figure 3a). Because an atom of hydrogen has just one proton and one electron, a positively charged hydrogen ion is simply a proton. This solitary proton is highly likely to participate in chemical reactions. Strong acids are compounds that release all of their $H^+$ in solution; that is, they ionize completely. Hydrochloric acid (HCl), which is released from cells in the lining of the stomach, is a strong acid because it releases all of its $H^+$ in the stomach's watery environment. This strong acid aids in digestion and kills ingested microbes. Weak acids do not ionize completely; that is, some of their hydrogenionsremain bonded within a compound in solution. An example of a weak acid is vinegar, or acetic acid; it is called acetate after it gives up a proton.

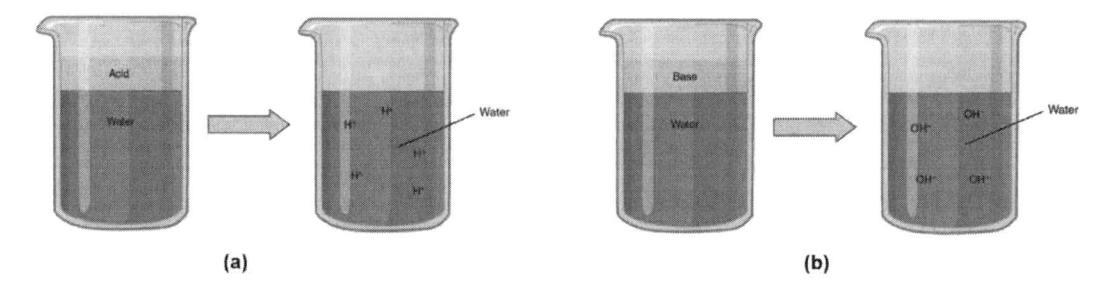

*Figure 3. Acids and Bases.* (a) *In aqueous solution, an acid dissociates into hydrogen ions ($H^+$) and anions. Nearly every molecule of a strong acid dissociates, producing a high concentration of $H^+$.* (b) *In aqueous solution, a base dissociates into hydroxyl ions ($OH^-$) and cations. Nearly every molecule of a strong base dissociates, producing a high concentration of $OH^-$.*

### Bases

A **base** is a substance that releases hydroxyl ions ($OH^-$) in solution, or one that accepts $H^+$ already present in solution (see Figure 3b). The hydroxyl ions or other base combine with $H^+$ present to form a water molecule, thereby removing $H^+$ and reducing the solution's acidity. Strong bases release most or all of their hydroxyl ions; weak bases release only some hydroxyl ions or absorb only a few $H^+$. Food mixed with hydrochloric acid from the stomach would burn the small intestine, the next portion of the digestive tract after the stomach, if it were not for the release of bicarbonate ($HCO_3^-$), a weak base that attracts $H^+$. Bicarbonate accepts some of the $H^+$ protons, thereby reducing the acidity of the solution.

## The Concept of pH

Hydrogen ions are spontaneously generated in pure water by the dissociation (ionization) of a small percentage of water molecules into equal numbers of hydrogen (H⁺) ions and hydroxide (OH⁻) ions. While the hydroxide ions are kept in solution by their hydrogen bonding with other water molecules, the hydrogen ions, consisting of naked protons, are immediately attracted to un-ionized water molecules, forming hydronium ions ($H_3O^+$). Still, by convention, scientists refer to hydrogen ions and their concentration as if they were free in this state in liquid water.

The relative acidity or alkalinity of a solution can be indicated by its pH. A solution's **pH** is the negative, base-10 logarithm of the hydrogen ion (H⁺) concentration of the solution. As an example, a pH 4 solution has an H⁺ concentration that is ten times greater than that of a pH 5 solution. That is, a solution with a pH of 4 is ten times more acidic than a solution with a pH of 5. The concept of pH will begin to make more sense when you study the pH scale, like that shown in Figure 4. The scale consists of a series of increments ranging from 0 to 14. A solution with a pH of 7 is considered neutral—neither acidic nor basic. Pure water has a pH of 7. The lower the number below 7, the more acidic the solution, or the greater the concentration of H⁺. The concentration of hydrogen ions at each pH value is 10 times different than the next pH. For instance, a pH value of 4 corresponds to a proton concentration of $10^{-4}$ M, or 0.0001M, while a pH value of 5 corresponds to a proton concentration of $10^{-5}$ M, or 0.00001M. The higher the number above 7, the more basic (alkaline) the solution, or the lower the concentration of H⁺. Human urine, for example, is ten times more acidic than pure water, and HCl is 10,000,000 times more acidic than water.

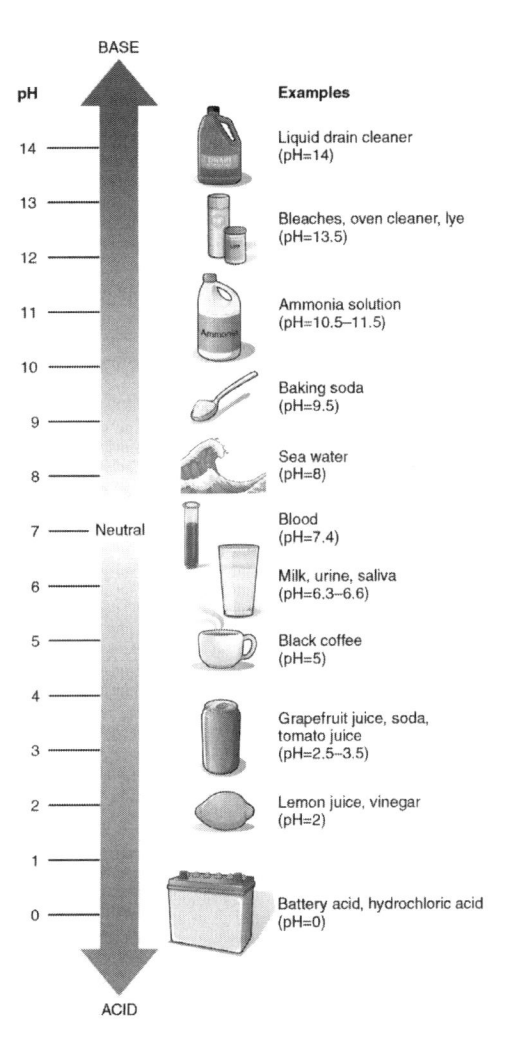

*Figure 4. The pH Scale*

## Buffers

So how can organisms whose bodies require a near-neutral pH ingest acidic and basic substances (a human drinking orange juice, for example) and survive? Buffers are the key. Buffers readily absorb excess H⁺ or OH⁻, keeping the pH of the body carefully maintained in the narrow range required for survival. Maintaining a constant blood pH is critical to a person's well-being. The buffer maintaining the pH of human blood involves carbonic acid ($H_2CO_3$), bicarbonate ion ($HCO_3^-$), and carbon dioxide ($CO_2$). When bicarbonate ions combine with free hydrogen ions and become carbonic acid, hydrogen ions are removed, moderating pH changes. Similarly, as shown in [Figure 8], excess carbonic acid can be converted to carbon dioxide gas and exhaled through the lungs. This prevents too many free hydrogen ions from building up in the blood and dangerously reducing the blood's pH. Likewise, if too much OH⁻ is introduced into the system, carbonic acid will combine with it to create bicarbonate, lowering the pH. Without this buffer system, the body's pH would fluctuate enough to put survival in jeopardy.

$$H^+ + HCO_3^- \longleftrightarrow H_2CO_3 \longleftrightarrow H_2O + CO_2$$

*Figure 8: This diagram shows the body's buffering of blood pH levels. The blue arrows show the process of raising pH as more CO2 is made. The purple arrows indicate the reverse process: the lowering of pH as more bicarbonate is created.*

Other examples of buffers are antacids used to combat excess stomach acid. Many of these over-the-counter medications work in the same way as blood buffers, usually with at least one ion capable of absorbing hydrogen and moderating pH, bringing relief to those that suffer "heartburn" after eating. The unique properties of water that contribute to this capacity to balance pH—as well as water's other characteristics—are essential to sustaining life on Earth.

The pH of human blood normally ranges from 7.35 to 7.45, although it is typically identified as pH 7.4. At this slightly basic pH, blood can reduce the acidity resulting from the carbon dioxide ($CO_2$) constantly being released into the bloodstream by the trillions of cells in the body. Homeostatic mechanisms (along with exhaling $CO_2$ while breathing) normally keep the pH of blood within this narrow range. This is critical, because fluctuations—either too acidic or too alkaline—can lead to life-threatening disorders.

All cells of the body depend on homeostatic regulation of acid–base balance at a pH of approximately 7.4. The body therefore has several mechanisms for this regulation, involving breathing, the excretion of chemicals in urine, and the internal release of chemicals collectively called buffers into body fluids.

A **buffer** is a solution of a weak acid and its conjugate base. A buffer can neutralize small amounts of acids or bases in body fluids. For example, if there is even a slight decrease below 7.35 in the pH of a bodily fluid, the buffer in the fluid—in this case, acting as a weak base—will bind the excess hydrogen ions. In contrast, if pH rises above 7.45, the buffer will act as a weak acid and contribute hydrogen ions.

### Homeostatic Imbalances: Acids and Bases

Excessive acidity of the blood and other body fluids is known as acidosis. Common causes of acidosis are situations and disorders that reduce the effectiveness of breathing, especially the person's ability to exhale fully, which causes a buildup of $CO_2$ (and $H^+$) in the bloodstream. Acidosis can also be caused by metabolic problems that reduce the level or function of buffers that act as bases, or that promote the production of acids. For instance, with severe diarrhea, too much bicarbonate can be lost from the body, allowing acids to build up in body fluids. In people with poorly managed diabetes (ineffective regulation of blood sugar), acids called ketones are produced as a form of body fuel. These can build up in the blood, causing a serious condition called diabetic ketoacidosis. Kidney failure, liver failure, heart failure, cancer, and other disorders also can prompt metabolic acidosis.

In contrast, alkalosis is a condition in which the blood and other body fluids are too alkaline (basic). As with acidosis, respiratory disorders are a major cause; however, in respiratory alkalosis, carbon dioxide levels fall too low. Lung disease, aspirin overdose, shock, and ordinary anxiety can cause respiratory alkalosis, which reduces the normal concentration of $H^+$.

Metabolic alkalosis often results from prolonged, severe vomiting, which causes a loss of hydrogen and chloride ions (as components of HCl). Medications also can prompt alkalosis. These include diuretics that cause the body to lose potassium ions, as well as antacids when taken in excessive amounts, for instance by someone with persistent heartburn or an ulcer.

### Footnotes

1. W. Humphrey W., A. Dalke, and K. Schulten, "VMD—Visual Molecular Dynamics," *Journal of Molecular Graphics* 14 (1996): 33-38.
2. W. Humphrey W., A. Dalke, and K. Schulten, "VMD—Visual Molecular Dynamics," *Journal of Molecular Graphics* 14 (1996): 33-38.

### Glossary

**acid:** molecule that donates hydrogen ions and increases the concentration of hydrogen ions in a solution

**base:** molecule that donates hydroxide ions or otherwise binds excess hydrogen ions and decreases the concentration of hydrogen ions in a solution

**buffer:** substance that prevents a change in pH by absorbing or releasing hydrogen or hydroxide ions

# Exercises

## PART 1

1. Explain what changes and what stays the same when 1.00 L of a solution of NaCl is diluted to 1.80 L.

2. What information do we need to calculate the molarity of a sulfuric acid solution?

3. What does it mean when we say that a 200-mL sample and a 400-mL sample of a solution of salt have the same molarity? In what ways are the two samples identical? In what ways are these two samples different

4. Determine the molarity of each of the following solutions:

   a. 1.457 mol KCl in 1.500 L of solution

   b. 0.515 g of $H_2SO_4$ in 1.00 L of solution

   c. 20.54 g of $Al(NO_3)_3$ in 1575 mL of solution

   d. 2.76 kg of $CuSO_4 \cdot 5H_2O$ in 1.45 L of solution

   e. 0.005653 mol of $Br_2$ in 10.00 mL of solution

   f. 0.000889 g of glycine, $C_2H_5NO_2$, in 1.05 mL of solution

5. Consider this question: What is the mass of the solute in 0.500 L of 0.30 $M$ glucose, $C_6H_{12}O_6$, used for intravenous injection?

   a. Outline the steps necessary to answer the question.

   b. Answer the question.

6. Consider this question: What is the mass of solute in 200.0 L of a 1.556-$M$ solution of KBr?

   a. Outline the steps necessary to answer the question.

   b. Answer the question.

7. Calculate the number of moles and the mass of the solute in each of the following solutions:

   a. 2.00 L of 18.5 $M$ $H_2SO_4$, concentrated sulfuric acid

   b. 100.0 mL of $3.8 \times 10^{-5}$ $M$ NaCN, the minimum lethal concentration of sodium cyanide in blood serum

   c. 5.50 L of 13.3 $M$ $H_2CO$, the formaldehyde used to "fix" tissue samples

   d. 325 mL of $1.8 \times 10^{-6}$ $M$ $FeSO_4$, the minimum concentration of iron sulfate detectable by taste in drinking water

8. Calculate the number of moles and the mass of the solute in each of the following solutions:

   a. 325 mL of $8.23 \times 10^{-5}$ $M$ KI a source of iodine in the diet

   b. 75.0 mL of $2.2 \times 10^{-5}$ $M$ $H_2SO_4$, a sample of acid rain

   c. 0.2500 L of 0.1135 $M$ $K_2CrO_4$, an analytical reagent used in iron assays

   d. 10.5 L of 3.716 $M$ $(NH_4)_2SO_4$, a liquid fertilizer

Exercises 187

9. Consider this question: What is the molarity of $KMnO_4$ in a solution of 0.0908 g of $KMnO_4$ in 0.500 L of solution?

    a. Outline the steps necessary to answer the question.

    b. Answer the question.

10. Consider this question: What is the molarity of HCl if 35.23 mL of a solution of HCl contain 0.3366 g of HCl?

    a. Outline the steps necessary to answer the question.

    b. Answer the question.

11. Calculate the molarity of each of the following solutions:

    a. 0.195 g of cholesterol, $C_{27}H_{46}O$, in 0.100 L of serum, the average concentration of cholesterol in human serum

    b. 4.25 g of $NH_3$ in 0.500 L of solution, the concentration of $NH_3$ in household ammonia

    c. 1.49 kg of isopropyl alcohol, $C_3H_7OH$, in 2.50 L of solution, the concentration of isopropyl alcohol in rubbing alcohol

    d. 0.029 g of $I_2$ in 0.100 L of solution, the solubility of $I_2$ in water at 20 °C

12. Calculate the molarity of each of the following solutions:

    a. 293 g HCl in 666 mL of solution, a concentrated HCl solution

    b. 2.026 g $FeCl_3$ in 0.1250 L of a solution used as an unknown in general chemistry laboratories

    c. 0.001 mg $Cd^{2+}$ in 0.100 L, the maximum permissible concentration of cadmium in drinking water

    d. 0.0079 g $C_7H_5SNO_3$ in one ounce (29.6 mL), the concentration of saccharin in a diet soft drink.

13. There is about 1.0 g of calcium, as $Ca^{2+}$, in 1.0 L of milk. What is the molarity of $Ca^{2+}$ in milk?

14. What volume of a 1.00-$M$ $Fe(NO_3)_3$ solution can be diluted to prepare 1.00 L of a solution with a concentration of 0.250 $M$?

15. If 0.1718 L of a 0.3556-$M$ $C_3H_7OH$ solution is diluted to a concentration of 0.1222 $M$, what is the volume of the resulting solution?

16. If 4.12 L of a 0.850 $M$-$H_3PO_4$ solution is be diluted to a volume of 10.00 L, what is the concentration the resulting solution?

17. What volume of a 0.33-$M$ $C_{12}H_{22}O_{11}$ solution can be diluted to prepare 25 mL of a solution with a concentration of 0.025 $M$?

18. What is the concentration of the NaCl solution that results when 0.150 L of a 0.556-$M$ solution is allowed to evaporate until the volume is reduced to 0.105 L?

19. A 2.00-L bottle of a solution of concentrated HCl was purchased for the general chemistry laboratory. The solution contained 868.8 g of HCl. What is the molarity of the solution?

20. An experiment in a general chemistry laboratory calls for a 2.00-$M$ solution of HCl. How many mL of 11.9 $M$ HCl would be required to make 250 mL of 2.00 $M$ HCl?

21. What volume of a 0.20-$M$ $K_2SO_4$ solution contains 57 g of $K_2SO_4$?

188  Water

22. The US Environmental Protection Agency (EPA) places limits on the quantities of toxic substances that may be discharged into the sewer system. Limits have been established for a variety of substances, including hexavalent chromium, which is limited to 0.50 mg/L. If an industry is discharging hexavalent chromium as potassium dichromate ($K_2Cr_2O_7$), what is the maximum permissible molarity of that substance?

## PART 2

1. Consider this question: What mass of a concentrated solution of nitric acid (68.0% $HNO_3$ by mass) is needed to prepare 400.0 g of a 10.0% solution of $HNO_3$ by mass?

    a. Outline the steps necessary to answer the question.

    b. Answer the question.

2. What mass of a 4.00% NaOH solution by mass contains 15.0 g of NaOH?

3. What mass of solid NaOH (97.0% NaOH by mass) is required to prepare 1.00 L of a 10.0% solution of NaOH by mass? The density of the 10.0% solution is 1.109 g/mL.

4. What mass of HCl is contained in 45.0 mL of an aqueous HCl solution that has a density of 1.19 g cm$^{-3}$ and contains 37.21% HCl by mass?

5. The hardness of water (hardness count) is usually expressed in parts per million (by mass) of $CaCO_3$, which is equivalent to milligrams of $CaCO_3$ per liter of water. What is the molar concentration of $Ca^{2+}$ ions in a water sample with a hardness count of 175 mg $CaCO_3$/L?

6. The level of mercury in a stream was suspected to be above the minimum considered safe (1 part per billion by weight). An analysis indicated that the concentration was 0.68 parts per billion. Assume a density of 1.0 g/mL and calculate the molarity of mercury in the stream.

7. In Canada and the United Kingdom, devices that measure blood glucose levels provide a reading in millimoles per liter. If a measurement of 5.3 m$M$ is observed, what is the concentration of glucose ($C_6H_{12}O_6$) in mg/dL?

8. A cough syrup contains 5.0% ethyl alcohol, $C_2H_5OH$, by mass. If the density of the solution is 0.9928 g/mL, determine the molarity of the alcohol in the cough syrup.

9. D5W is a solution used as an intravenous fluid. It is a 5.0% by mass solution of dextrose ($C_6H_{12}O_6$) in water. If the density of D5W is 1.029 g/mL, calculate the molarity of dextrose in the solution.

10. Find the molarity of a 40.0% by mass aqueous solution of sulfuric acid, $H_2SO_4$, for which the density is 1.3057 g/mL.

## PART 3

1. Explain why solutions of HBr in benzene (a nonpolar solvent) are nonconductive, while solutions in water (a polar solvent) are conductive.

2. Consider the solutions presented:

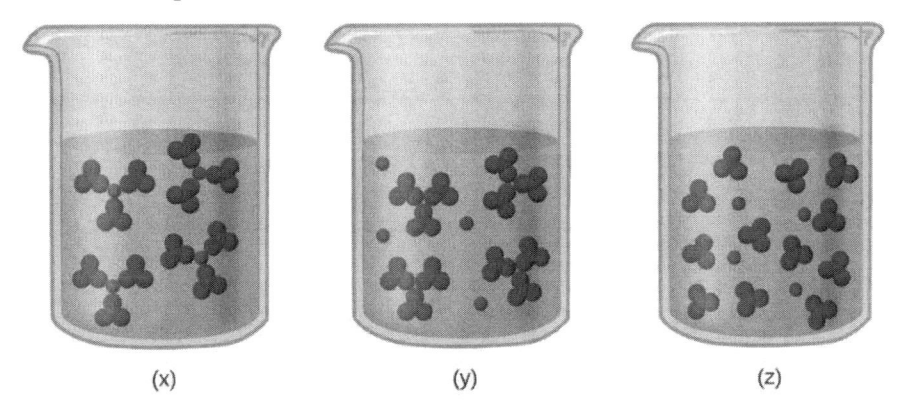

(x)　　　　　　(y)　　　　　　(z)

   a. Which of the following sketches best represents the ions in a solution of Fe(NO$_3$)$_3$(aq)?

   b. Write a balanced chemical equation showing the products of the dissolution of Fe(NO$_3$)$_3$.

3. Compare the processes that occur when methanol (CH$_3$OH), hydrogen chloride (HCl), and sodium hydroxide (NaOH) dissolve in water. Write equations and prepare sketches showing the form in which each of these compounds is present in its respective solution.

4. What is the expected electrical conductivity of the following solutions?

   a. NaOH(aq)

   b. HCl(aq)

   c. C$_6$H$_{12}$O$_6$(aq) (glucose)

   d. NH$_3$(l)

5. Why are most *solid* ionic compounds electrically nonconductive, whereas aqueous solutions of ionic compounds are good conductors? Would you expect a *liquid* (molten) ionic compound to be electrically conductive or nonconductive? Explain.

6. Indicate the most important type of intermolecular attraction responsible for solvation in each of the following solutions:

   a. the solutions in Figure 2

   b. methanol, CH$_3$OH, dissolved in ethanol, C$_2$H$_5$OH

   c. methane, CH$_4$, dissolved in benzene, C$_6$H$_6$

   d. the polar halocarbon CF$_2$Cl$_2$ dissolved in the polar halocarbon CF$_2$ClCFCl$_2$

   e. O$_2$(l) in N$_2$(l)

# Check Your Knowledge: Self-Test

1. Determine the molarity for each of the following solutions:

   a. 0.444 mol of $CoCl_2$ in 0.654 L of solution

   b. 98.0 g of phosphoric acid, $H_3PO_4$, in 1.00 L of solution

   c. 0.2074 g of calcium hydroxide, $Ca(OH)_2$, in 40.00 mL of solution

   d. 10.5 kg of $Na_2SO_4 \cdot 10H_2O$ in 18.60 L of solution

   e. $7.0 \times 10^{-3}$ mol of $I_2$ in 100.0 mL of solution

   f. $1.8 \times 10^4$ mg of HCl in 0.075 L of solution

2. What is the molarity of the diluted solution when each of the following solutions is diluted to the given final volume?

   a. 1.00 L of a 0.250-$M$ solution of $Fe(NO_3)_3$ is diluted to a final volume of 2.00 L

   b. 0.5000 L of a 0.1222-$M$ solution of $C_3H_7OH$ is diluted to a final volume of 1.250 L

   c. 2.35 L of a 0.350-$M$ solution of $H_3PO_4$ is diluted to a final volume of 4.00 L

   d. 22.50 mL of a 0.025-$M$ solution of $C_{12}H_{22}O_{11}$ is diluted to 100.0 mL

3. What is the final concentration of the solution produced when 225.5 mL of a 0.09988-$M$ solution of $Na_2CO_3$ is allowed to evaporate until the solution volume is reduced to 45.00 mL?

4. A throat spray is 1.40% by mass phenol, $C_6H_5OH$, in water. If the solution has a density of 0.9956 g/mL, calculate the molarity of the solution.

5. Copper(I) iodide (CuI) is often added to table salt as a dietary source of iodine. How many moles of CuI are contained in 1.00 lb (454 g) of table salt containing 0.0100% CuI by mass?

6. Explain why the ions $Na^+$ and $Cl^-$ are strongly solvated in water but not in hexane, a solvent composed of nonpolar molecules.

7. Which of the following statements is not true?

   a. Water is polar.

   b. Water stabilizes temperature.

   c. Water is essential for life.

   d. Water is the most abundant molecule in the Earth's atmosphere.

8. When acids are added to a solution, the pH should _____.

   1. decrease

   2. increase

   3. stay the same

   4. cannot tell without testing

9. A molecule that binds up excess hydrogen ions in a solution is called a(n) _____.

   1. acid
   2. isotope
   3. base
   4. donator

10. Which of the following statements is true?

    1. Acids and bases cannot mix together.
    2. Acids and bases will neutralize each other.
    3. Acids, but not bases, can change the pH of a solution.
    4. Acids donate hydroxide ions (OH$^-$); bases donate hydrogen ions (H$^+$).

11. Discuss how buffers help prevent drastic swings in pH.
12. Why can some insects walk on water?

# Chapter 6: Organic Molecules

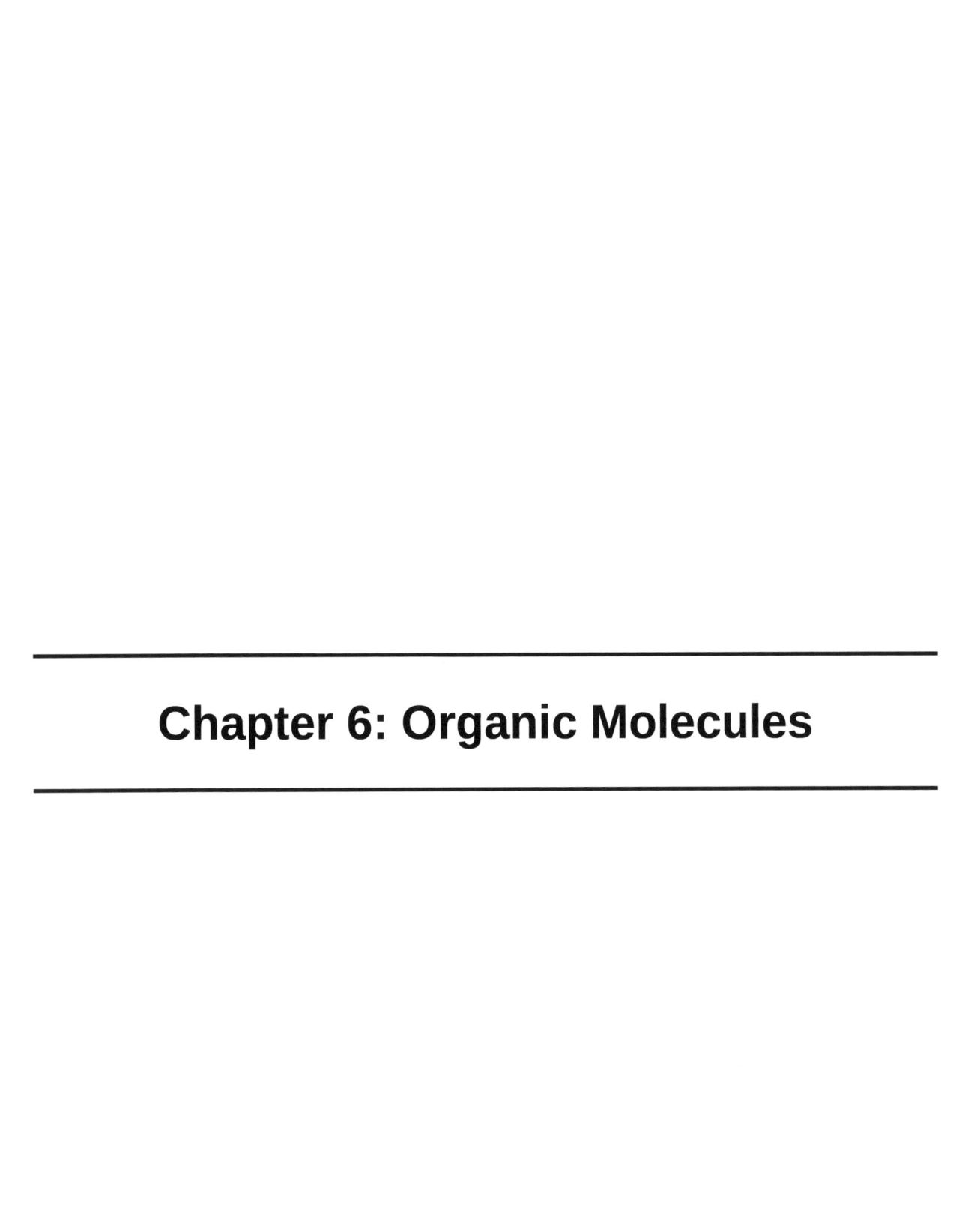

# Introduction

The elements carbon, hydrogen, nitrogen, oxygen, sulfur, and phosphorus are the key building blocks of the chemicals found in living things. They form the carbohydrates, nucleic acids, proteins, and lipids (all of which will be defined later in this chapter) that are the fundamental molecular components of all organisms. In this chapter, we will discuss these important building blocks and learn how the unique properties of the atoms of different elements affect their interactions with other atoms to form the molecules of life.

*Figure 1. Foods such as bread, fruit, and cheese are rich sources of biological macromolecules. (credit: modification of work by Bengt Nyman)*

Food provides an organism with nutrients—the matter it needs to survive. Many of these critical nutrients come in the form of biological macromolecules, or large molecules necessary for life. These macromolecules are built from different combinations of smaller organic molecules. What specific types of biological macromolecules do living things require? How are these molecules formed? What functions do they serve? In this chapter, we will explore these questions.

# Chemical Formulas

### Learning Objectives

By the end of this section, you will be able to:
- Symbolize the composition of molecules using molecular formulas and empirical formulas
- Represent the bonding arrangement of atoms within molecules using structural formulas

A **molecular formula** is a representation of a molecule that uses chemical symbols to indicate the types of atoms followed by subscripts to show the number of atoms of each type in the molecule. (A subscript is used only when more than one atom of a given type is present.) Molecular formulas are also used as abbreviations for the names of compounds.

The **structural formula** for a compound gives the same information as its molecular formula (the types and numbers of atoms in the molecule) but also shows how the atoms are connected in the molecule. The structural formula for methane contains symbols for one C atom and four H atoms, indicating the number of atoms in the molecule (Figure 1). The lines represent bonds that hold the atoms together. (A chemical bond is an attraction between atoms or ions that holds them together in a molecule or a crystal.) We will discuss chemical bonds and see how to predict the arrangement of atoms in a molecule later. For now, simply know that the lines are an indication of how the atoms are connected in a molecule. A ball-and-stick model shows the geometric arrangement of the atoms with atomic sizes not to scale, and a space-filling model shows the relative sizes of the atoms.

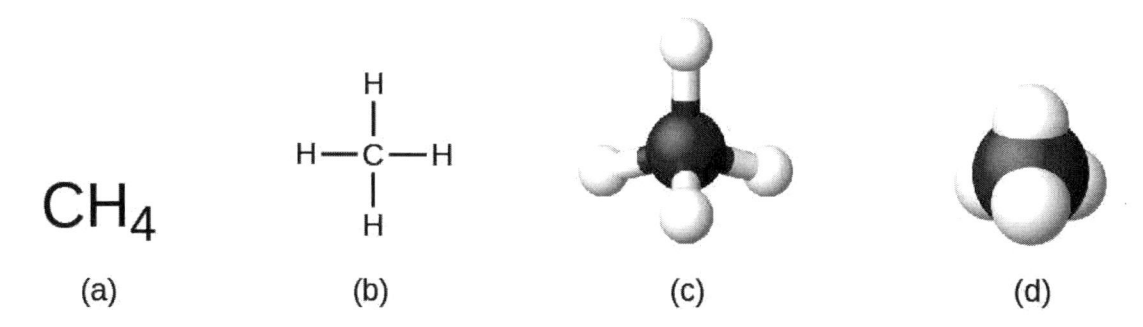

*Figure 1. A methane molecule can be represented as (a) a molecular formula, (b) a structural formula, (c) a ball-and-stick model, and (d) a space-filling model. Carbon and hydrogen atoms are represented by black and white spheres, respectively.*

Although many elements consist of discrete, individual atoms, some exist as molecules made up of two or more atoms of the element chemically bonded together. For example, most samples of the elements hydrogen, oxygen, and nitrogen are composed of molecules that contain two atoms each (called diatomic molecules) and thus have the molecular formulas $H_2$, $O_2$, and $N_2$, respectively. Other elements commonly found as diatomic molecules are fluorine ($F_2$), chlorine ($Cl_2$), bromine ($Br_2$), and iodine ($I_2$).

The most common form of the element sulfur is composed of molecules that consist of eight atoms of sulfur; its molecular formula is $S_8$ (Figure 2).

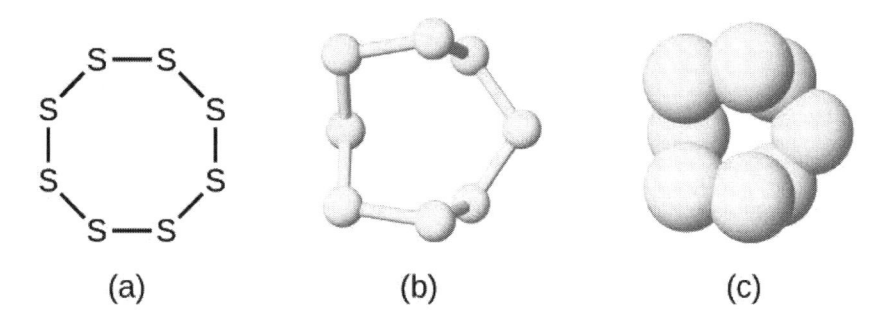

*Figure 2. A molecule of sulfur is composed of eight sulfur atoms and is therefore written as $S_8$. It can be represented as (a) a structural formula, (b) a ball-and-stick model, and (c) a space-filling model. Sulfur atoms are represented by yellow spheres.*

It is important to note that a subscript following a symbol and a number in front of a symbol do not represent the same thing; for example, $H_2$ and 2H represent distinctly different species. $H_2$ is a molecular formula; it represents a diatomic molecule of hydrogen, consisting of two atoms of the element that are chemically bonded together. The expression 2H, on the other hand, indicates two separate hydrogen atoms that are not combined as a unit. The expression $2H_2$ represents two molecules of diatomic hydrogen (Figure 3).

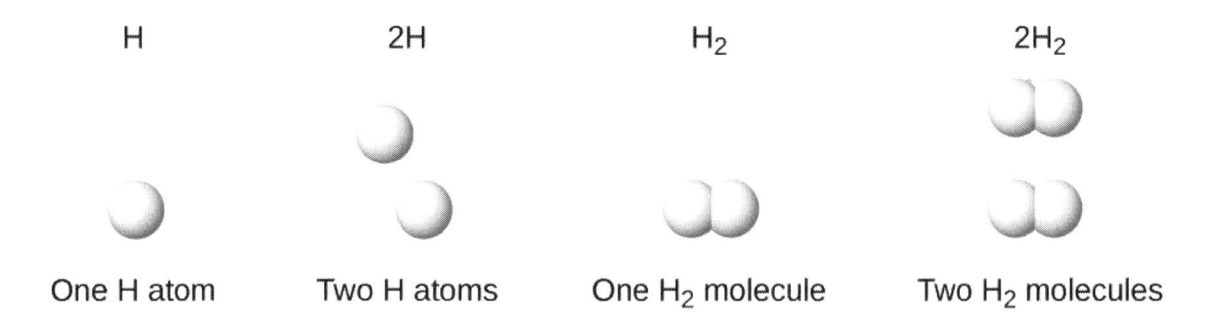

*Figure 3. The symbols H, 2H, $H_2$, and $2H_2$ represent very different entities.*

Compounds are formed when two or more elements chemically combine, resulting in the formation of bonds. For example, hydrogen and oxygen can react to form water, and sodium and chlorine can react to form table salt. We sometimes describe the composition of these compounds with an **empirical formula**, which indicates the types of atoms present and *the simplest whole-number ratio of the number of atoms (or ions) in the compound.* For example, titanium dioxide (used as pigment in white paint and in the thick, white, blocking type of sunscreen) has an empirical formula of $TiO_2$. This identifies the elements titanium (Ti) and oxygen (O) as the constituents of titanium dioxide, and indicates the presence of twice as many atoms of the element oxygen as atoms of the element titanium (Figure 4).

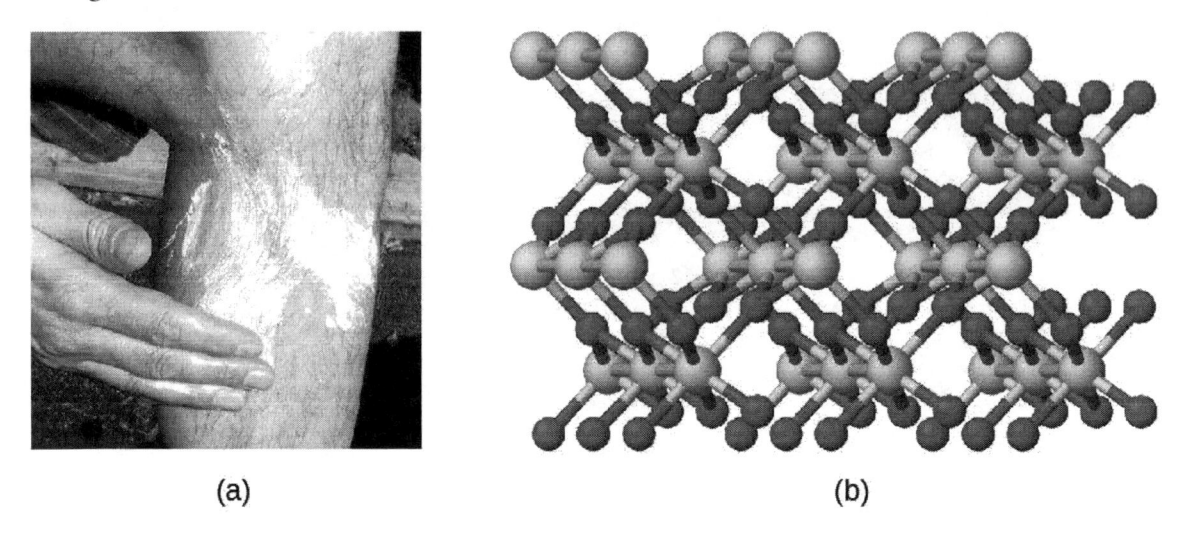

*Figure 4.* (a) *The white compound titanium dioxide provides effective protection from the sun.* (b) *A crystal of titanium dioxide, TiO$_2$, contains titanium and oxygen in a ratio of 1 to 2. The titanium atoms are gray and the oxygen atoms are red. (credit a: modification of work by "osseous"/Flickr)*

As discussed previously, we can describe a compound with a molecular formula, in which the subscripts indicate the *actual numbers of atoms* of each element in a molecule of the compound. In many cases, the molecular formula of a substance is derived from experimental determination of both its empirical formula and its **molecular mass** (the sum of atomic masses for all atoms composing the molecule). For example, it can be determined experimentally that benzene contains two elements, carbon (C) and hydrogen (H), and that for every carbon atom in benzene, there is one hydrogen atom. Thus, the empirical formula is CH. An experimental determination of the molecular mass reveals that a molecule of benzene contains six carbon atoms and six hydrogen atoms, so the molecular formula for benzene is C$_6$H$_6$ (Figure 5).

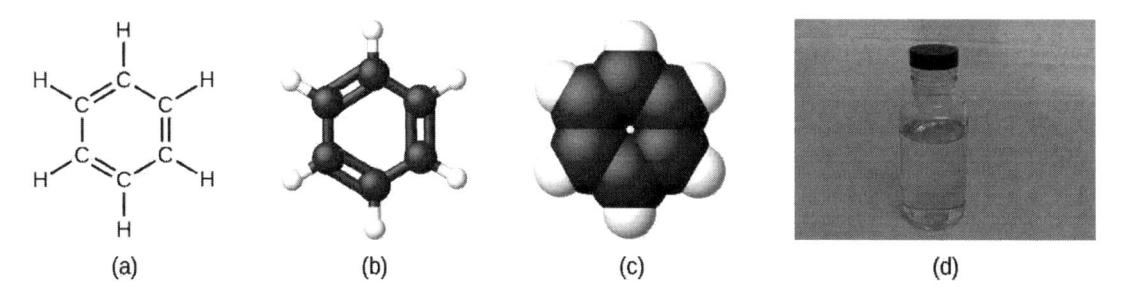

*Figure 5. Benzene, C$_6$H$_6$, is produced during oil refining and has many industrial uses. A benzene molecule can be represented as (a) a structural formula, (b) a ball-and-stick model, and (c) a space-filling model. (d) Benzene is a clear liquid. (credit d: modification of work by Sahar Atwa)*

If we know a compound's formula, we can easily determine the empirical formula. (This is somewhat of an academic exercise; the reverse chronology is generally followed in actual practice.) For example, the molecular formula for acetic acid, the component that gives vinegar its sharp taste, is C$_2$H$_4$O$_2$. This formula indicates that a molecule of acetic acid (Figure 6) contains two carbon atoms, four hydrogen atoms, and two oxygen atoms. The ratio of atoms is 2:4:2. Dividing by the lowest common denominator (2) gives the simplest, whole-number ratio of atoms, 1:2:1, so the empirical formula is CH$_2$O. Note that a molecular formula is always a whole-number multiple of an empirical formula.

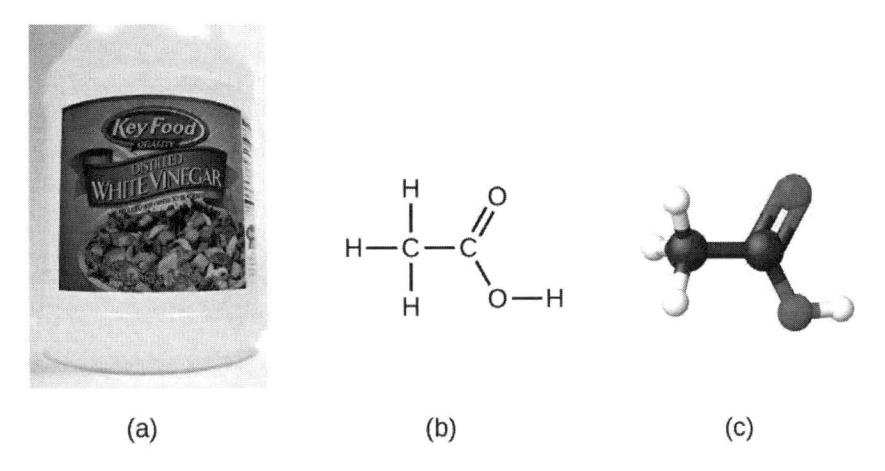

*Figure 6. (a) Vinegar contains acetic acid, $C_2H_4O_2$, which has an empirical formula of $CH_2O$. It can be represented as (b) a structural formula and (c) as a ball-and-stick model. (credit a: modification of work by "HomeSpot HQ"/Flickr)*

### Example 1: Empirical and Molecular Formulas

Molecules of glucose (blood sugar) contain 6 carbon atoms, 12 hydrogen atoms, and 6 oxygen atoms. What are the molecular and empirical formulas of glucose?

---

You can explore PhET's molecule building using an online simulation: https://phet.colorado.edu/en/simulation/build-a-molecule.

---

### Portrait of a Chemist: Lee Cronin

What is it that chemists do? According to Lee **Cronin** (Figure 7), chemists make very complicated molecules by "chopping up" small molecules and "reverse engineering" them. He wonders if we could "make a really cool universal chemistry set" by what he calls "app-ing" chemistry. Could we "app" chemistry?

In a 2012 TED talk, Lee describes one fascinating possibility: combining a collection of chemical "inks" with a 3D printer capable of fabricating a reaction apparatus (tiny test tubes, beakers, and the like) to fashion a "universal toolkit of chemistry." This toolkit could be used to create custom-tailored drugs to fight a new superbug or to "print" medicine personally configured to your genetic makeup, environment, and health situation. Says Cronin, "What Apple did for music, I'd like to do for the discovery and distribution of prescription drugs."[1] View his full talk from the TED website.

*Figure 7. Chemist Lee Cronin has been named one of the UK's 10 most inspirational scientists. (credit: image courtesy of Lee Cronin)*

---

[1] Lee Cronin, "Print Your Own Medicine," Talk presented at TED Global 2012, Edinburgh, Scotland, June 2012.

## 200  Organic Molecules

[Video: TED — Lee Cronin: Print your own medicine]
https://youtu.be/mAEqvn7B2Qg

It is important to be aware that it may be possible for the same atoms to be arranged in different ways: Compounds with the same molecular formula may have different atom-to-atom bonding and therefore different structures. For example, could there be another compound with the same formula as acetic acid, $C_2H_4O_2$? And if so, what would be the structure of its molecules?

If you predict that another compound with the formula $C_2H_4O_2$ could exist, then you demonstrated good chemical insight and are correct. Two C atoms, four H atoms, and two O atoms can also be arranged to form a methyl formate, which is used in manufacturing, as an insecticide, and for quick-drying finishes. Methyl formate molecules have one of the oxygen atoms between the two carbon atoms, differing from the arrangement in acetic acid molecules. Acetic acid and methyl formate are examples of **isomers**—compounds with the same chemical formula but different molecular structures (Figure 8). Note that this small difference in the arrangement of the atoms has a major effect on their respective chemical properties. You would certainly not want to use a solution of methyl formate as a substitute for a solution of acetic acid (vinegar) when you make salad dressing.

Acetic acid
$C_2H_4O_2$
(a)

Methyl formate
$C_2H_4O_2$
(b)

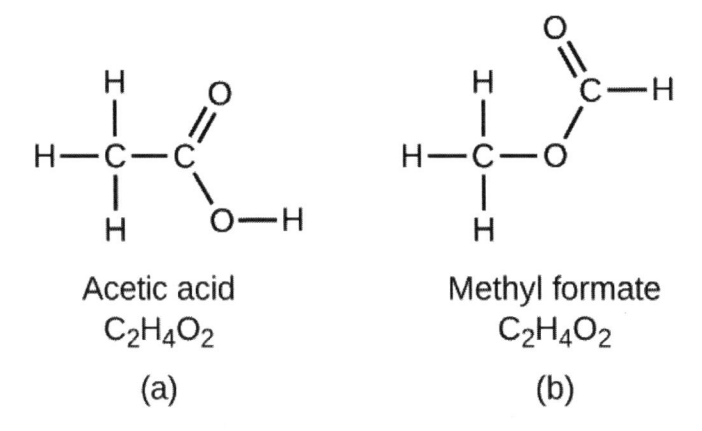

*Figure 8. Molecules of (a) acetic acid and methyl formate (b) are structural isomers; they have the same formula ($C_2H_4O_2$) but different structures (and therefore different chemical properties).*

Many types of isomers exist (Figure 9). Acetic acid and methyl formate are **structural isomers**, compounds in which the molecules differ in how the atoms are connected to each other. There are also various types of **spatial isomers**, in which the relative orientations of the atoms in space can be different. For example, the compound carvone (found in caraway seeds, spearmint, and mandarin orange peels) consists of two isomers that are mirror images of each other. *S*-(+)-carvone smells like caraway, and *R*-(−)-carvone smells like spearmint.

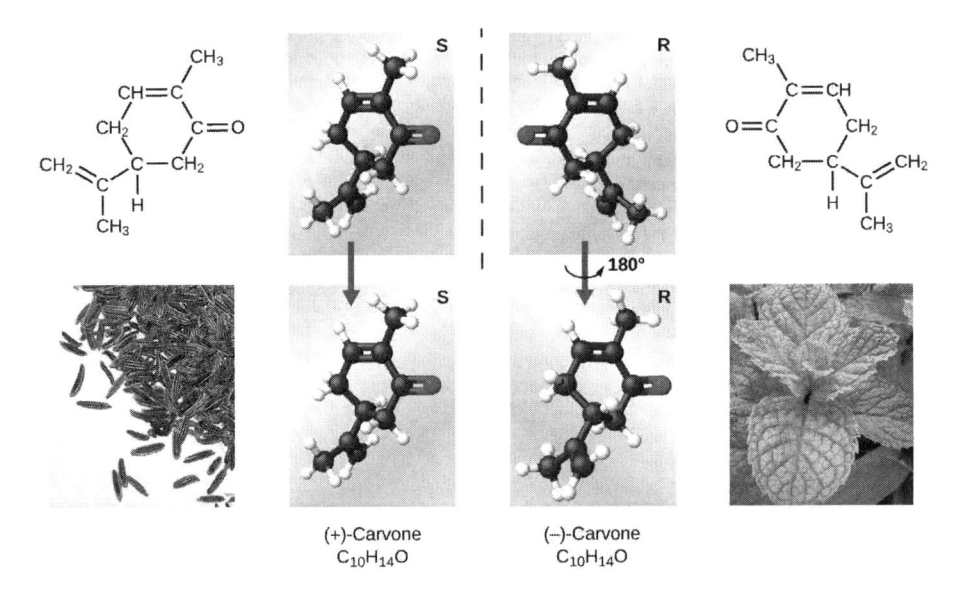

*Figure 9. Molecules of carvone are spatial isomers; they only differ in the relative orientations of the atoms in space. (credit bottom left: modification of work by "Miansari66"/Wikimedia Commons; credit bottom right: modification of work by Forest & Kim Starr)*

### Key Concepts and Summary

A molecular formula uses chemical symbols and subscripts to indicate the exact numbers of different atoms in a molecule or compound. An empirical formula gives the simplest, whole-number ratio of atoms in a compound. A structural formula indicates the bonding arrangement of the atoms in the molecule. Ball-and-stick and space-filling models show the geometric arrangement of atoms in a molecule. Isomers are compounds with the same molecular formula but different arrangements of atoms.

### Glossary

**empirical formula:** formula showing the composition of a compound given as the simplest whole-number ratio of atoms

**isomers:** compounds with the same chemical formula but different structures

**molecular formula:** formula indicating the composition of a molecule of a compound and giving the actual number of atoms of each element in a molecule of the compound.

**spatial isomers:** compounds in which the relative orientations of the atoms in space differ

**structural formula:** shows the atoms in a molecule and how they are connected

# Carbon

### Learning Objectives

By the end of this section, you will be able to:

- Explain why carbon is important for life
- Describe the role of functional groups in biological molecules

Cells are made of many complex molecules called macromolecules, such as proteins, nucleic acids (RNA and DNA), carbohydrates, and lipids. The macromolecules are a subset of organic molecules (any carbon-containing liquid, solid, or gas) that are especially important for life. The fundamental component for all of these macromolecules is carbon. The carbon atom has unique properties that allow it to form covalent bonds to as many as four different atoms, making this versatile element ideal to serve as the basic structural component, or "backbone," of the macromolecules.

Individual carbon atoms have an incomplete outermost electron shell. With an atomic number of 6 (six electrons and six protons), the first two electrons fill the inner shell, leaving four in the second shell. Therefore, carbon atoms can form up to four covalent bonds with other atoms to satisfy the octet rule. The methane molecule provides an example: it has the chemical formula $CH_4$. Each of its four hydrogen atoms forms a single covalent bond with the carbon atom by sharing a pair of electrons. This results in a filled outermost shell.

## HYDROCARBONS

Hydrocarbons are organic molecules consisting entirely of carbon and hydrogen, such as methane ($CH_4$) described above. We often use hydrocarbons in our daily lives as fuels—like the propane in a gas grill or the butane in a lighter. The many covalent bonds between the atoms in hydrocarbons store a great amount of energy, which is released when these molecules are burned (oxidized). Methane, an excellent fuel, is the simplest hydrocarbon molecule, with a central carbon atom bonded to four different hydrogen atoms, as illustrated in [link]. The geometry of the methane molecule, where the atoms reside in three dimensions, is determined by the shape of its electron orbitals. The carbons and the four hydrogen atoms form a shape known as a tetrahedron, with four triangular faces; for this reason, methane is described as having tetrahedral geometry.

Methane has a tetrahedral geometry, with each of the four hydrogen atoms spaced 109.5° apart.

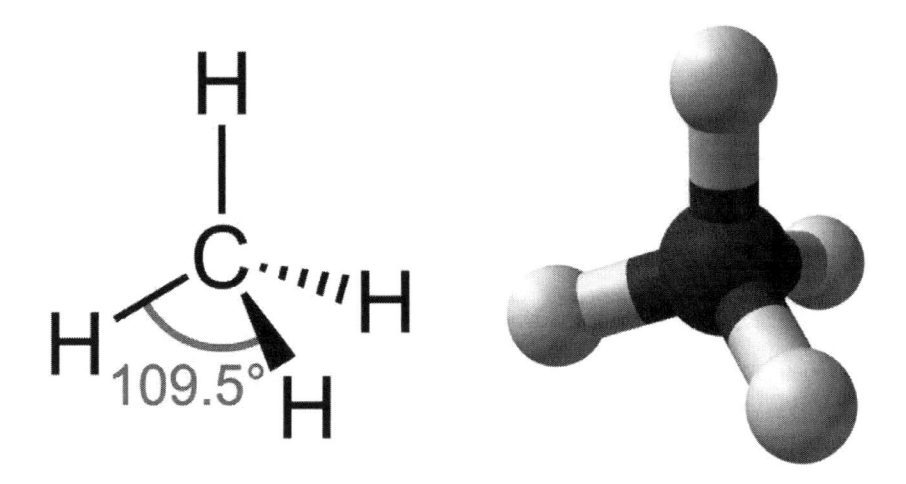

As the backbone of the large molecules of living things, hydrocarbons may exist as linear carbon chains, carbon rings, or combinations of both. Furthermore, individual carbon-to-carbon bonds may be single, double, or triple covalent bonds, and each type of bond affects the geometry of the molecule in a specific way. This three-dimensional shape or conformation of the large molecules of life (macromolecules) is critical to how they function.

*Hydrocarbon Chains*

Hydrocarbon chains are formed by successive bonds between carbon atoms and may be branched or unbranched. Furthermore, the overall geometry of the molecule is altered by the different geometries of single, double, and triple covalent bonds, illustrated in **[link]**. The hydrocarbons ethane, ethene, and ethyne serve as examples of how different carbon-to-carbon bonds affect the geometry of the molecule. The names of all three molecules start with the prefix "eth-," which is the prefix for two carbon hydrocarbons. The suffixes "-ane," "-ene," and "-yne" refer to the presence of single, double, or triple carbon-carbon bonds, respectively. Thus, propane, propene, and propyne follow the same pattern with three carbon molecules, butane, butane, and butyne for four carbon molecules, and so on. Double and triple bonds change the geometry of the molecule: single bonds allow rotation along the axis of the bond, whereas double bonds lead to a planar configuration and triple bonds to a linear one. These geometries have a significant impact on the shape a particular molecule can assume.

When carbon forms single bonds with other atoms, the shape is tetrahedral. When two carbon atoms form a double bond, the shape is planar, or flat. Single bonds, like those found in ethane, are able to rotate. Double bonds, like those found in ethene cannot rotate, so the atoms on either side are locked in place.

| Methane (CH$_4$) | Ethane (C$_2$H$_6$) | Ethene (C$_2$H$_4$) |
|---|---|---|
| Tetrahedral (single bond) | Tetrahedral (single bond) | Planar (double bond) |

### Hydrocarbon Rings

So far, the hydrocarbons we have discussed have been aliphatic hydrocarbons, which consist of linear chains of carbon atoms. Another type of hydrocarbon, aromatic hydrocarbons, consists of closed rings of carbon atoms. Ring structures are found in hydrocarbons, sometimes with the presence of double bonds, which can be seen by comparing the structure of cyclohexane to benzene in **[link]**. Examples of biological molecules that incorporate the benzene ring include some amino acids and cholesterol and its derivatives, including the hormones estrogen and testosterone. The benzene ring is also found in the herbicide 2,4-D. Benzene is a natural component of crude oil and has been classified as a carcinogen. Some hydrocarbons have both aliphatic and aromatic portions; beta-carotene is an example of such a hydrocarbon.

Carbon can form five-and six membered rings. Single or double bonds may connect the carbons in the ring, and nitrogen may be substituted for carbon.

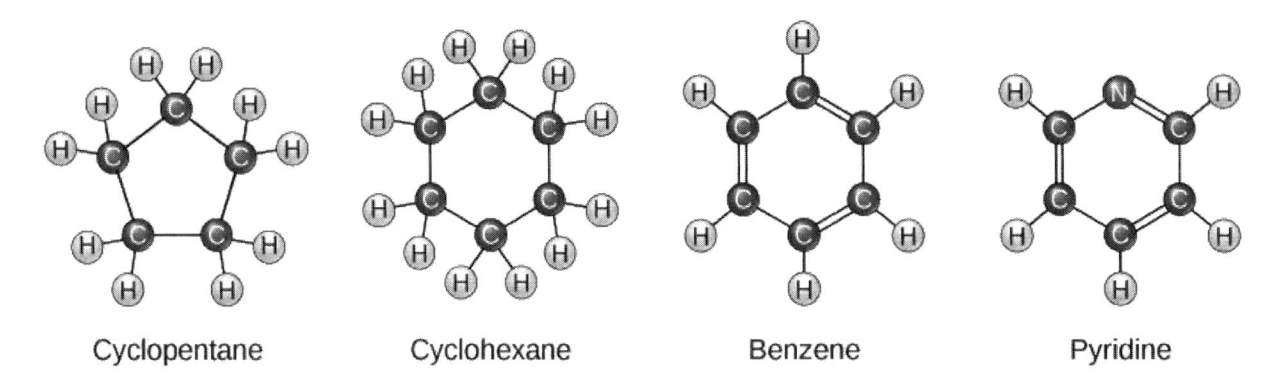

Cyclopentane    Cyclohexane    Benzene    Pyridine

## ISOMERS

The three-dimensional placement of atoms and chemical bonds within organic molecules is central to understanding their chemistry. Molecules that share the same chemical formula but differ in the placement (structure) of their atoms and/or chemical bonds are known as isomers. Structural isomers (like butane and isobutene shown in **[link]** a) differ in the placement of their covalent bonds: both molecules have four carbons and ten hydrogens ($C_4H_{10}$), but the different arrangement of the atoms within the molecules leads to differences in their chemical properties. For example, due to their different chemical properties, butane is suited for use as a fuel for cigarette lighters and torches, whereas isobutene is suited for use as a refrigerant and a propellant in spray cans.

Geometric isomers, on the other hand, have similar placements of their covalent bonds but differ in how these bonds are made to the surrounding atoms, especially in carbon-to-carbon double bonds. In the simple molecule butene ($C_4H_8$), the two methyl groups ($CH_3$) can be on either side of the double covalent bond central to the molecule, as illustrated in **[link]** b. When the carbons are bound on the same side of the double bond, this is the *cis* configuration; if they are on opposite sides of the double bond, it is a *trans* configuration. In the *trans* configuration, the carbons form a more or less linear structure, whereas the carbons in the *cis* configuration make a bend (change in direction) of the carbon backbone.

### Art Connection

Molecules that have the same number and type of atoms arranged differently are called isomers. (a) Structural isomers have a different covalent arrangement of atoms. (b) Geometric isomers have a different arrangement of atoms around a double bond. (c) Enantiomers are mirror images of each other.

**(a) Structural isomers**

Butane

Isobutane

**(b) Geometric isomers**

*cis*-2-butene

*trans*-2-butene

methyl groups on same side of double bond

methyl groups on opposite sides of double bond

**(c) Enantiomers**

L-isomer

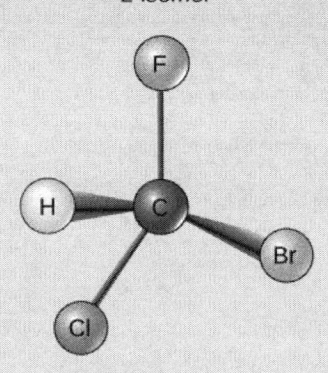

D-isomer

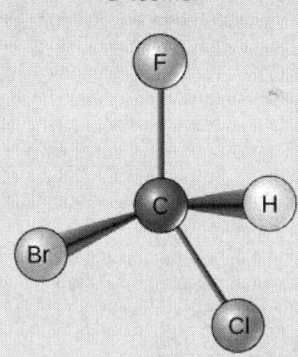

Which of the following statements is false?

1. Molecules with the formulas $CH_3CH_2COOH$ and $C_3H_6O_2$ could be structural isomers.
2. Molecules must have a double bond to be *cis–trans* isomers.
3. To be enantiomers, a molecule must have at least three different atoms or groups connected to a central carbon.
4. To be enantiomers, a molecule must have at least four different atoms or groups connected to a central carbon.

206   Organic Molecules

In triglycerides (fats and oils), long carbon chains known as fatty acids may contain double bonds, which can be in either the *cis* or *trans* configuration, illustrated in **[link]**. Fats with at least one double bond between carbon atoms are unsaturated fats. When some of these bonds are in the *cis* configuration, the resulting bend in the carbon backbone of the chain means that triglyceride molecules cannot pack tightly, so they remain liquid (oil) at room temperature. On the other hand, triglycerides with *trans* double bonds (popularly called trans fats), have relatively linear fatty acids that are able to pack tightly together at room temperature and form solid fats. In the human diet, trans fats are linked to an increased risk of cardiovascular disease, so many food manufacturers have reduced or eliminated their use in recent years. In contrast to unsaturated fats, triglycerides without double bonds between carbon atoms are called saturated fats, meaning that they contain all the hydrogen atoms available. Saturated fats are a solid at room temperature and usually of animal origin.

These space-filling models show a *cis* (oleic acid) and a *trans* (eliadic acid) fatty acid. Notice the bend in the molecule cause by the *cis* configuration.

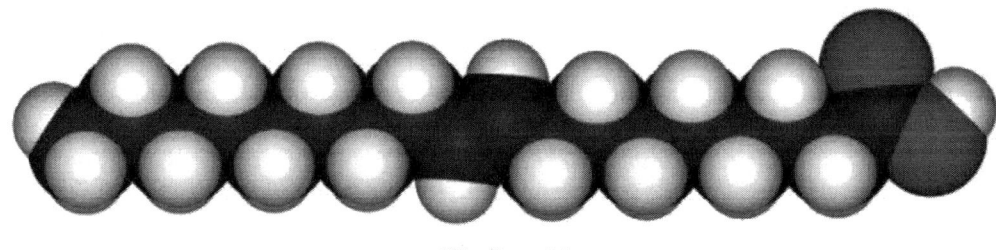

Eliadic acid

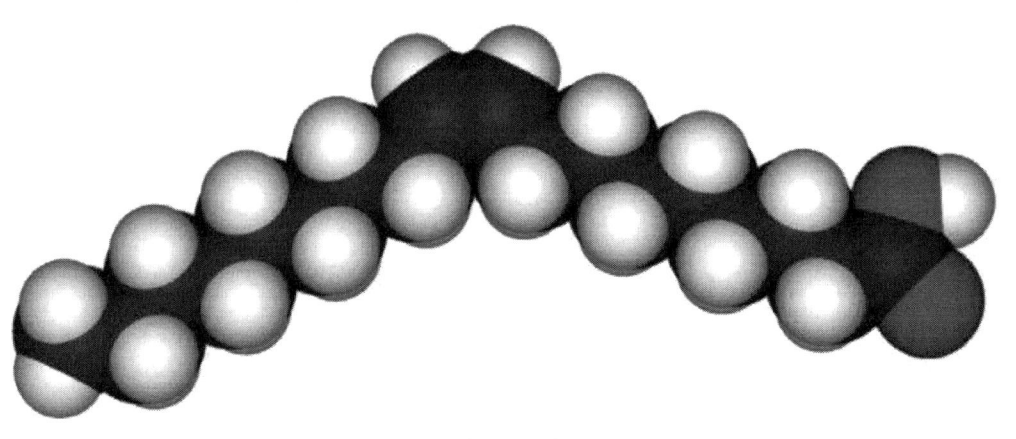

Oleic acid

## ENANTIOMERS

Enantiomers are molecules that share the same chemical structure and chemical bonds but differ in the three-dimensional placement of atoms so that they are mirror images. As shown in **[link]**, an amino acid alanine example, the two structures are non-superimposable. In nature, only the L-forms of amino acids are used to make proteins. Some D forms of amino acids are seen in the cell walls of bacteria, but never in their proteins. Similarly, the D-form of glucose is the main product of photosynthesis and the L-form of the molecule is rarely seen in nature.

D-alanine and L-alanine are examples of enantiomers or mirror images. Only the L-forms of amino acids are used to make proteins.

D-alanine    L-alanine

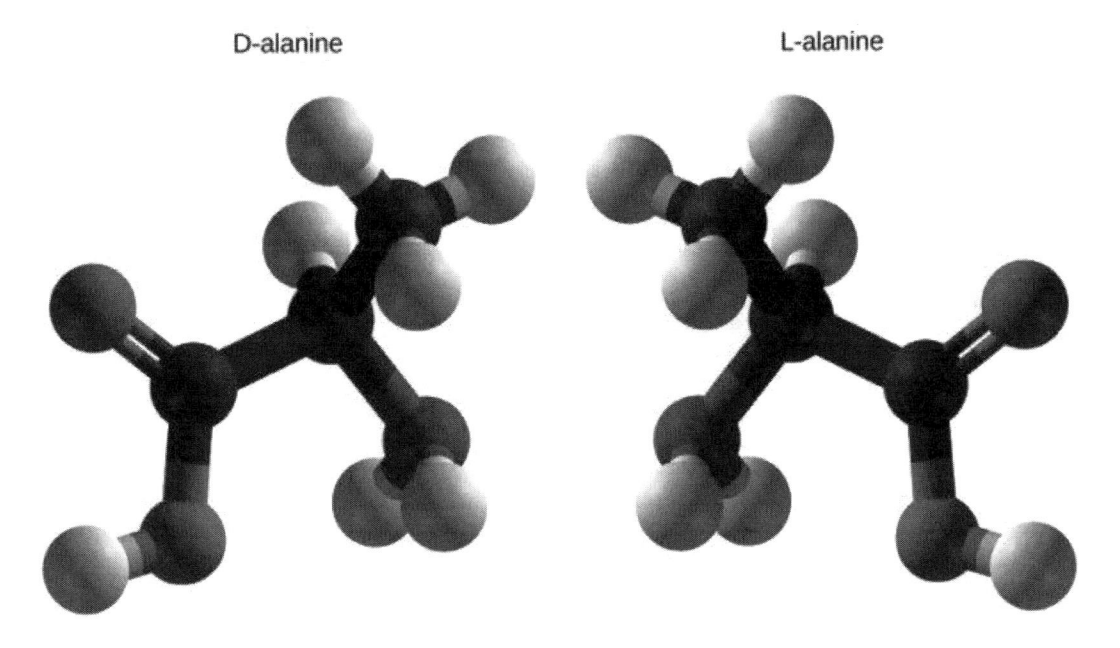

## *FUNCTIONAL GROUPS*

Functional groups are groups of atoms that occur within molecules and confer specific chemical properties to those molecules. They are found along the "carbon backbone" of macromolecules. This carbon backbone is formed by chains and/or rings of carbon atoms with the occasional substitution of an element such as nitrogen or oxygen. Molecules with other elements in their carbon backbone are substituted hydrocarbons.

The functional groups in a macromolecule are usually attached to the carbon backbone at one or several different places along its chain and/or ring structure. Each of the four types of macromolecules—proteins, lipids, carbohydrates, and nucleic acids—has its own characteristic set of functional groups that contributes greatly to its differing chemical properties and its function in living organisms.

A functional group can participate in specific chemical reactions. Some of the important functional groups in biological molecules are shown in [link]; they include: hydroxyl, methyl, carbonyl, carboxyl, amino, phosphate, and sulfhydryl. These groups play an important role in the formation of molecules like DNA, proteins, carbohydrates, and lipids. Functional groups are usually classified as hydrophobic or hydrophilic depending on their charge or polarity characteristics. An example of a hydrophobic group is the non-polar methane molecule. Among the hydrophilic functional groups is the carboxyl group found in amino acids, some amino acid side chains, and the fatty acids that form triglycerides and phospholipids. This carboxyl group ionizes to release hydrogen ions ($H^+$) from the COOH group resulting in the negatively charged $COO^-$ group; this contributes to the hydrophilic nature of whatever molecule it is found on. Other functional groups, such as the carbonyl group, have a partially negatively charged oxygen atom that may form hydrogen bonds with water molecules, again making the molecule more hydrophilic.

The functional groups shown here are found in many different biological molecules.

| Functional Group | Structure | Properties |
|---|---|---|
| Hydroxyl | R—O—H | Polar |
| Methyl | R—CH₃ | Nonpolar |
| Carbonyl | R—C(=O)—R' | Polar |
| Carboxyl | R—C(=O)—OH | Charged, ionizes to release $H^+$. Since carboxyl groups can release $H^+$ ions into solution, they are considered acidic. |
| Amino | R—NH₂ | Charged, accepts $H^+$ to form $NH_3^+$. Since amino groups can remove $H^+$ from solution, they are considered basic. |
| Phosphate | R—O—P(=O)(OH)—OH | Charged, ionizes to release $H^+$. Since phosphate groups can release $H^+$ ions into solution, they are considered acidic. |
| Sulfhydryl | R—S—H | Polar |

Hydrogen bonds between functional groups (within the same molecule or between different molecules) are important to the function of many macromolecules and help them to fold properly into and maintain the appropriate shape for functioning. Hydrogen bonds are also involved in various recognition processes, such as DNA complementary base pairing and the binding of an enzyme to its substrate, as illustrated in [link].

Hydrogen bonds connect two strands of DNA together to create the double-helix structure.

## SECTION SUMMARY

The unique properties of carbon make it a central part of biological molecules. Carbon binds to oxygen, hydrogen, and nitrogen covalently to form the many molecules important for cellular function. Carbon has four electrons in its outermost shell and can form four bonds. Carbon and hydrogen can form hydrocarbon chains or rings. Functional groups are groups of atoms that confer specific properties to hydrocarbon (or substituted hydrocarbon) chains or rings that define their overall chemical characteristics and function.

## ART CONNECTIONS

Which of the following statements is false?

1. Molecules with the formulas $CH_3CH_2COOH$ and $C_3H_6O_2$ could be structural isomers.
2. Molecules must have a double bond to be *cis–trans* isomers.
3. To be enantiomers, a molecule must have at least three different atoms or groups connected to a central carbon.
4. To be enantiomers, a molecule must have at least four different atoms or groups connected to a central carbon.

## Glossary

**aliphatic hydrocarbon:** hydrocarbon consisting of a linear chain of carbon atoms

**aromatic hydrocarbon:** hydrocarbon consisting of closed rings of carbon atoms

**enantiomers:** molecules that share overall structure and bonding patterns, but differ in how the atoms are three dimensionally placed such that they are mirror images of each other

**functional group:** group of atoms that provides or imparts a specific function to a carbon skeleton

**geometric isomer:** isomer with similar bonding patterns differing in the placement of atoms alongside a double covalent bond

**hydrocarbon:** molecule that consists only of carbon and hydrogen

**isomers:** molecules that differ from one another even though they share the same chemical formula

**organic molecule:** any molecule containing carbon (except carbon dioxide)

**structural isomers:** molecules that share a chemical formula but differ in the placement of their chemical bonds

**substituted hydrocarbon:** hydrocarbon chain or ring containing an atom of another element in place of one of the backbone carbons

# Synthesis of Biological Macromolecules

**Learning Objectives**

By the end of this section, you will be able to:

- Understand the synthesis of macromolecules
- Explain dehydration (or condensation) and hydrolysis reactions

As you've learned, biological macromolecules are large molecules, necessary for life, that are built from smaller organic molecules. There are four major classes of biological macromolecules (carbohydrates, lipids, proteins, and nucleic acids); each is an important cell component and performs a wide array of functions. Combined, these molecules make up the majority of a cell's dry mass (recall that water makes up the majority of its complete mass). Biological macromolecules are organic, meaning they contain carbon. In addition, they may contain hydrogen, oxygen, nitrogen, and additional minor elements.

## DEHYDRATION SYNTHESIS

Most macromolecules are made from single subunits, or building blocks, called monomers. The monomers combine with each other using covalent bonds to form larger molecules known as polymers. In doing so, monomers release water molecules as byproducts. This type of reaction is known as dehydration synthesis, which means "to put together while losing water."

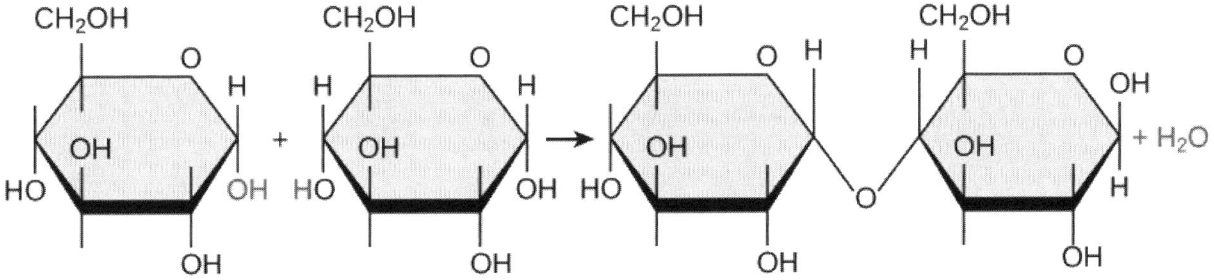

*Figure 1: In the dehydration synthesis reaction depicted above, two molecules of glucose are linked together to form the disaccharide maltose. In the process, a water molecule is formed.*

In a dehydration synthesis reaction ([Figure 1]), the hydrogen of one monomer combines with the hydroxyl group of another monomer, releasing a molecule of water. At the same time, the monomers share electrons and form covalent bonds. As additional monomers join, this chain of repeating monomers forms a polymer. Different types of monomers can combine in many configurations, giving rise to a diverse group of macromolecules. Even one kind of monomer can combine in a variety of ways to form several different polymers: for example, glucose monomers are the constituents of starch, glycogen, and cellulose.

## HYDROLYSIS

Polymers are broken down into monomers in a process known as hydrolysis, which means "to split water," a reaction in which a water molecule is used during the breakdown ([Figure 2]). During these reactions, the polymer is broken into two components: one part gains a hydrogen atom (H+) and the other gains a hydroxyl molecule (OH−) from a split water molecule.

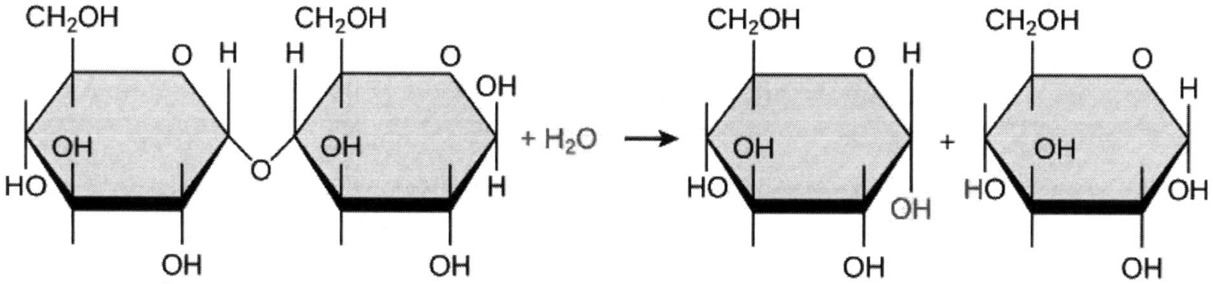

*Figure 2: In the hydrolysis reaction shown here, the disaccharide maltose is broken down to form two glucose monomers with the addition of a water molecule. Note that this reaction is the reverse of the synthesis reaction shown in [Figure 1].*

Dehydration and hydrolysis reactions are catalyzed, or "sped up," by specific enzymes; dehydration reactions involve the formation of new bonds, requiring energy, while hydrolysis reactions break bonds and release energy. These reactions are similar for most macromolecules, but each monomer and polymer reaction is specific for its class. For example, in our bodies, food is hydrolyzed, or broken down, into smaller molecules by catalytic enzymes in the digestive system. This allows for easy absorption of nutrients by cells in the intestine. Each macromolecule is broken down by a specific enzyme. For instance, carbohydrates are broken down by amylase, sucrase, lactase, or maltase. Proteins are broken down by the enzymes pepsin and peptidase, and by hydrochloric acid. Lipids are broken down by lipases. Breakdown of these macromolecules provides energy for cellular activities.

> Visit this site to see visual representations of dehydration synthesis and hydrolysis: https://www.youtube.com/watch?v=ZMTeqZLXBSo.

## SECTION SUMMARY

Proteins, carbohydrates, nucleic acids, and lipids are the four major classes of biological macromolecules—large molecules necessary for life that are built from smaller organic molecules. Macromolecules are made up of single units known as monomers that are joined by covalent bonds to form larger polymers. The polymer is more than the sum of its parts: it acquires new characteristics, and leads to an osmotic pressure that is much lower than that formed by its ingredients; this is an important advantage in the maintenance of cellular osmotic conditions. A monomer joins with another monomer with the release of a water molecule, leading to the formation of a covalent bond. These types of reactions are known as dehydration or condensation reactions. When polymers are broken down into smaller units (monomers), a molecule of water is used for each bond broken by these reactions; such reactions are known as hydrolysis reactions. Dehydration and hydrolysis reactions are similar for all macromolecules, but each monomer and polymer reaction is specific to its class. Dehydration reactions typically require an investment of energy for new bond formation, while hydrolysis reactions typically release energy by breaking bonds.

## Glossary

**biological macromolecule:** large molecule necessary for life that is built from smaller organic molecules

**dehydration synthesis:** (also, condensation) reaction that links monomer molecules together, releasing a molecule of water for each bond formed

**hydrolysis:** reaction causes breakdown of larger molecules into smaller molecules with the utilization of water

**monomer:** smallest unit of larger molecules called polymers

**polymer:** chain of monomer residues that is linked by covalent bonds; polymerization is the process of polymer formation from monomers by condensation

# Exercises

## PART 1

1. Explain why the symbol for an atom of the element oxygen and the formula for a molecule of oxygen differ.

2. Explain why the symbol for the element sulfur and the formula for a molecule of sulfur differ.

3. Write the molecular and empirical formulas of the following compounds:

   a. 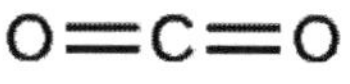 O=C=O

   b. H—C≡C—H

   c. H₂C=CH₂ (with H's on each carbon)

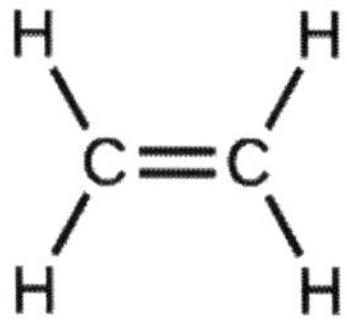

   d. 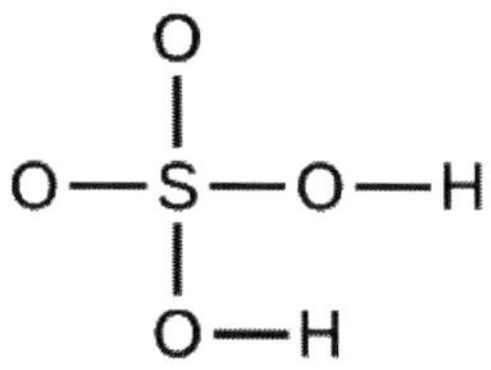 
   ```
        O
        ‖
   O—S—O—H
        |
        O—H
   ```

4. Write the molecular and empirical formulas of the following compounds:

   a. 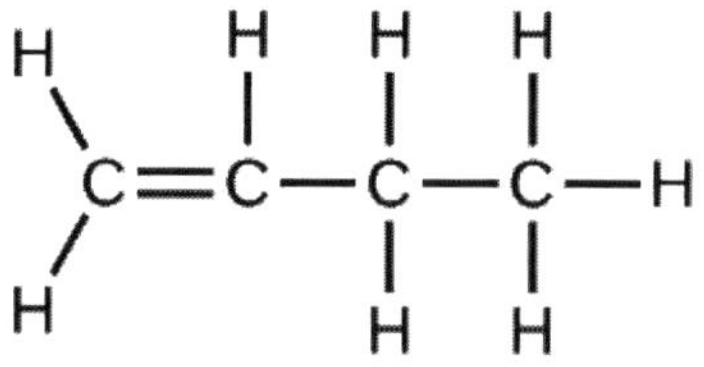 H₂C=CH—CH₂—CH₃ (shown with all H's)

   b. 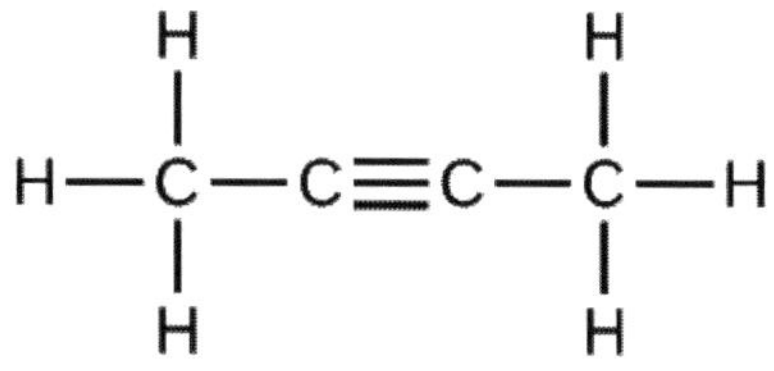 H₃C—C≡C—CH₃ (shown with all H's)

   c.
   ```
       Cl  Cl
       |   |
   Cl—Si—Si—Cl
       |   |
       H   H
   ```

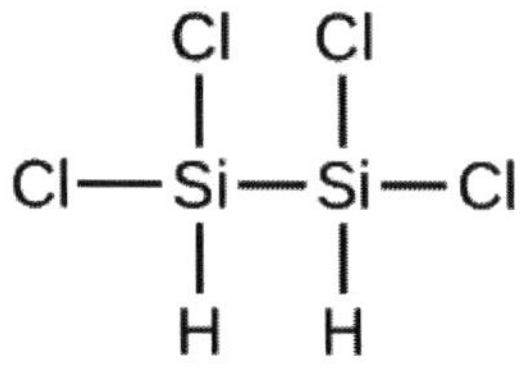

   d. 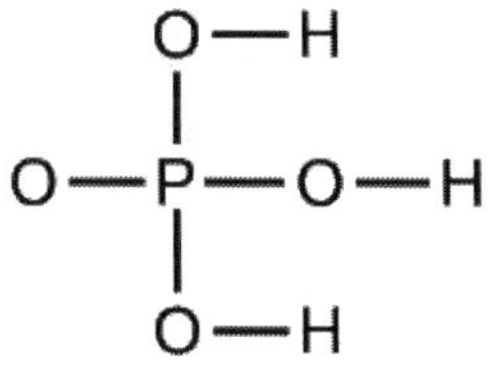
   ```
         O—H
         |
     O—P—O—H
         |
         O—H
   ```

214

5. Determine the empirical formulas for the following compounds:

   a. caffeine, $C_8H_{10}N_4O_2$

   b. fructose, $C_{12}H_{22}O_{11}$

   c. hydrogen peroxide, $H_2O_2$

   d. glucose, $C_6H_{12}O_6$

   e. ascorbic acid (vitamin C), $C_6H_8O_6$

6. Determine the empirical formulas for the following compounds:

   a. acetic acid, $C_2H_4O_2$

   b. citric acid, $C_6H_8O_7$

   c. hydrazine, $N_2H_4$

   d. nicotine, $C_{10}H_{14}N_2$

   e. butane, $C_4H_{10}$

7. Write the empirical formulas for the following compounds:

   a.
   $$\begin{array}{c} H \quad\; O \\ | \quad\;\; \| \\ H-C-C-O-H \\ | \\ H \end{array}$$

   b.
   $$\begin{array}{c} H \quad\; O \quad\quad H \quad H \\ | \quad\;\; \| \quad\quad\; | \quad\;\; | \\ H-C-C-O-C-C-H \\ | \quad\quad\quad\quad\;\; | \quad\;\; | \\ H \quad\quad\quad\quad H \quad H \end{array}$$

## PART 2

1. Open the **Build a Molecule simulation** and select the "Playground" tab. Select an appropriate atoms "Kit" to build a molecule with two carbon and six hydrogen atoms. Drag atoms into the space above the "Kit" to make a molecule. A name will appear when you have made an actual molecule that exists (even if it is not the one you want). You can use the scissors tool to separate atoms if you would like to change the connections.

   1. Draw the structural formula of this molecule and state its name.

   2. Can you arrange these atoms in any way to make a different compound?

2. Use the **Build a Molecule simulation** to repeat question 1, but build a molecule with two carbons, six hydrogens, and one oxygen.

   1. Draw the structural formula of this molecule and state its name.

   2. Can you arrange these atoms to make a different molecule? If so, draw its structural formula and state its name.

   3. How are the molecules drawn in (a) and (b) the same? How do they differ? What are they called (the type of relationship between these molecules, not their names).

216   Organic Molecules

3. Use the **Build a Molecule simulation** to repeat question 1, but build a molecule with three carbons, seven hydrogens, and one chlorine.

   a. Draw the structural formula of this molecule and state its name.

   b. Can you arrange these atoms to make a different molecule? If so, draw its structural formula and state its name.

   c. How are the molecules drawn in (a) and (b) the same? How do they differ? What are they called (the type of relationship between these molecules, not their names)?

## PART 3

1. What property of carbon makes it essential for organic life?

2. Compare and contrast saturated and unsaturated triglycerides:

## PART 4

1. Why are biological macromolecules considered organic?

2. What role do electrons play in dehydration synthesis and hydrolysis?

3. Name the reaction in which macromolecules are produced:

4. Name the reaction in which macromolecules are broken down:

# Check Your Knowledge: Self-Test

1. A molecule of metaldehyde (a pesticide used for snails and slugs) contains 8 carbon atoms, 16 hydrogen atoms, and 4 oxygen atoms. What are the molecular and empirical formulas of metaldehyde?

2. Each carbon molecule can bond with as many as _____ other atom(s) or molecule(s).

    a) one

    b) two

    c) six

    d) four

3. Which of the following is not a functional group that can bond with carbon?

    a) sodium

    b) hydroxyl

    c) phosphate

    d) carbonyl

4. Dehydration synthesis leads to the formation of:

    a) monomers

    b) polymers

    c) water and polymers

    d) none of the above

5. During the breakdown of polymers, which of the following reactions takes place?

    a) hydrolysis

    b) dehydration

    c) condensation

    d) covalent bond

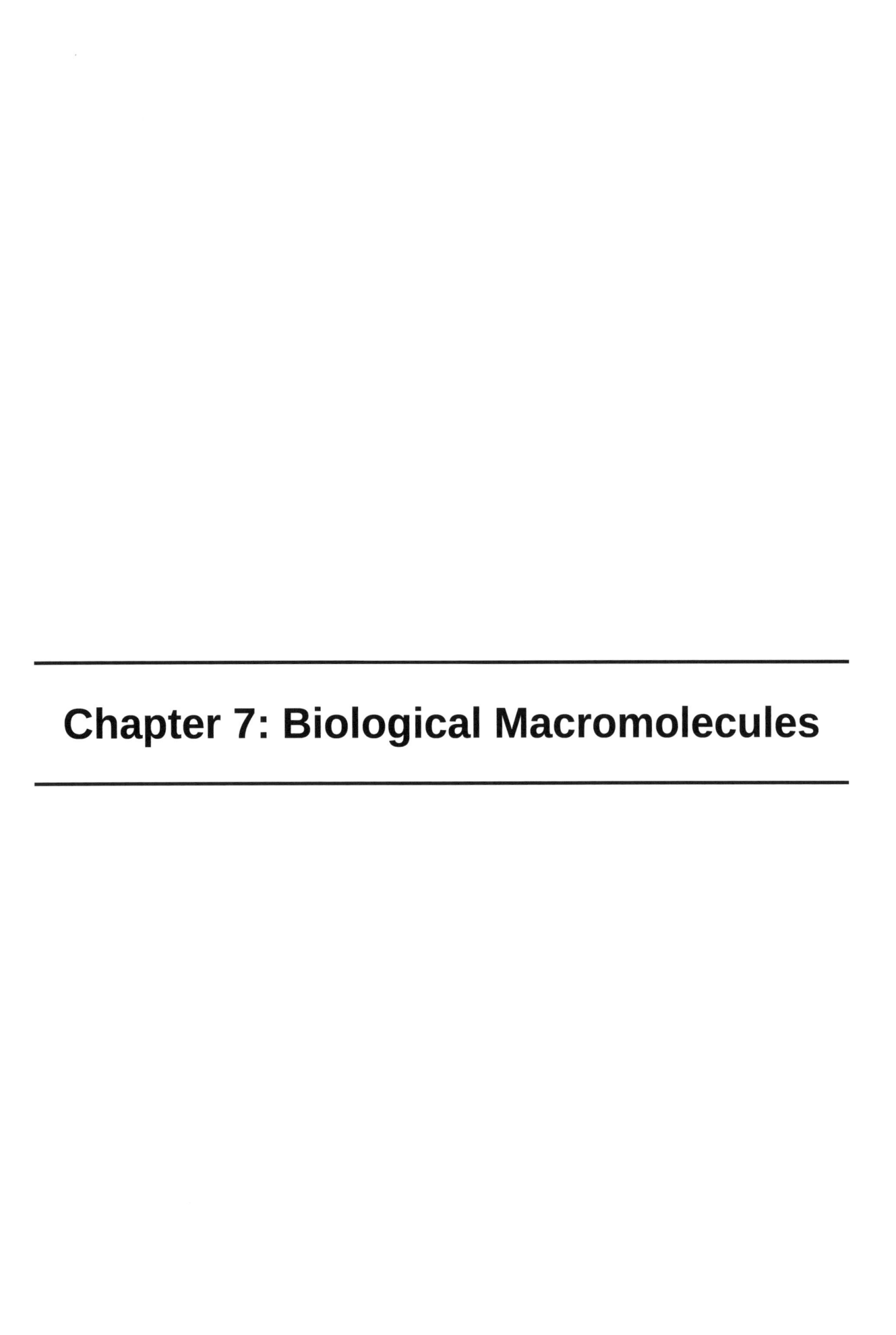

# Chapter 7: Biological Macromolecules

# Carbohydrates

### Learning Objectives

By the end of this section, you will be able to:

- Discuss the role of carbohydrates in cells and in the extracellular materials of animals and plants
- Explain the classifications of carbohydrates
- List common monosaccharides, disaccharides, and polysaccharides

Most people are familiar with carbohydrates, one type of macromolecule, especially when it comes to what we eat. To lose weight, some individuals adhere to "low-carb" diets. Athletes, in contrast, often "carb-load" before important competitions to ensure that they have enough energy to compete at a high level. Carbohydrates are, in fact, an essential part of our diet; grains, fruits, and vegetables are all natural sources of carbohydrates. Carbohydrates provide energy to the body, particularly through glucose, a simple sugar that is a component of starch and an ingredient in many staple foods. Carbohydrates also have other important functions in humans, animals, and plants.

## MOLECULAR STRUCTURES

Carbohydrates can be represented by the stoichiometric formula $(CH_2O)_n$, where n is the number of carbons in the molecule. In other words, the ratio of carbon to hydrogen to oxygen is 1:2:1 in carbohydrate molecules. This formula also explains the origin of the term "carbohydrate": the components are carbon ("carbo") and the components of water (hence, "hydrate"). Carbohydrates are classified into three subtypes: monosaccharides, disaccharides, and polysaccharides.

### Monosaccharides

Monosaccharides (mono- = "one"; sacchar- = "sweet") are simple sugars, the most common of which is glucose. In monosaccharides, the number of carbons usually ranges from three to seven. Most monosaccharide names end with the suffix -ose. If the sugar has an aldehyde group (the functional group with the structure R-CHO), it is known as an aldose, and if it has a ketone group (the functional group with the structure RC(=O)R'), it is known as a ketose. Depending on the number of carbons in the sugar, they also may be known as trioses (three carbons), pentoses (five carbons), and or hexoses (six carbons). See **[Figure 1]** for an illustration of the monosaccharides.

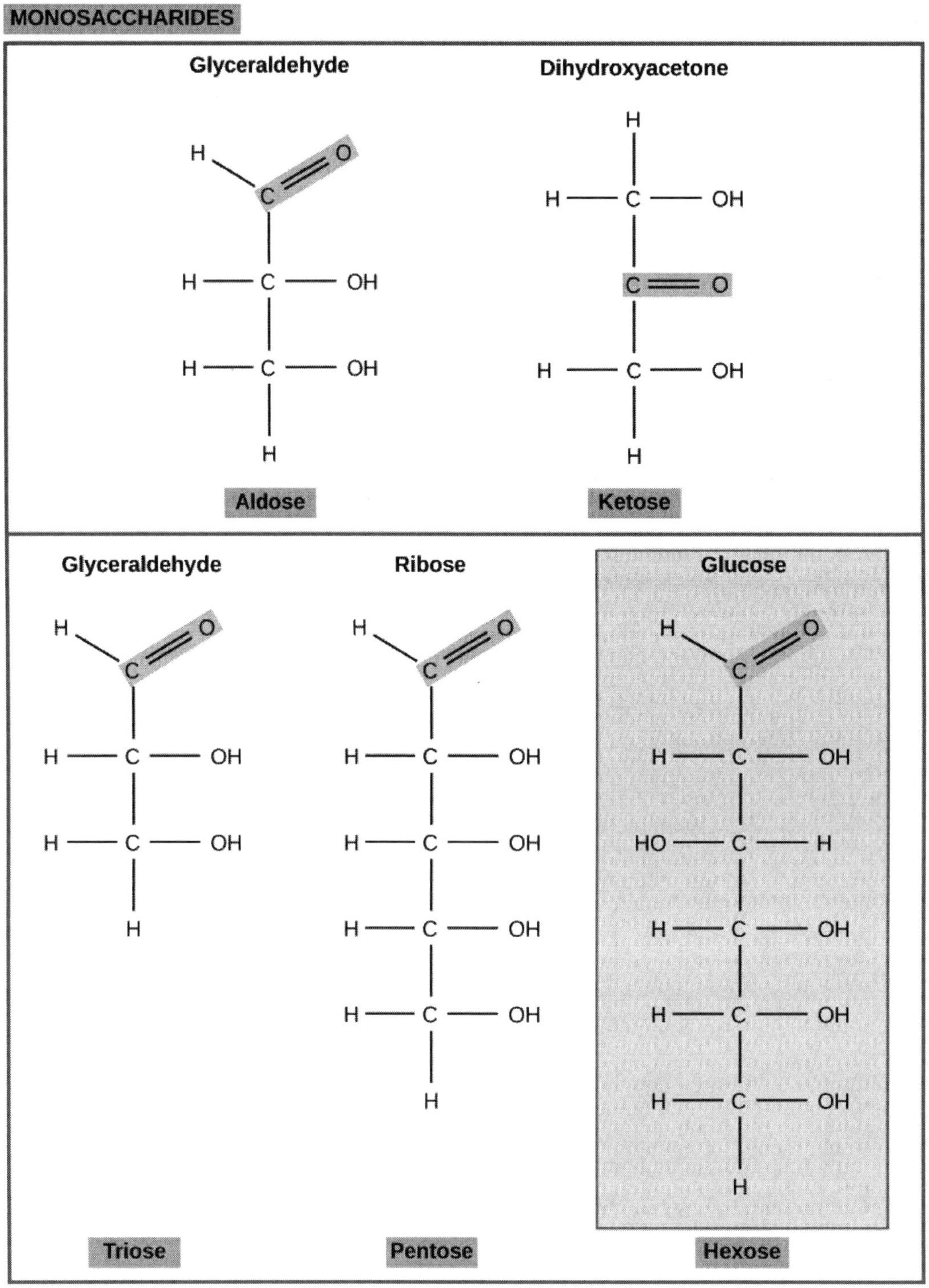

222  Biological Macromolecules

**MONOSACCHARIDES**

*Figure 1: Monosaccharides are classified based on the position of their carbonyl group and the number of carbons in the backbone. Aldoses have a carbonyl group (indicated in green) at the end of the carbon chain, and ketoses have a carbonyl group in the middle of the carbon chain. Trioses, pentoses, and hexoses have three, five, and six carbon backbones, respectively.*

The chemical formula for glucose is $C_6H_{12}O_6$. In humans, glucose is an important source of energy. During cellular respiration, energy is released from glucose, and that energy is used to help make adenosine triphosphate (ATP). Plants synthesize glucose using carbon dioxide and water, and glucose in turn is used for energy requirements for the plant. Excess glucose is often stored as starch that is catabolized (the breakdown of larger molecules by cells) by humans and other animals that feed on plants.

Carbohydrates 223

Galactose (part of lactose, or milk sugar) and fructose (found in sucrose, in fruit) are other common monosaccharides. Although glucose, galactose, and fructose all have the same chemical formula ($C_6H_{12}O_6$), they differ structurally and chemically (and are known as isomers) because of the different arrangement of functional groups around the asymmetric carbon; all of these monosaccharides have more than one asymmetric carbon ([**Figure 2**]).

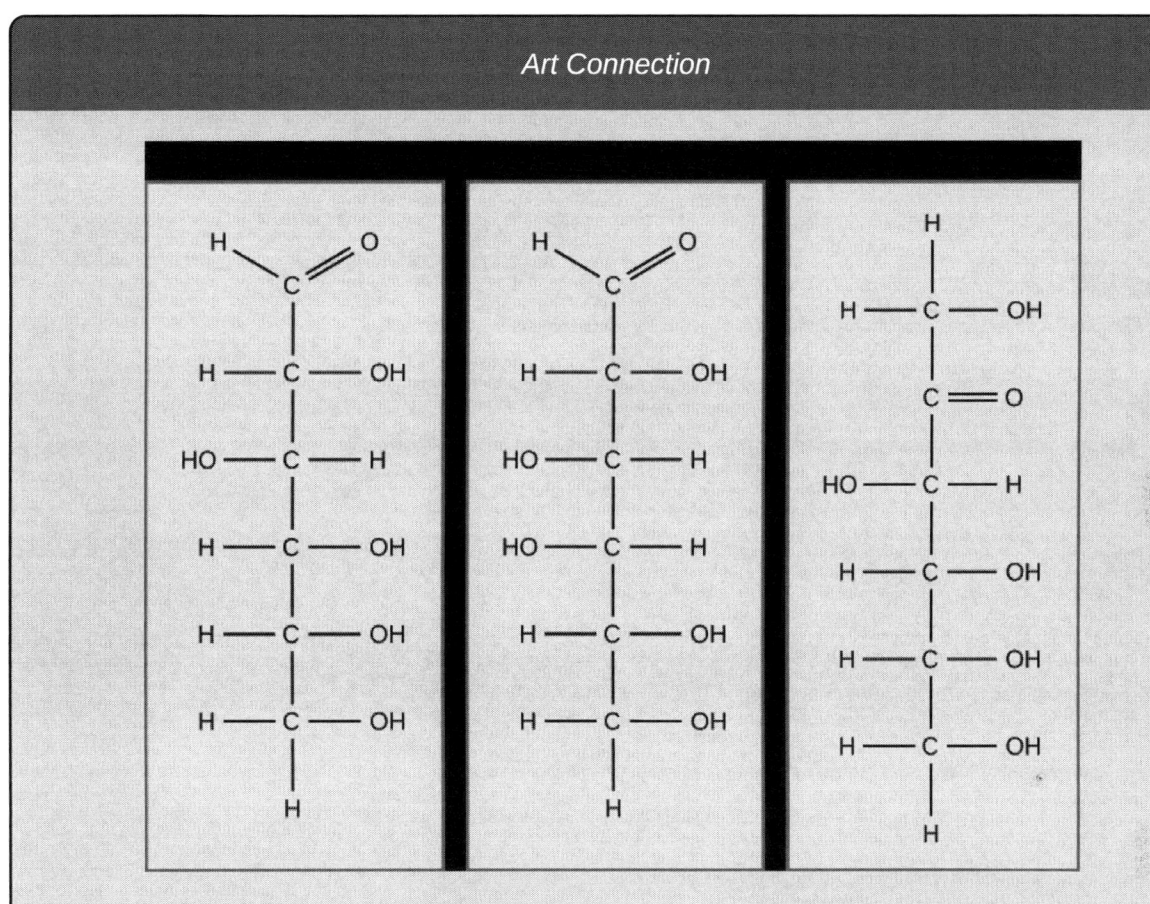

### Art Connection

*Figure 2: Glucose, galactose, and fructose are all hexoses. They are structural isomers, meaning they have the same chemical formula (C6H12O6) but a different arrangement of atoms.*

What kind of sugars are these, aldose or ketose?

Glucose, galactose, and fructose are isomeric monosaccharides (hexoses), meaning they have the same chemical formula but have slightly different structures. Glucose and galactose are aldoses, and fructose is a ketose.

Monosaccharides can exist as a linear chain or as ring-shaped molecules; in aqueous solutions they are usually found in ring forms ([**Figure 3**]). Glucose in a ring form can have two different arrangements of the hydroxyl group (OH) around the anomeric carbon (carbon 1 that becomes asymmetric in the process of ring formation). If the hydroxyl group is below carbon number 1 in the sugar, it is said to be in the alpha (α) position, and if it is above the plane, it is said to be in the beta (β) position.

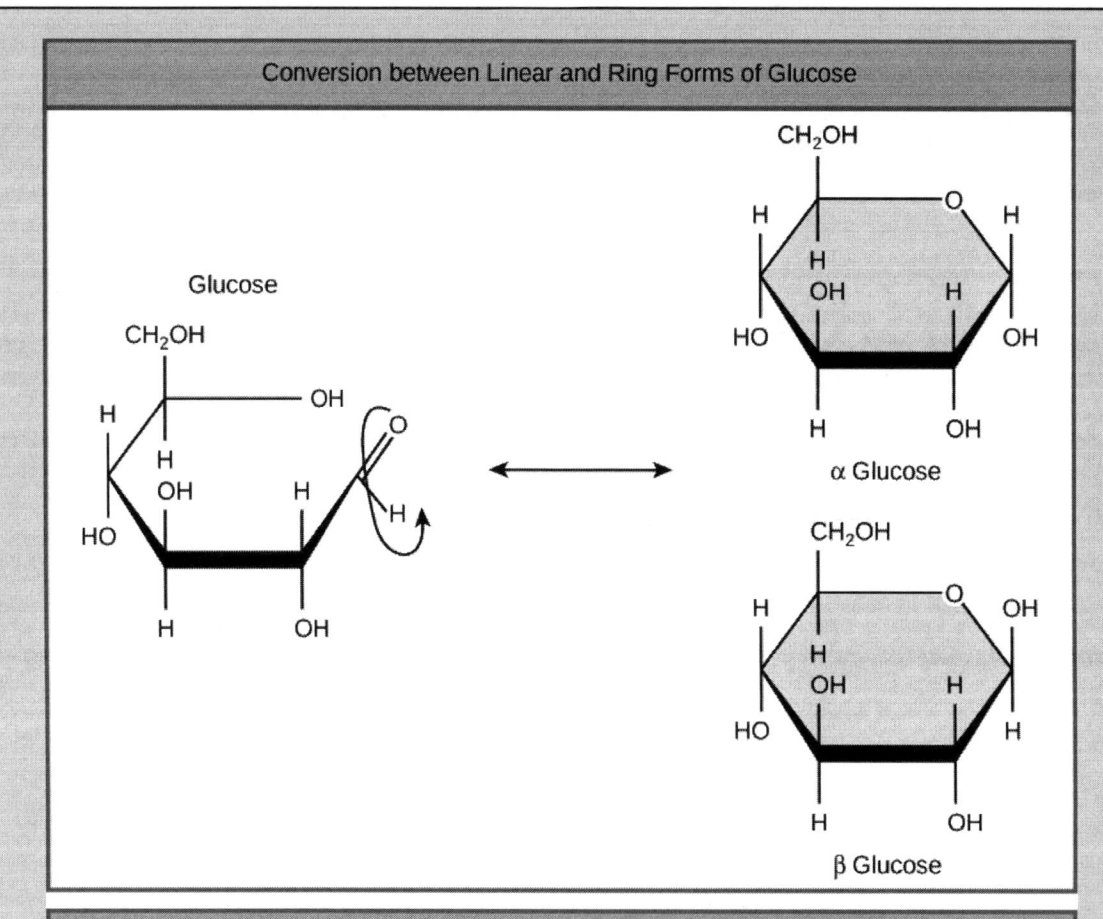

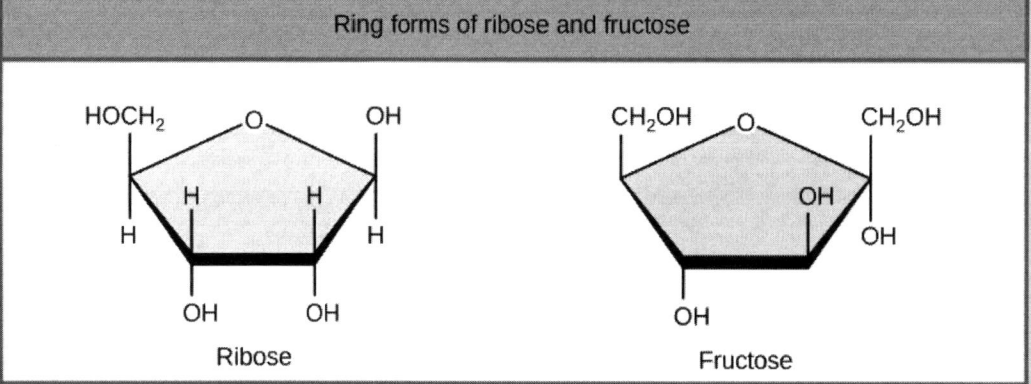

*Figure 3: Five and six carbon monosaccharides exist in equilibrium between linear and ring forms. When the ring forms, the side chain it closes on is locked into an α or β position. Fructose and ribose also form rings, although they form five-membered rings as opposed to the six-membered ring of glucose.*

### Disaccharides

Disaccharides (di- = "two") form when two monosaccharides undergo a dehydration reaction (also known as a condensation reaction or dehydration synthesis). During this process, the hydroxyl group of one monosaccharide combines with the hydrogen of another monosaccharide, releasing a molecule of water and forming a covalent bond. A covalent bond formed between a carbohydrate molecule and another molecule (in this case, between two monosaccharides) is known as a glycosidic bond ([Figure 4]). Glycosidic bonds (also called glycosidic linkages) can be of the alpha or the beta type.

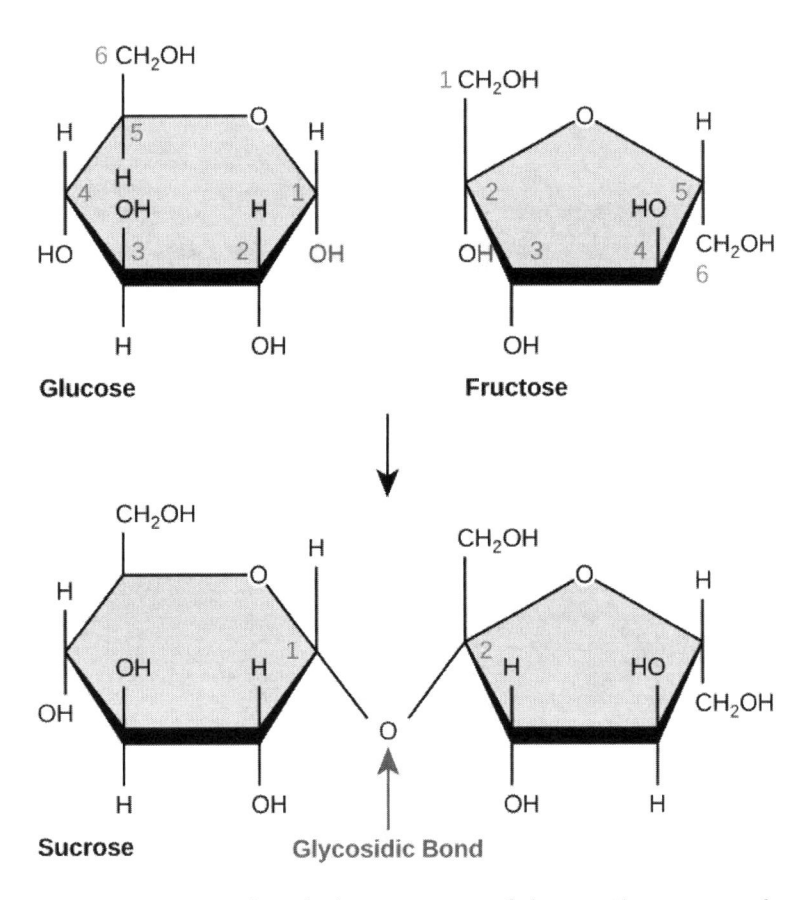

*Figure 4: Sucrose is formed when a monomer of glucose and a monomer of fructose are joined in a dehydration reaction to form a glycosidic bond. In the process, a water molecule is lost. By convention, the carbon atoms in a monosaccharide are numbered from the terminal carbon closest to the carbonyl group. In sucrose, a glycosidic linkage is formed between carbon 1 in glucose and carbon 2 in fructose.*

Common disaccharides include lactose, maltose, and sucrose ([Figure 5]). Lactose is a disaccharide consisting of the monomers glucose and galactose. It is found naturally in milk. Maltose, or malt sugar, is a disaccharide formed by a dehydration reaction between two glucose molecules. The most common disaccharide is sucrose, or table sugar, which is composed of the monomers glucose and fructose.

226  Biological Macromolecules

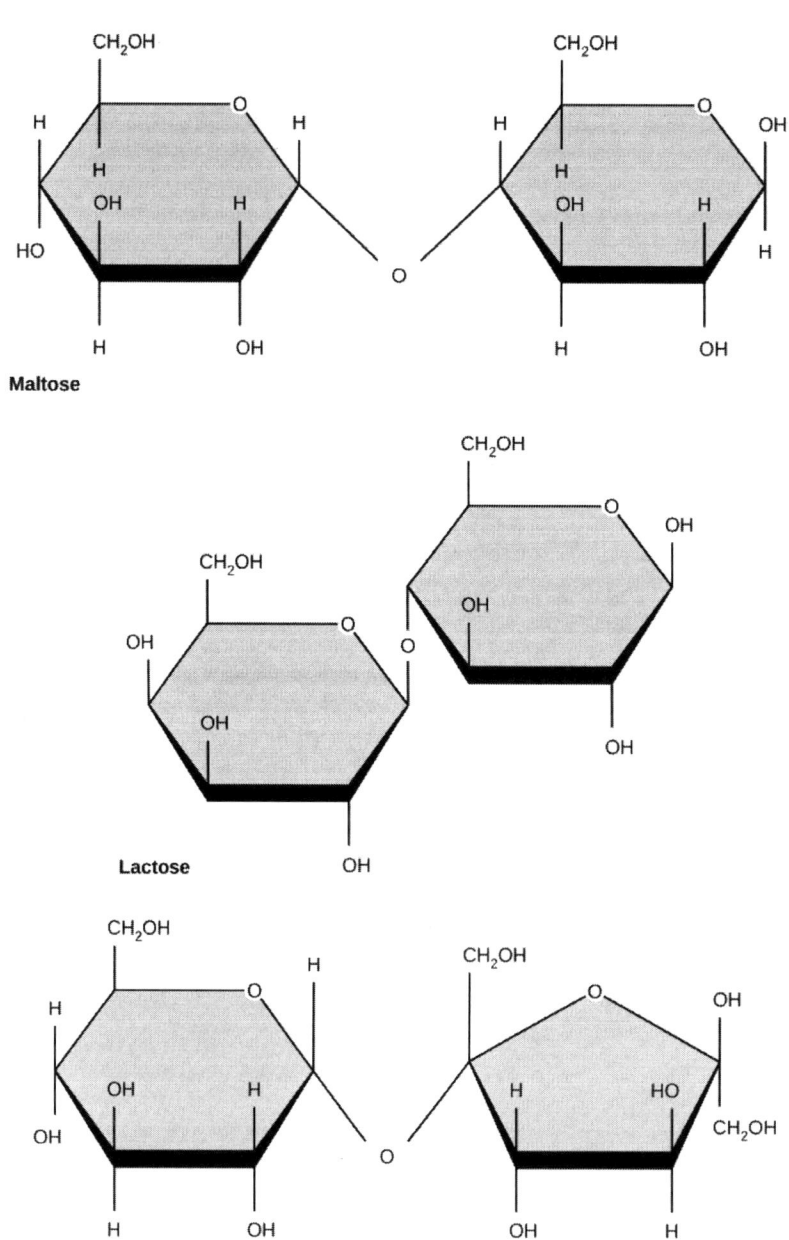

*Figure 5: Common disaccharides include maltose (grain sugar), lactose (milk sugar), and sucrose (table sugar).*

## Polysaccharides

A long chain of monosaccharides linked by glycosidic bonds is known as a polysaccharide (poly- = "many"). The chain may be branched or unbranched, and it may contain different types of monosaccharides. The molecular weight may be 100,000 daltons or more depending on the number of monomers joined. Starch, glycogen, cellulose, and chitin are primary examples of polysaccharides.

Starch is the stored form of sugars in plants and is made up of a mixture of amylose and amylopectin (both polymers of glucose). Plants are able to synthesize glucose, and the excess glucose, beyond the plant's immediate energy needs, is stored as starch in different plant parts, including roots and seeds. The starch in the seeds provides food for the embryo as it germinates and can also act as a source of food for humans and animals. The starch that is consumed by humans is broken down by enzymes, such as salivary amylases, into smaller molecules, such as maltose and glucose. The cells can then absorb the glucose.

Starch is made up of glucose monomers that are joined by α 1-4 or α 1-6 glycosidic bonds. The numbers 1-4 and 1-6 refer to the carbon number of the two residues that have joined to form the bond. As illustrated in **[Figure 6]**, amylose is starch formed by unbranched chains of glucose monomers (only α 1-4 linkages), whereas amylopectin is a branched polysaccharide (α 1-6 linkages at the branch points).

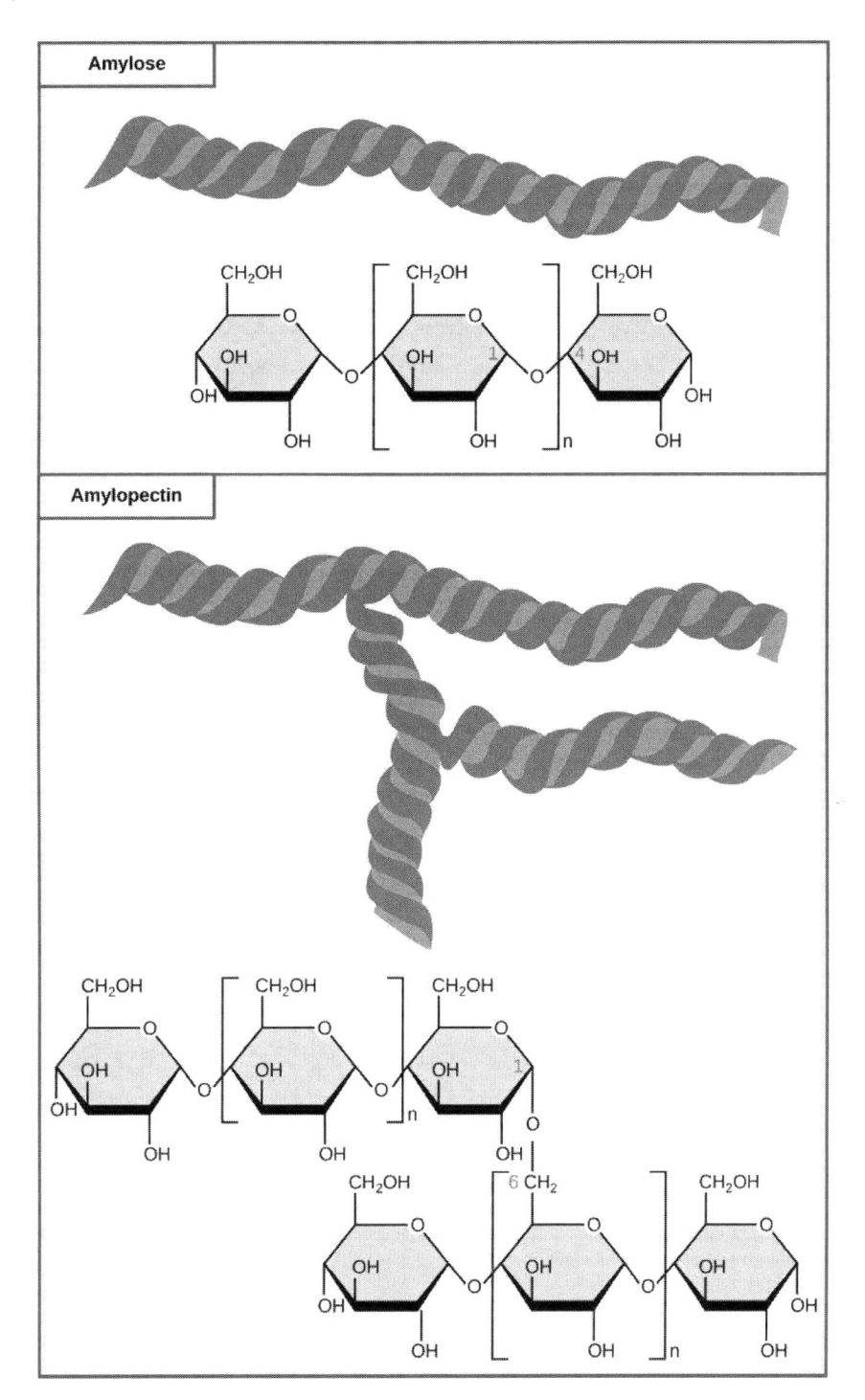

*Figure 6: Amylose and amylopectin are two different forms of starch. Amylose is composed of unbranched chains of glucose monomers connected by α 1,4 glycosidic linkages. Amylopectin is composed of branched chains of glucose monomers connected by α 1,4 and α 1,6 glycosidic linkages. Because of the way the subunits are joined, the glucose chains have a helical structure. Glycogen (not shown) is similar in structure to amylopectin but more highly branched.*

Glycogen is the storage form of glucose in humans and other vertebrates and is made up of monomers of glucose. Glycogen is the animal equivalent of starch and is a highly branched molecule usually stored in liver and muscle cells. Whenever blood glucose levels decrease, glycogen is broken down to release glucose in a process known as glycogenolysis.

Cellulose is the most abundant natural biopolymer. The cell wall of plants is mostly made of cellulose; this provides structural support to the cell. Wood and paper are mostly cellulosic in nature. Cellulose is made up of glucose monomers that are linked by β 1-4 glycosidic bonds ([Figure 7]).

**Cellulose fibers**

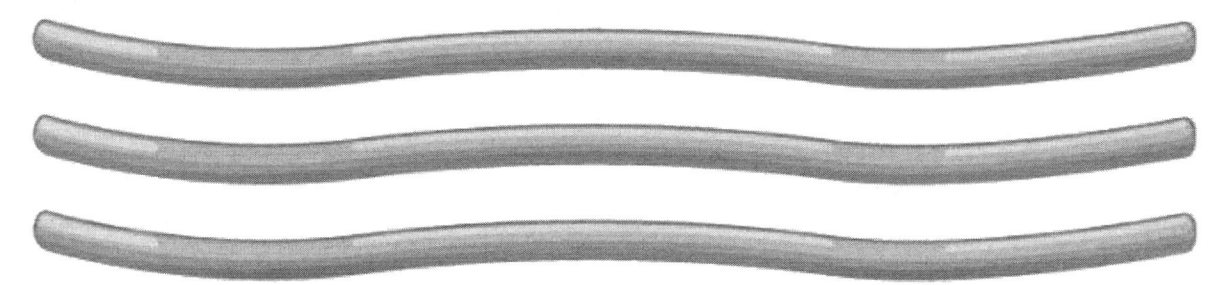

**Cellulose structure**

As shown in [Figure 7], every other glucose monomer in cellulose is flipped over, and the monomers are packed tightly as extended long chains. This gives cellulose its rigidity and high tensile strength—which is so important to plant cells. While the β 1-4 linkage cannot be broken down by human digestive enzymes, herbivores such as cows, koalas, and buffalos are able, with the help of the specialized flora in their stomach, to digest plant material that is rich in cellulose and use it as a food source. In these animals, certain species of bacteria and protists reside in the rumen (part of the digestive system of herbivores) and secrete the enzyme cellulase. The appendix of grazing animals also contains bacteria that digest cellulose, giving it an important role in the digestive systems of ruminants. Cellulases can break down cellulose into glucose monomers that can be used as an energy source by the animal. Termites are also able to break down cellulose because of the presence of other organisms in their bodies that secrete cellulases.

Carbohydrates serve various functions in different animals. Arthropods (insects, crustaceans, and others) have an outer skeleton, called the exoskeleton, which protects their internal body parts (as seen in the bee in [Figure 8]). This exoskeleton is made of the biological macromolecule chitin, which is a polysaccharide-containing nitrogen. It is made of repeating units of N-acetyl-β-d-glucosamine, a modified sugar. Chitin is also a major component of fungal cell walls; fungi are neither animals nor plants and form a kingdom of their own in the domain Eukarya.

Career Connections

Registered DietitianObesity is a worldwide health concern, and many diseases such as diabetes and heart disease are becoming more prevalent because of obesity. This is one of the reasons why registered dietitians are increasingly sought after for advice. Registered dietitians help plan nutrition programs for individuals in various settings. They often work with patients in health care facilities, designing nutrition plans to treat and prevent diseases. For example, dietitians may teach a patient with diabetes how to manage blood sugar levels by eating the correct types and amounts of carbohydrates. Dietitians may also work in nursing homes, schools, and private practices.

To become a registered dietitian, one needs to earn at least a bachelor's degree in dietetics, nutrition, food technology, or a related field. In addition, registered dietitians must complete a supervised internship program and pass a national exam. Those who pursue careers in dietetics take courses in nutrition, chemistry, biochemistry, biology, microbiology, and human physiology. Dietitians must become experts in the chemistry and physiology (biological functions) of food (proteins, carbohydrates, and fats).

*Figure 8: Insects have a hard outer exoskeleton made of chitin, a type of polysaccharide. (credit: Louise Docker)*

## BENEFITS OF CARBOHYDRATES

Are carbohydrates good for you? People who wish to lose weight are often told that carbohydrates are bad for them and should be avoided. Some diets completely forbid carbohydrate consumption, claiming that a low-carbohydrate diet helps people to lose weight faster. However, carbohydrates have been an important part of the human diet for thousands of years; artifacts from ancient civilizations show the presence of wheat, rice, and corn in our ancestors' storage areas.

Carbohydrates should be supplemented with proteins, vitamins, and fats to be parts of a well-balanced diet. Calorie-wise, a gram of carbohydrate provides 4.3 Kcal. For comparison, fats provide 9 Kcal/g, a less desirable ratio. Carbohydrates contain soluble and insoluble elements; the insoluble part is known as fiber, which is mostly cellulose. Fiber has many uses; it promotes regular bowel movement by adding bulk, and it regulates the rate of consumption of blood glucose. Fiber also helps to remove excess cholesterol from the body: fiber binds to the cholesterol in the small intestine, then attaches to the cholesterol and prevents the cholesterol particles from entering the bloodstream, and then cholesterol exits the body via the feces. Fiber-rich diets also have a protective role in reducing the occurrence of colon cancer. In addition, a meal containing whole grains and vegetables gives a feeling of fullness. As an immediate source of energy, glucose is broken down during the process of cellular respiration, which produces ATP, the energy currency of the cell. Without the consumption of carbohydrates, the availability of "instant energy" would be reduced. Eliminating carbohydrates from the diet is not the best way to lose weight. A low-calorie diet that is rich in whole grains, fruits, vegetables, and lean meat, together with plenty of exercise and plenty of water, is the more sensible way to lose weight.

For an additional perspective on carbohydrates, explore "Biomolecules: the Carbohydrates" through this interactive animation: https://www.wisc-online.com/learn/general-education/anatomy-and-physiology2/ap21316/biomolecules-the-carbohydrates-video

## SECTION SUMMARY

Carbohydrates are a group of macromolecules that are a vital energy source for the cell and provide structural support to plant cells, fungi, and all of the arthropods that include lobsters, crabs, shrimp, insects, and spiders. Carbohydrates are classified as monosaccharides, disaccharides, and polysaccharides depending on the number of monomers in the molecule. Monosaccharides are linked by glycosidic bonds that are formed as a result of dehydration reactions, forming disaccharides and polysaccharides with the elimination of a water molecule for each bond formed. Glucose, galactose, and fructose are common monosaccharides, whereas common disaccharides include lactose, maltose, and sucrose. Starch and glycogen, examples of polysaccharides, are the storage forms of glucose in plants and animals, respectively. The long polysaccharide chains may be branched or unbranched. Cellulose is an example of an unbranched polysaccharide, whereas amylopectin, a constituent of starch, is a highly branched molecule. Storage of glucose, in the form of polymers like starch of glycogen, makes it slightly less accessible for metabolism; however, this prevents it from leaking out of the cell or creating a high osmotic pressure that could cause excessive water uptake by the cell.

### Glossary

**carbohydrate:** biological macromolecule in which the ratio of carbon to hydrogen and to oxygen is 1:2:1; carbohydrates serve as energy sources and structural support in cells and form the a cellular exoskeleton of arthropods

**cellulose:** polysaccharide that makes up the cell wall of plants; provides structural support to the cell

**chitin:** type of carbohydrate that forms the outer skeleton of all arthropods that include crustaceans and insects; it also forms the cell walls of fungi

**disaccharide:** two sugar monomers that are linked together by a glycosidic bond

**glycogen:** storage carbohydrate in animals

**glycosidic bond:** bond formed by a dehydration reaction between two monosaccharides with the elimination of a water molecule

**monosaccharide:** single unit or monomer of carbohydrates

**polysaccharide:** long chain of monosaccharides; may be branched or unbranched

**starch:** storage carbohydrate in plants

# Lipids

## Learning Objectives

By the end of this section, you will be able to:

- Describe the four major types of lipids
- Explain the role of fats in storing energy
- Differentiate between saturated and unsaturated fatty acids
- Describe phospholipids and their role in cells
- Define the basic structure of a steroid and some functions of steroids
- Explain the how cholesterol helps to maintain the fluid nature of the plasma membrane

Lipids include a diverse group of compounds that are largely nonpolar in nature. This is because they are hydrocarbons that include mostly nonpolar carbon–carbon or carbon–hydrogen bonds. Nonpolar molecules are hydrophobic ("water fearing"), or insoluble in water. Lipids perform many different functions in a cell. Cells store energy for long-term use in the form of fats. Lipids also provide insulation from the environment for plants and animals ([**Figure 1:** ]). For example, they help keep aquatic birds and mammals dry when forming a protective layer over fur or feathers because of their water-repellant hydrophobic nature. Lipids are also the building blocks of many hormones and are an important constituent of all cellular membranes. Lipids include fats, oils, waxes, phospholipids, and steroids.

*Figure 1: Hydrophobic lipids in the fur of aquatic mammals, such as this river otter, protect them from the elements. (credit: Ken Bosma)*

## FATS AND OILS

A fat molecule consists of two main components—glycerol and fatty acids. Glycerol is an organic compound (alcohol) with three carbons, five hydrogens, and three hydroxyl (OH) groups. Fatty acids have a long chain of hydrocarbons to which a carboxyl group is attached, hence the name "fatty acid." The number of carbons in the fatty acid may range from 4 to 36; most common are those containing 12–18 carbons. In a fat molecule, the fatty acids are attached to each of the three carbons of the glycerol molecule with an ester bond through an oxygen atom ([**Figure 2**]).

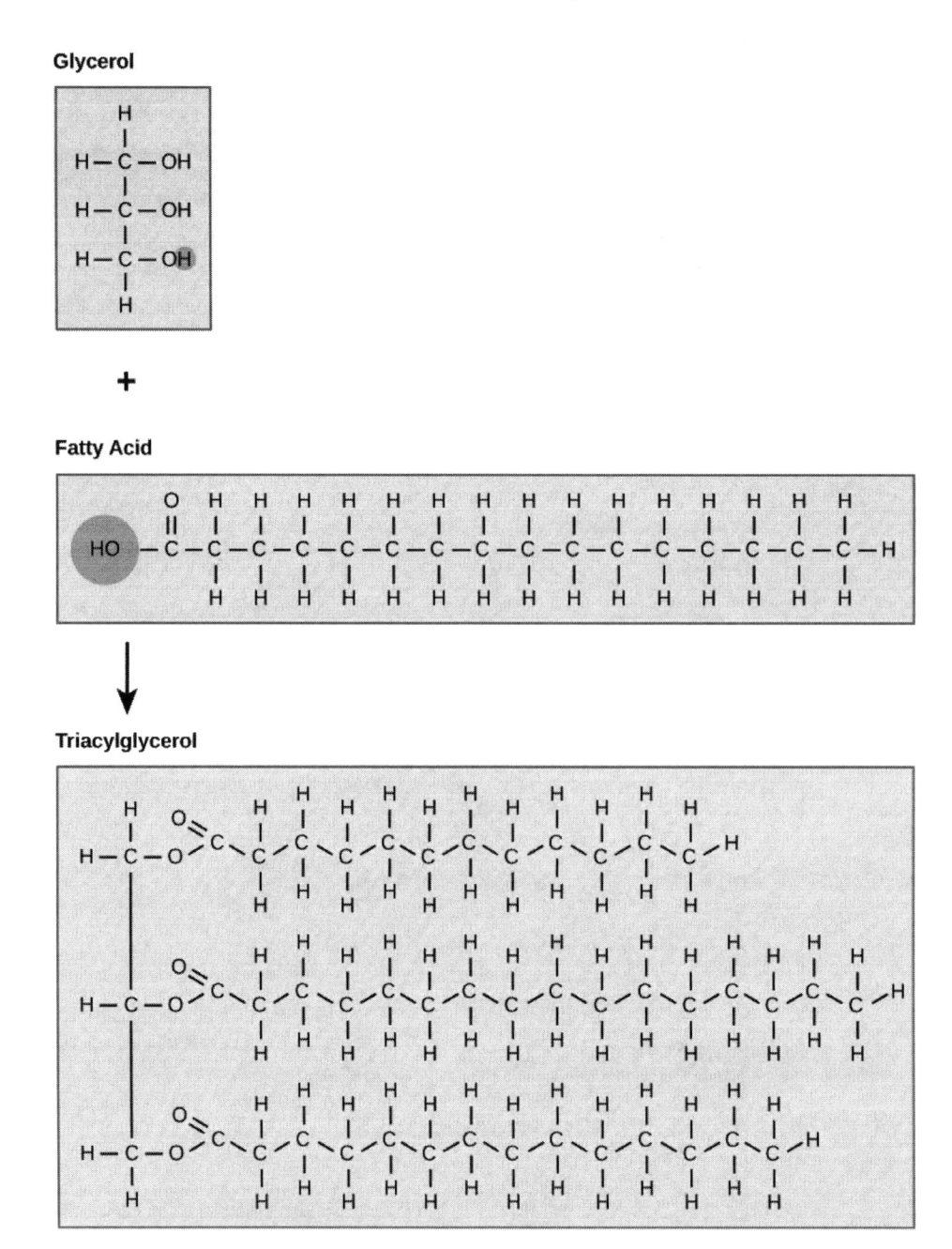

*Figure 2: Triacylglycerol is formed by the joining of three fatty acids to a glycerol backbone in a dehydration reaction. Three molecules of water are released in the process.*

During this ester bond formation, three water molecules are released. The three fatty acids in the triacylglycerol may be similar or dissimilar. Fats are also called triacylglycerols or triglycerides because of their chemical structure. Some fatty acids have common names that specify their origin. For example, palmitic acid, a saturated fatty acid, is derived from the palm tree. Arachidic acid is derived from *Arachis hypogea,* the scientific name for groundnuts or peanuts.

Fatty acids may be saturated or unsaturated. In a fatty acid chain, if there are only single bonds between neighboring carbons in the hydrocarbon chain, the fatty acid is said to be saturated. Saturated fatty acids are saturated with hydrogen; in other words, the number of hydrogen atoms attached to the carbon skeleton is maximized. Stearic acid is an example of a saturated fatty acid ([Figure 3])

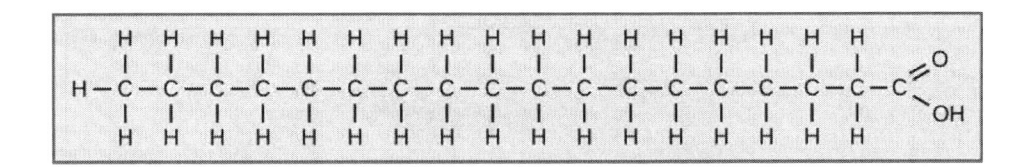

*Figure 3: Stearic acid is a common saturated fatty acid.*

When the hydrocarbon chain contains a double bond, the fatty acid is said to be unsaturated. Oleic acid is an example of an unsaturated fatty acid ([Figure 4]).

*Figure 4: Oleic acid is a common unsaturated fatty acid.*

Most unsaturated fats are liquid at room temperature and are called oils. If there is one double bond in the molecule, then it is known as a monounsaturated fat (e.g., olive oil), and if there is more than one double bond, then it is known as a polyunsaturated fat (e.g., canola oil).

When a fatty acid has no double bonds, it is known as a saturated fatty acid because no more hydrogen may be added to the carbon atoms of the chain. A fat may contain similar or different fatty acids attached to glycerol. Long straight fatty acids with single bonds tend to get packed tightly and are solid at room temperature. Animal fats with stearic acid and palmitic acid (common in meat) and the fat with butyric acid (common in butter) are examples of saturated fats. Mammals store fats in specialized cells called adipocytes, where globules of fat occupy most of the cell's volume. In plants, fat or oil is stored in many seeds and is used as a source of energy during seedling development. Unsaturated fats or oils are usually of plant origin and contain *cis* unsaturated fatty acids. *Cis* and *trans* indicate the configuration of the molecule around the double bond. If hydrogens are present in the same plane, it is referred to as a cis fat; if the hydrogen atoms are on two different planes, it is referred to as a trans fat. The *cis* double bond causes a bend or a "kink" that prevents the fatty acids from packing tightly, keeping them liquid at room temperature ([Figure 5]). Olive oil, corn oil, canola oil, and cod liver oil are examples of unsaturated fats. Unsaturated fats help to lower blood cholesterol levels whereas saturated fats contribute to plaque formation in the arteries.

## 234   Biological Macromolecules

**Saturated fatty acid**

Stearic acid

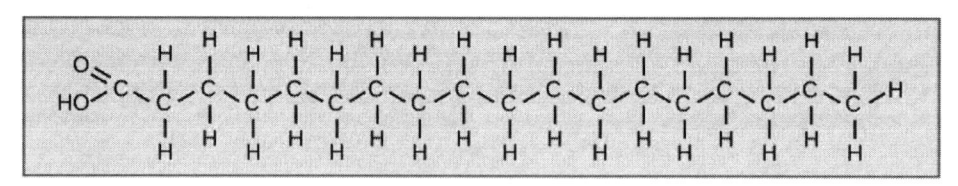

**Unsaturated fatty acids**

*Cis* oleic acid

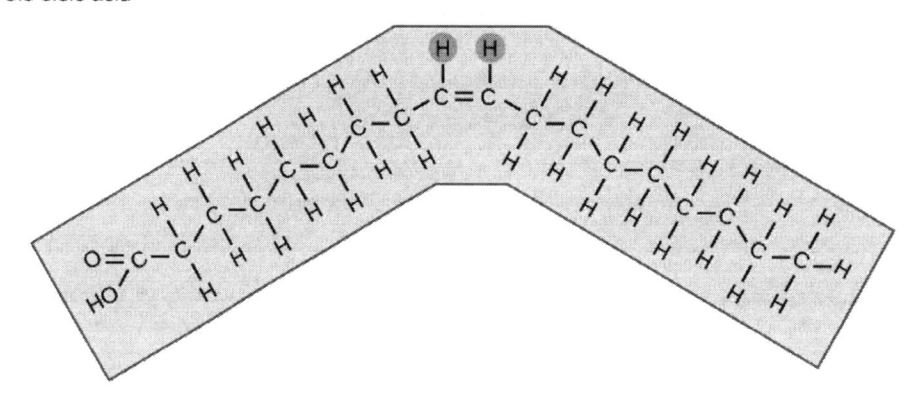

*Trans* oleic acid

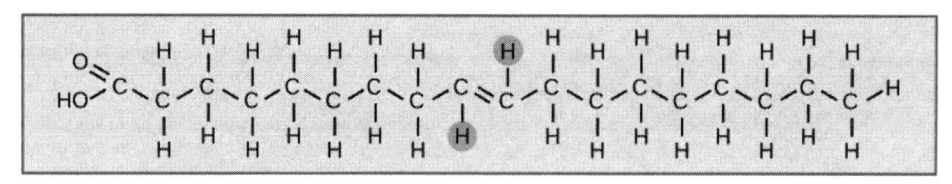

*Figure 5: Saturated fatty acids have hydrocarbon chains connected by single bonds only. Unsaturated fatty acids have one or more double bonds. Each double bond may be in a cis or trans configuration. In the cis configuration, both hydrogens are on the same side of the hydrocarbon chain. In the trans configuration, the hydrogens are on opposite sides. A cis double bond causes a kink in the chain.*

### Trans Fats

In the food industry, oils are artificially hydrogenated to make them semi-solid and of a consistency desirable for many processed food products. Simply speaking, hydrogen gas is bubbled through oils to solidify them. During this hydrogenation process, double bonds of the *cis-* conformation in the hydrocarbon chain may be converted to double bonds in the trans- conformation.

Margarine, some types of peanut butter, and shortening are examples of artificially hydrogenated trans fats. Recent studies have shown that an increase in trans fats in the human diet may lead to an increase in levels of low-density lipoproteins (LDL), or "bad" cholesterol, which in turn may lead to plaque deposition in the arteries, resulting in heart disease. Many fast food restaurants have recently banned the use of trans fats, and food labels are required to display the trans fat content.

### Omega Fatty Acids

Essential fatty acids are fatty acids required but not synthesized by the human body. Consequently, they have to be supplemented through ingestion via the diet. Omega-3 fatty acids (like that shown in **[Figure 6]**) fall into this category and are one of only two known for humans (the other being omega-6 fatty acid). These are polyunsaturated fatty acids and are called omega-3 because the third carbon from the end of the hydrocarbon chain is connected to its neighboring carbon by a double bond.

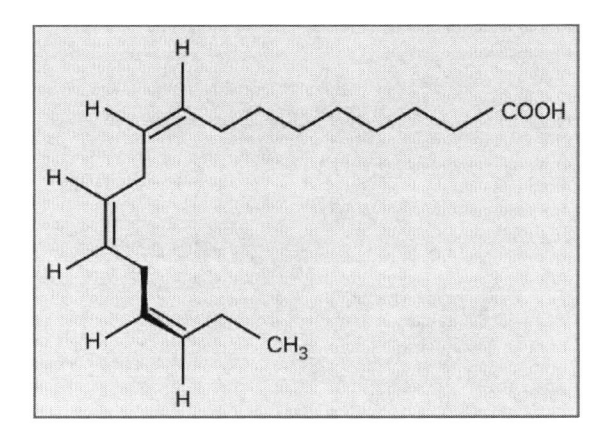

*Figure 6: Alpha-linolenic acid is an example of an omega-3 fatty acid. It has three cis double bonds and, as a result, a curved shape. For clarity, the carbons are not shown. Each singly bonded carbon has two hydrogens associated with it, also not shown.*

The farthest carbon away from the carboxyl group is numbered as the omega ($\omega$) carbon, and if the double bond is between the third and fourth carbon from that end, it is known as an omega-3 fatty acid. Nutritionally important because the body does not make them, omega-3 fatty acids include alpha-linoleic acid (ALA), eicosapentaenoic acid (EPA), and docosahexaenoic acid (DHA), all of which are polyunsaturated. Salmon, trout, and tuna are good sources of omega-3 fatty acids. Research indicates that omega-3 fatty acids reduce the risk of sudden death from heart attacks, reduce triglycerides in the blood, lower blood pressure, and prevent thrombosis by inhibiting blood clotting. They also reduce inflammation, and may help reduce the risk of some cancers in animals.

Like carbohydrates, fats have received a lot of bad publicity. It is true that eating an excess of fried foods and other "fatty" foods leads to weight gain. However, fats do have important functions. Many vitamins are fat soluble, and fats serve as a long-term storage form of fatty acids: a source of energy. They also provide insulation for the body. Therefore, "healthy" fats in moderate amounts should be consumed on a regular basis.

## WAXES

Wax covers the feathers of some aquatic birds and the leaf surfaces of some plants. Because of the hydrophobic nature of waxes, they prevent water from sticking on the surface ([Figure 7]). Waxes are made up of long fatty acid chains esterified to long-chain alcohols.

## PHOSPHOLIPIDS

Phospholipids are major constituents of the plasma membrane, the outermost layer of animal cells. Like fats, they are composed of fatty acid chains attached to a glycerol or sphingosine backbone. Instead of three fatty acids attached

*Figure 7: Waxy coverings on some leaves are made of lipids. (credit: Roger Griffith)*

## 236 Biological Macromolecules

as in triglycerides, however, there are two fatty acids forming diacylglycerol, and the third carbon of the glycerol backbone is occupied by a modified phosphate group ([**Figure 8**]). A phosphate group alone attached to a diaglycerol does not qualify as a phospholipid; it is phosphatidate (diacylglycerol 3-phosphate), the precursor of phospholipids. The phosphate group is modified by an alcohol. Phosphatidylcholine and phosphatidylserine are two important phospholipids that are found in plasma membranes.

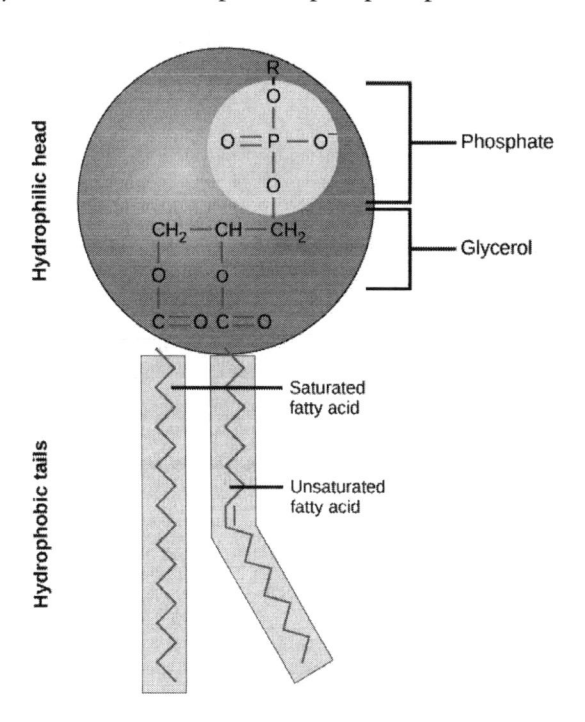

*Figure 8: A phospholipid is a molecule with two fatty acids and a modified phosphate group attached to a glycerol backbone. The phosphate may be modified by the addition of charged or polar chemical groups. Two chemical groups that may modify the phosphate, choline and serine, are shown here. Both choline and serine attach to the phosphate group at the position labeled R via the hydroxyl group indicated in green.*

A phospholipid is an amphipathic molecule, meaning it has a hydrophobic and a hydrophilic part. The fatty acid chains are hydrophobic and cannot interact with water, whereas the phosphate-containing group is hydrophilic and interacts with water ([**Figure 9**]).

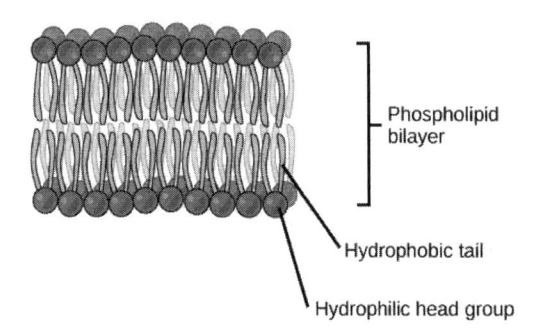

*Figure 9: The phospholipid bilayer is the major component of all cellular membranes. The hydrophilic head groups of the phospholipids face the aqueous solution. The hydrophobic tails are sequestered in the middle of the bilayer.*

The head is the hydrophilic part, and the tail contains the hydrophobic fatty acids. In a membrane, a bilayer of phospholipids forms the matrix of the structure, the fatty acid tails of phospholipids face inside, away from water, whereas the phosphate group faces the outside, aqueous side ([Figure 9]).

Phospholipids are responsible for the dynamic nature of the plasma membrane. If a drop of phospholipids is placed in water, it spontaneously forms a structure known as a micelle, where the hydrophilic phosphate heads face the outside and the fatty acids face the interior of this structure.

## STEROIDS

Unlike the phospholipids and fats discussed earlier, steroids have a fused ring structure. Although they do not resemble the other lipids, they are grouped with them because they are also hydrophobic and insoluble in water. All steroids have four linked carbon rings and several of them, like cholesterol, have a short tail ([Figure 10]). Many steroids also have the –OH functional group, which puts them in the alcohol classification (sterols).

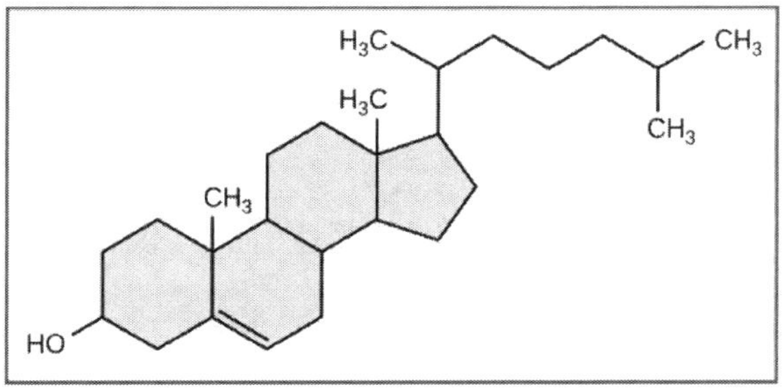

*Figure 10: Steroids such as cholesterol and cortisol are composed of four fused hydrocarbon rings.*

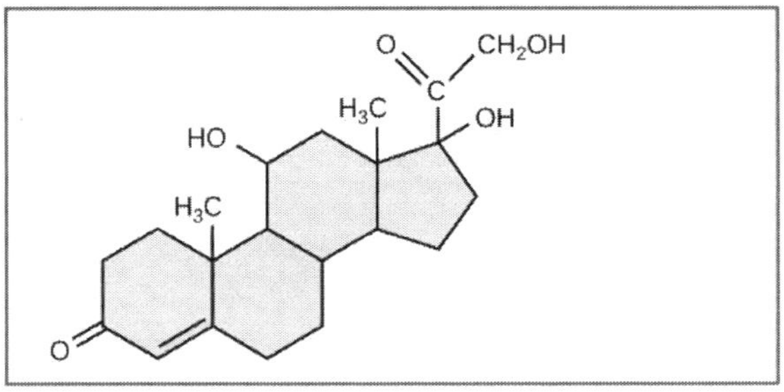

Cholesterol is the most common steroid. Cholesterol is mainly synthesized in the liver and is the precursor to many steroid hormones such as testosterone and estradiol, which are secreted by the gonads and endocrine glands. It is also the precursor to Vitamin D. Cholesterol is also the precursor of bile salts, which help in the emulsification of fats and their subsequent absorption by cells. Although cholesterol is often spoken of in negative terms by lay people, it is necessary for proper functioning of the body. It is a component of the plasma membrane of animal cells and is found within the phospholipid bilayer. Being the outermost structure in animal cells, the plasma membrane is responsible for the transport of materials and cellular recognition and it is involved in cell-to-cell communication.

For an additional perspective on lipids, explore the interactive animation "Biomolecules: The Lipids": http://openstaxcollege.org/l/lipids.

## SECTION SUMMARY

Lipids are a class of macromolecules that are nonpolar and hydrophobic in nature. Major types include fats and oils, waxes, phospholipids, and steroids. Fats are a stored form of energy and are also known as triacylglycerols or triglycerides. Fats are made up of fatty acids and either glycerol or sphingosine. Fatty acids may be unsaturated or saturated, depending on the presence or absence of double bonds in the hydrocarbon chain. If only single bonds are present, they are known as saturated fatty acids. Unsaturated fatty acids may have one or more double bonds in the hydrocarbon chain. Phospholipids make up the matrix of membranes. They have a glycerol or sphingosine backbone to which two fatty acid chains and a phosphate-containing group are attached. Steroids are another class of lipids. Their basic structure has four fused carbon rings. Cholesterol is a type of steroid and is an important constituent of the plasma membrane, where it helps to maintain the fluid nature of the membrane. It is also the precursor of steroid hormones such as testosterone.

### Glossary

**lipid:** macromolecule that is nonpolar and insoluble in water

**omega fat:** type of polyunsaturated fat that is required by the body; the numbering of the carbon omega starts from the methyl end or the end that is farthest from the carboxylic end

**phospholipid:** major constituent of the membranes; composed of two fatty acids and a phosphate-containing group attached to a glycerol backbone

**saturated fatty acid:** long-chain of hydrocarbon with single covalent bonds in the carbon chain; the number of hydrogen atoms attached to the carbon skeleton is maximized

**steroid:** type of lipid composed of four fused hydrocarbon rings forming a planar structure

**trans fat:** fat formed artificially by hydrogenating oils, leading to a different arrangement of double bond(s) than those found in naturally occurring lipids

**triacylglycerol (also, triglyceride):** fat molecule; consists of three fatty acids linked to a glycerol molecule

**unsaturated fatty acid:** long-chain hydrocarbon that has one or more double bonds in the hydrocarbon chain

**wax:** lipid made of a long-chain fatty acid that is esterified to a long-chain alcohol; serves as a protective coating on some feathers, aquatic mammal fur, and leaves

# Proteins

### Learning Objectives

By the end of this section, you will be able to:

- Describe the functions proteins perform in the cell and in tissues
- Discuss the relationship between amino acids and proteins
- Explain the four levels of protein organization
- Describe the ways in which protein shape and function are linked

Proteins are one of the most abundant organic molecules in living systems and have the most diverse range of functions of all macromolecules. Proteins may be structural, regulatory, contractile, or protective; they may serve in transport, storage, or membranes; or they may be toxins or enzymes. Each cell in a living system may contain thousands of proteins, each with a unique function. Their structures, like their functions, vary greatly. They are all, however, polymers of amino acids, arranged in a linear sequence.

## TYPES AND FUNCTIONS OF PROTEINS

Enzymes, which are produced by living cells, are catalysts in biochemical reactions (like digestion) and are usually complex or conjugated proteins. Each enzyme is specific for the substrate (a reactant that binds to an enzyme) it acts on. The enzyme may help in breakdown, rearrangement, or synthesis reactions. Enzymes that break down their substrates are called catabolic enzymes, enzymes that build more complex molecules from their substrates are called anabolic enzymes, and enzymes that affect the rate of reaction are called catalytic enzymes. It should be noted that all enzymes increase the rate of reaction and, therefore, are considered to be organic catalysts. An example of an enzyme is salivary amylase, which hydrolyzes its substrate amylose, a component of starch.

Hormones are chemical-signaling molecules, usually small proteins or steroids, secreted by endocrine cells that act to control or regulate specific physiological processes, including growth, development, metabolism, and reproduction. For example, insulin is a protein hormone that helps to regulate the blood glucose level. The primary types and functions of proteins are listed in [Figure 1].

240  Biological Macromolecules

*Protein Types and Functions*

| Type | Examples | Functions |
|---|---|---|
| Digestive Enzymes | Amylase, lipase, pepsin, trypsin | Help in digestion of food by catabolizing nutrients into monomeric units |
| Transport | Hemoglobin, albumin | Carry substances in the blood or lymph throughout the body |
| Structural | Actin, tubulin, keratin | Construct different structures, like the cytoskeleton |
| Hormones | Insulin, thyroxine | Coordinate the activity of different body systems |
| Defense | Immunoglobulins | Protect the body from foreign pathogens |
| Contractile | Actin, myosin | Effect muscle contraction |
| Storage | Legume storage proteins, egg white (albumin) | Provide nourishment in early development of the embryo and the seedling |

Proteins have different shapes and molecular weights; some proteins are globular in shape whereas others are fibrous in nature. For example, hemoglobin is a globular protein, but collagen, found in our skin, is a fibrous protein. Protein shape is critical to its function, and this shape is maintained by many different types of chemical bonds. Changes in temperature, pH, and exposure to chemicals may lead to permanent changes in the shape of the protein, leading to loss of function, known as denaturation. All proteins are made up of different arrangements of the same 20 types of amino acids.

## AMINO ACIDS

Amino acids are the monomers that make up proteins. Each amino acid has the same fundamental structure, which consists of a central carbon atom, also known as the alpha (α) carbon, bonded to an amino group (NH$_2$), a carboxyl group (COOH), and to a hydrogen atom. Every amino acid also has another atom or group of atoms bonded to the central atom known as the R group ([Figure 1]).

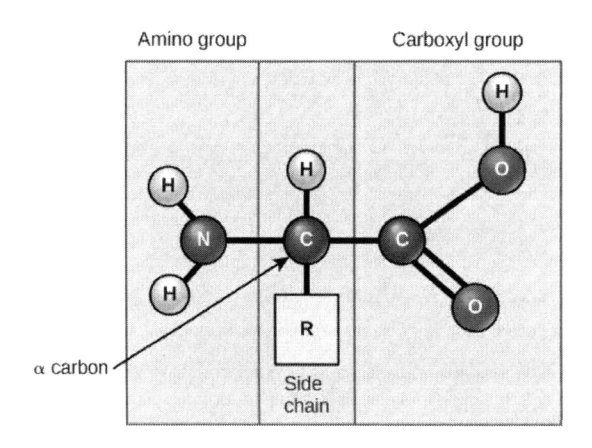

*Figure 1: Amino acids have a central asymmetric carbon to which an amino group, a carboxyl group, a hydrogen atom, and a side chain (R group) are attached.*

The name "amino acid" is derived from the fact that they contain both amino group and carboxyl-acid-group in their basic structure. As mentioned, there are 20 amino acids present in proteins. Ten of these are considered essential amino acids in humans because the human body cannot produce them and they are obtained from the diet. For each amino acid, the R group (or side chain) is different ([Figure 2]).

### Art Connection

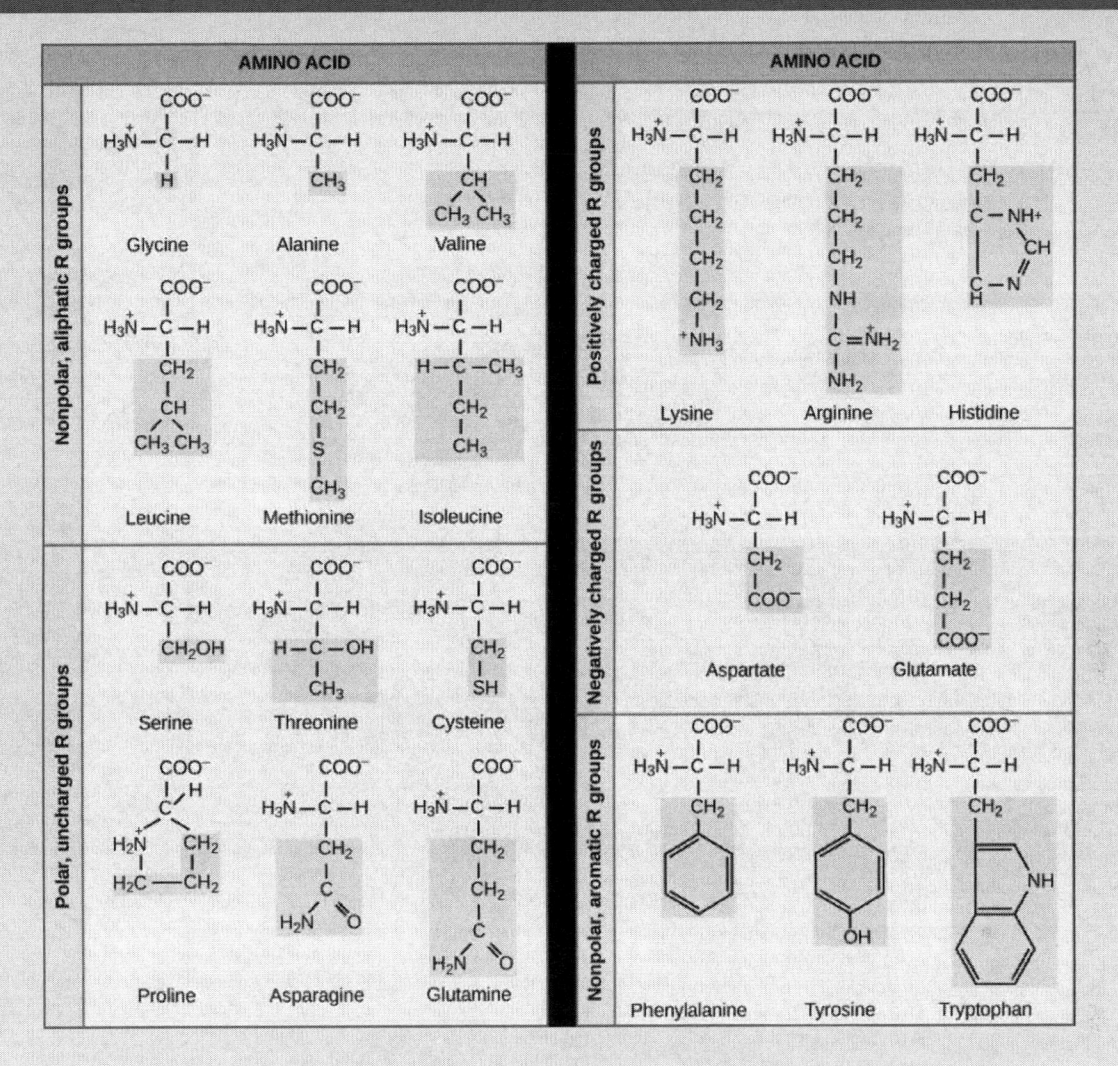

*Figure 2: There are 20 common amino acids commonly found in proteins, each with a different R group (variant group) that determines its chemical nature.*

Which categories of amino acid would you expect to find on the surface of a soluble protein, and which would you expect to find in the interior? What distribution of amino acids would you expect to find in a protein embedded in a lipid bilayer?

The chemical nature of the side chain determines the nature of the amino acid (that is, whether it is acidic, basic, polar, or nonpolar). For example, the amino acid glycine has a hydrogen atom as the R group. Amino acids such as valine, methionine, and alanine are nonpolar or hydrophobic in nature, while amino acids such as serine, threonine, and cysteine are polar and have hydrophilic side chains. The side chains of lysine and arginine are positively charged, and therefore these amino acids are also known as basic amino acids. Proline has an R group that is linked to the amino group, forming a ring-like structure. Proline is an exception to the standard structure of an animo acid since its amino group is not separate from the side chain (**[Figure 2]**).

Amino acids are represented by a single upper case letter or a three-letter abbreviation. For example, valine is known by the letter V or the three-letter symbol val. Just as some fatty acids are essential to a diet, some amino acids are necessary as well. They are known as essential amino acids, and in humans they include isoleucine, leucine, and cysteine. Essential amino acids refer to those necessary for construction of proteins in the body, although not produced by the body; which amino acids are essential varies from organism to organism.

The sequence and the number of amino acids ultimately determine the protein's shape, size, and function. Each amino acid is attached to another amino acid by a covalent bond, known as a peptide bond, which is formed by a dehydration reaction. The carboxyl group of one amino acid and the amino group of the incoming amino acid combine, releasing a molecule of water. The resulting bond is the peptide bond ([Figure 3]).

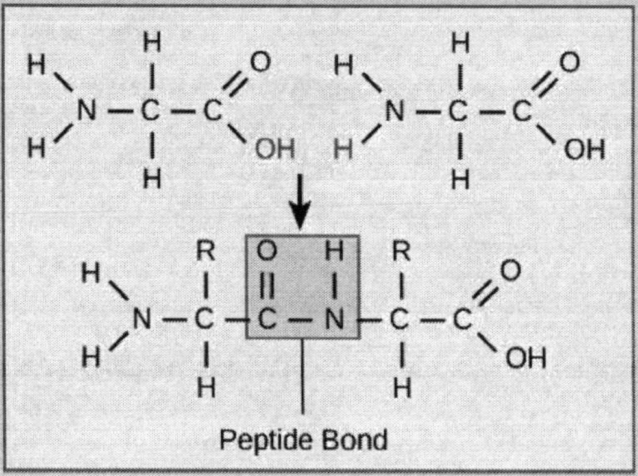

*Figure 3: Peptide bond formation is a dehydration synthesis reaction. The carboxyl group of one amino acid is linked to the amino group of the incoming amino acid. In the process, a molecule of water is released.*

The products formed by such linkages are called peptides. As more amino acids join to this growing chain, the resulting chain is known as a polypeptide. Each polypeptide has a free amino group at one end. This end is called the N terminal, or the amino terminal, and the other end has a free carboxyl group, also known as the C or carboxyl terminal. While the terms polypeptide and protein are sometimes used interchangeably, a polypeptide is technically a polymer of amino acids, whereas the term protein is used for a polypeptide or polypeptides that have combined together, often have bound non-peptide prosthetic groups, have a distinct shape, and have a unique function. After protein synthesis (translation), most proteins are modified. These are known as post-translational modifications. They may undergo cleavage, phosphorylation, or may require the addition of other chemical groups. Only after these modifications is the protein completely functional.

Click through the steps of protein synthesis in this interactive tutorial: http://openstaxcollege.org/l/protein_synth.

Evolution Connection

The Evolutionary Significance of Cytochrome cCytochrome c is an important component of the electron transport chain, a part of cellular respiration, and it is normally found in the cellular organelle, the mitochondrion. This protein has a heme prosthetic group, and the central ion of the heme gets alternately reduced and oxidized during electron transfer. Because this essential protein's role in producing cellular energy is crucial, it has changed very little over millions of years. Protein sequencing has shown that there is a considerable amount of cytochrome c amino acid sequence homology among different species; in other words, evolutionary kinship can be assessed by measuring the similarities or differences among various species' DNA or protein sequences.

Scientists have determined that human cytochrome c contains 104 amino acids. For each cytochrome c molecule from different organisms that has been sequenced to date, 37 of these amino acids appear in the same position in all samples of cytochrome c. This indicates that there may have been a common ancestor. On comparing the human and chimpanzee protein sequences, no sequence difference was found. When human and rhesus monkey sequences were compared, the single difference found was in one amino acid. In another comparison, human to yeast sequencing shows a difference in the 44th position.

## PROTEIN STRUCTURE

As discussed earlier, the shape of a protein is critical to its function. For example, an enzyme can bind to a specific substrate at a site known as the active site. If this active site is altered because of local changes or changes in overall protein structure, the enzyme may be unable to bind to the substrate. To understand how the protein gets its final shape or conformation, we need to understand the four levels of protein structure: primary, secondary, tertiary, and quaternary.

### Primary Structure

The unique sequence of amino acids in a polypeptide chain is its primary structure. For example, the pancreatic hormone insulin has two polypeptide chains, A and B, and they are linked together by disulfide bonds. The N terminal amino acid of the A chain is glycine, whereas the C terminal amino acid is asparagine (**[Figure 4]**). The sequences of amino acids in the A and B chains are unique to insulin.

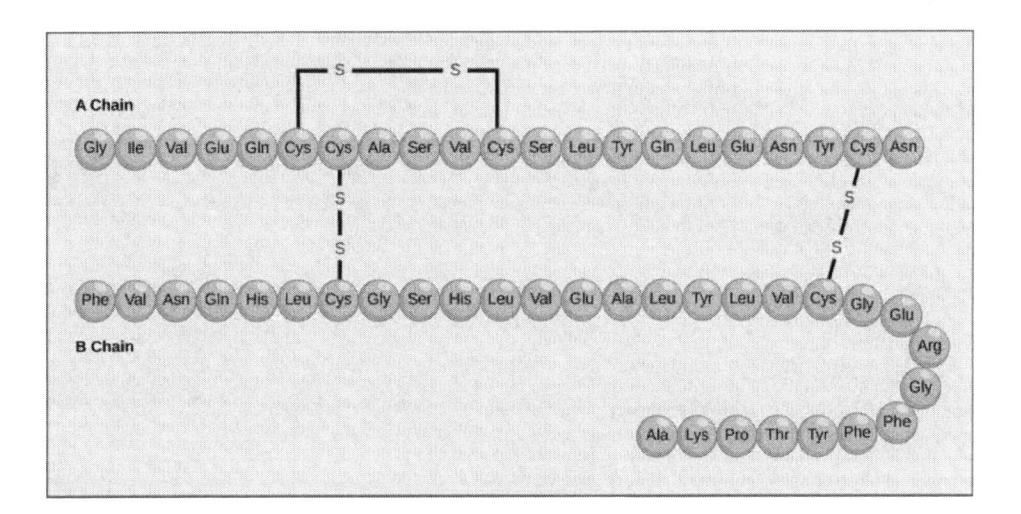

*Figure 4: Bovine serum insulin is a protein hormone made of two peptide chains, A (21 amino acids long) and B (30 amino acids long). In each chain, primary structure is indicated by three-letter abbreviations that represent the names of the amino acids in the order they are present. The amino acid cysteine (cys) has a sulfhydryl (SH) group as a side chain. Two sulfhydryl groups can react in the presence of oxygen to form a disulfide (S-S) bond. Two disulfide bonds connect the A and B chains together, and a third helps the A chain fold into the correct shape. Note that all disulfide bonds are the same length, but are drawn different sizes for clarity.*

## 244 Biological Macromolecules

The unique sequence for every protein is ultimately determined by the gene encoding the protein. A change in nucleotide sequence of the gene's coding region may lead to a different amino acid being added to the growing polypeptide chain, causing a change in protein structure and function. In sickle cell anemia, the hemoglobin β chain (a small portion of which is shown in **[Figure 5]**) has a single amino acid substitution, causing a change in protein structure and function. Specifically, the amino acid glutamic acid is substituted by valine in the β chain. What is most remarkable to consider is that a hemoglobin molecule is made up of two alpha chains and two beta chains that each consist of about 150 amino acids. The molecule, therefore, has about 600 amino acids. The structural difference between a normal hemoglobin molecule and a sickle cell molecule—which dramatically decreases life expectancy—is a single amino acid of the 600. What is even more remarkable is that those 600 amino acids are encoded by three nucleotides each, and the mutation is caused by a single base change (point mutation), 1 in 1800 bases.

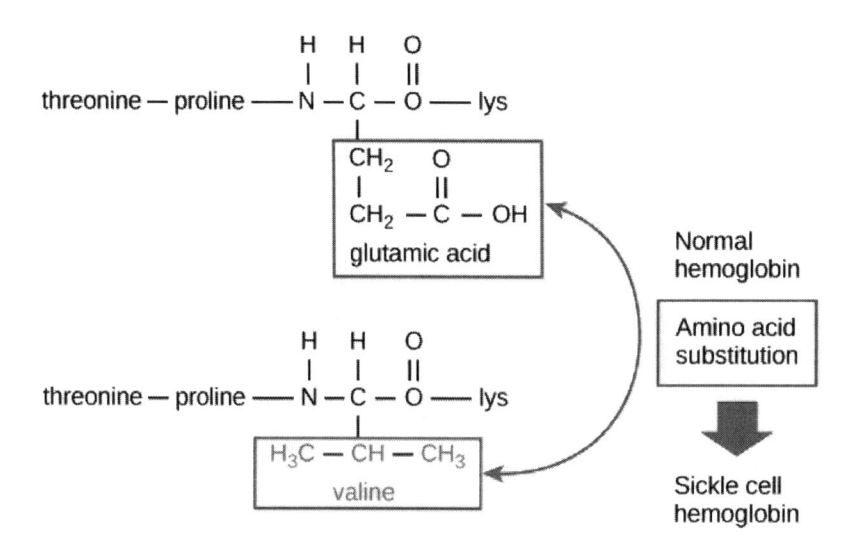

*Figure 5: The beta chain of hemoglobin is 147 residues in length, yet a single amino acid substitution leads to sickle cell anemia. In normal hemoglobin, the amino acid at position seven is glutamate. In sickle cell hemoglobin, this glutamate is replaced by a valine.*

Because of this change of one amino acid in the chain, hemoglobin molecules form long fibers that distort the biconcave, or disc-shaped, red blood cells and assume a crescent or "sickle" shape, which clogs arteries (**[Figure 6]**). This can lead to myriad serious health problems such as breathlessness, dizziness, headaches, and abdominal pain for those affected by this disease.

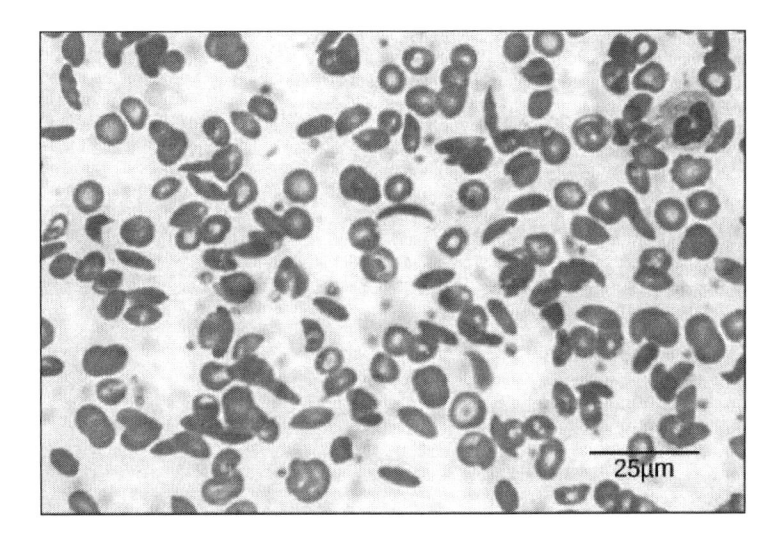

*Figure 6: In this blood smear, visualized at 535x magnification using bright field microscopy, sickle cells are crescent shaped, while normal cells are disc-shaped. (credit: modification of work by Ed Uthman; scale-bar data from Matt Russell)*

## Secondary Structure

The local folding of the polypeptide in some regions gives rise to the secondary structure of the protein. The most common are the α-helix and β-pleated sheet structures ([**Figure 7**]). Both structures are the α-helix structure—the helix held in shape by hydrogen bonds. The hydrogen bonds form between the oxygen atom in the carbonyl group in one amino acid and another amino acid that is four amino acids farther along the chain.

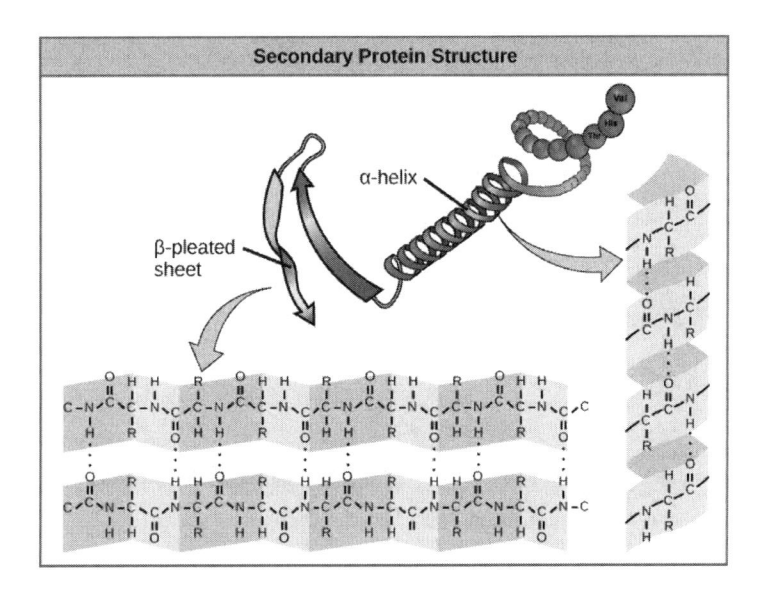

*Figure 7: The α-helix and β-pleated sheet are secondary structures of proteins that form because of hydrogen bonding between carbonyl and amino groups in the peptide backbone. Certain amino acids have a propensity to form an α-helix, while others have a propensity to form a β-pleated sheet.*

## 246 Biological Macromolecules

Every helical turn in an alpha helix has 3.6 amino acid residues. The R groups (the variant groups) of the polypeptide protrude out from the α-helix chain. In the β-pleated sheet, the "pleats" are formed by hydrogen bonding between atoms on the backbone of the polypeptide chain. The R groups are attached to the carbons and extend above and below the folds of the pleat. The pleated segments align parallel or antiparallel to each other, and hydrogen bonds form between the partially positive nitrogen atom in the amino group and the partially negative oxygen atom in the carbonyl group of the peptide backbone. The α-helix and β-pleated sheet structures are found in most globular and fibrous proteins and they play an important structural role.

### Tertiary Structure

The unique three-dimensional structure of a polypeptide is its tertiary structure ([Figure 8). This structure is in part due to chemical interactions at work on the polypeptide chain. Primarily, the interactions among R groups creates the complex three-dimensional tertiary structure of a protein. The nature of the R groups found in the amino acids involved can counteract the formation of the hydrogen bonds described for standard secondary structures. For example, R groups with like charges are repelled by each other and those with unlike charges are attracted to each other (ionic bonds). When protein folding takes place, the hydrophobic R groups of nonpolar amino acids lay in the interior of the protein, whereas the hydrophilic R groups lay on the outside. The former types of interactions are also known as hydrophobic interactions. Interaction between cysteine side chains forms disulfide linkages in the presence of oxygen, the only covalent bond forming during protein folding.

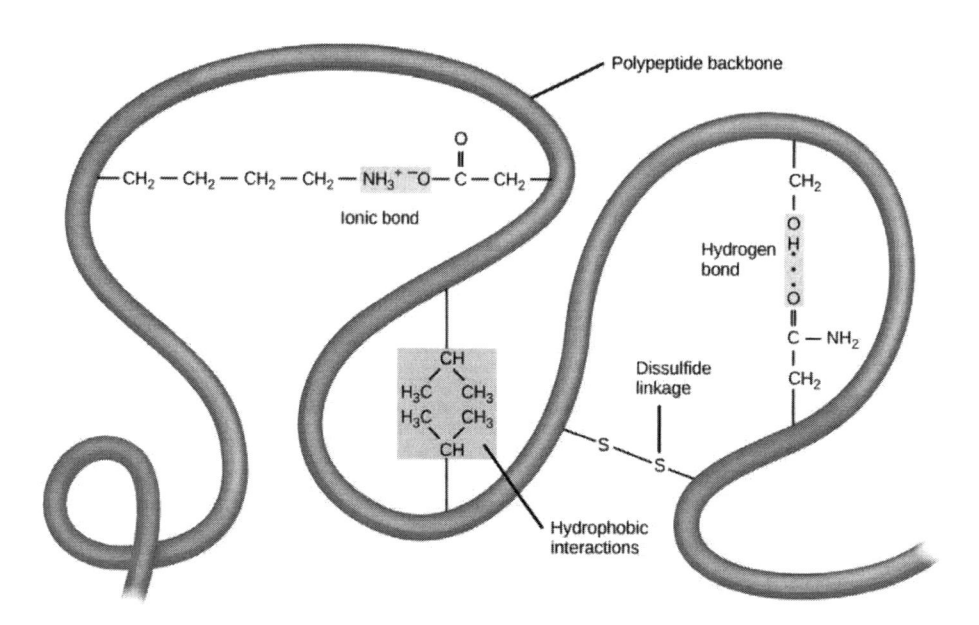

*Figure 8: The tertiary structure of proteins is determined by a variety of chemical interactions. These include hydrophobic interactions, ionic bonding, hydrogen bonding and disulfide linkages.*

All of these interactions, weak and strong, determine the final three-dimensional shape of the protein. When a protein loses its three-dimensional shape, it may no longer be functional.

### Quaternary Structure

In nature, some proteins are formed from several polypeptides, also known as subunits, and the interaction of these subunits forms the quaternary structure. Weak interactions between the subunits help to stabilize the overall structure. For example, insulin (a globular protein) has a combination of hydrogen bonds and disulfide bonds that cause it to be mostly clumped into a ball shape. Insulin starts out as a single polypeptide and loses some internal sequences in the presence of post-translational modification after the formation of the disulfide linkages that hold the remaining chains together. Silk (a fibrous protein), however, has a β-pleated sheet structure that is the result of hydrogen bonding between different chains.

The four levels of protein structure (primary, secondary, tertiary, and quaternary) are illustrated in [**Figure 9**].

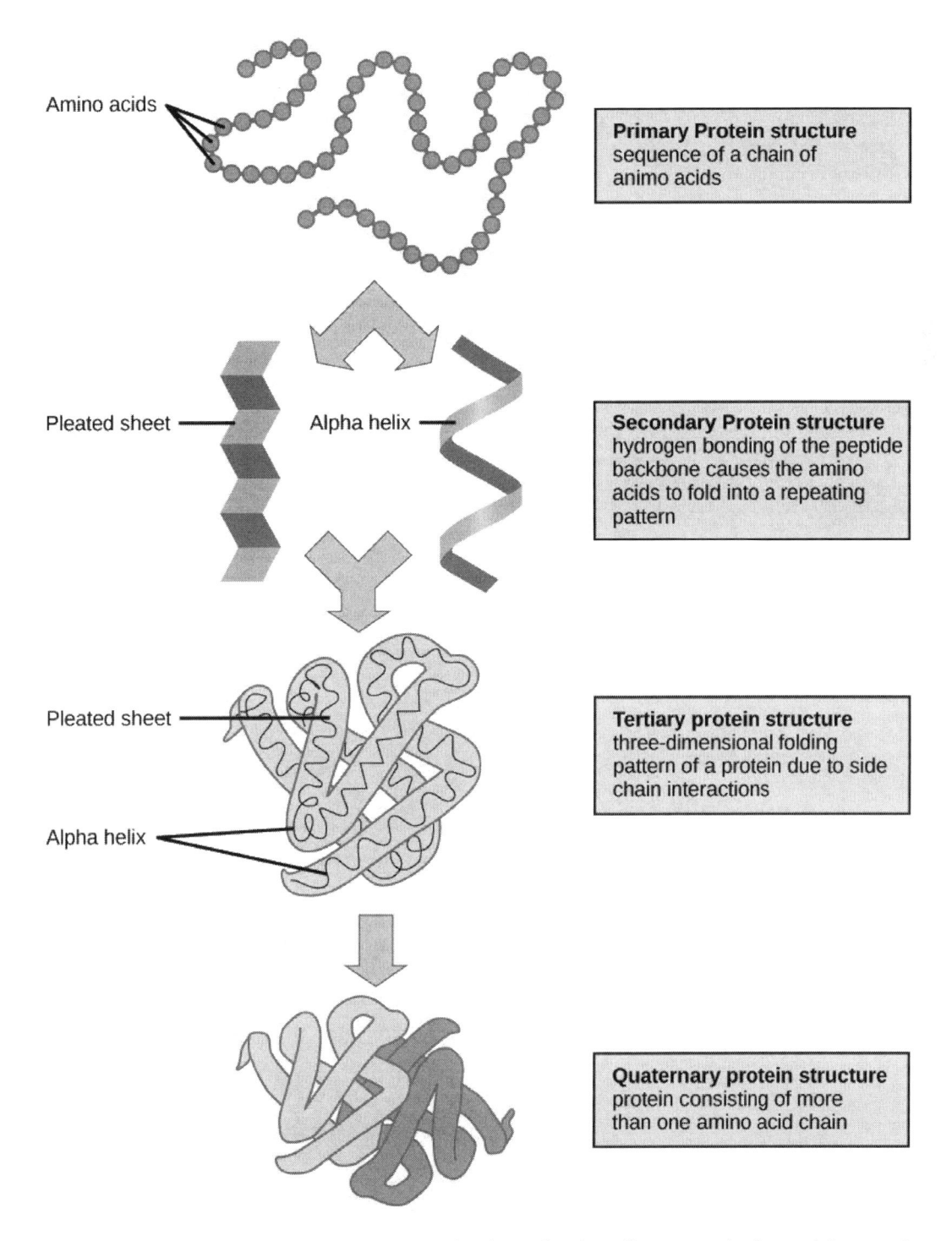

*Figure 9: The four levels of protein structure can be observed in these illustrations. (credit: modification of work by National Human Genome Research Institute)*

## DENATURATION AND PROTEIN FOLDING

Each protein has its own unique sequence and shape that are held together by chemical interactions. If the protein is subject to changes in temperature, pH, or exposure to chemicals, the protein structure may change, losing its shape without losing its primary sequence in what is known as denaturation. Denaturation is often reversible because the primary structure of the polypeptide is conserved in the process if the denaturing agent is removed, allowing the protein to resume its function. Sometimes denaturation is irreversible, leading to loss of function. One example of irreversible protein denaturation is when an egg is fried. The albumin protein in the liquid egg white is denatured when placed in a hot pan. Not all proteins are denatured at high temperatures; for instance, bacteria that survive in hot springs have proteins that function at temperatures close to boiling. The stomach is also very acidic, has a low pH, and denatures proteins as part of the digestion process; however, the digestive enzymes of the stomach retain their activity under these conditions.

Protein folding is critical to its function. It was originally thought that the proteins themselves were responsible for the folding process. Only recently was it found that often they receive assistance in the folding process from protein helpers known as chaperones (or chaperonins) that associate with the target protein during the folding process. They act by preventing aggregation of polypeptides that make up the complete protein structure, and they disassociate from the protein once the target protein is folded.

---

For an additional perspective on proteins, view this animation called "Biomolecules: The Proteins": http://openstaxcollege.org/l/proteins.

---

## SECTION SUMMARY

Proteins are a class of macromolecules that perform a diverse range of functions for the cell. They help in metabolism by providing structural support and by acting as enzymes, carriers, or hormones. The building blocks of proteins (monomers) are amino acids. Each amino acid has a central carbon that is linked to an amino group, a carboxyl group, a hydrogen atom, and an R group or side chain. There are 20 commonly occurring amino acids, each of which differs in the R group. Each amino acid is linked to its neighbors by a peptide bond. A long chain of amino acids is known as a polypeptide.

Proteins are organized at four levels: primary, secondary, tertiary, and (optional) quaternary. The primary structure is the unique sequence of amino acids. The local folding of the polypeptide to form structures such as the α helix and β-pleated sheet constitutes the secondary structure. The overall three-dimensional structure is the tertiary structure. When two or more polypeptides combine to form the complete protein structure, the configuration is known as the quaternary structure of a protein. Protein shape and function are intricately linked; any change in shape caused by changes in temperature or pH may lead to protein denaturation and a loss in function.

## Glossary

**alpha-helix structure (α-helix):** type of secondary structure of proteins formed by folding of the polypeptide into a helix shape with hydrogen bonds stabilizing the structure

**amino acid:** monomer of a protein; has a central carbon or alpha carbon to which an amino group, a carboxyl group, a hydrogen, and an R group or side chain is attached; the R group is different for all 20 amino acids

**beta-pleated sheet (β-pleated):** secondary structure found in proteins in which "pleats" are formed by hydrogen bonding between atoms on the backbone of the polypeptide chain

**chaperone:** (also, chaperonin) protein that helps nascent protein in the folding process

**denaturation:** loss of shape in a protein as a result of changes in temperature, pH, or exposure to chemicals

**enzyme:** catalyst in a biochemical reaction that is usually a complex or conjugated protein

**hormone:** chemical signaling molecule, usually protein or steroid, secreted by endocrine cells that act to control or regulate specific physiological processes

**peptide bond:** bond formed between two amino acids by a dehydration reaction

**polypeptide:** long chain of amino acids linked by peptide bonds

**primary structure:** linear sequence of amino acids in a protein

**protein:** biological macromolecule composed of one or more chains of amino acids

**quaternary structure:** association of discrete polypeptide subunits in a protein

**secondary structure:** regular structure formed by proteins by intramolecular hydrogen bonding between the oxygen atom of one amino acid residue and the hydrogen attached to the nitrogen atom of another amino acid residue

**tertiary structure:** three-dimensional conformation of a protein, including interactions between secondary structural elements; formed from interactions between amino acid side chains

# Enzymes

## Learning Objectives

By the end of this section, you will be able to:
- Describe the role of enzymes in metabolic pathways
- Explain how enzymes function as molecular catalysts
- Discuss enzyme regulation by various factors

A substance that helps a chemical reaction to occur is a catalyst, and the special molecules that catalyze biochemical reactions are called enzymes. Almost all enzymes are proteins, made up of chains of amino acids, and they perform the critical task of lowering the activation energies of chemical reactions inside the cell. Enzymes do this by binding to the reactant molecules, and holding them in such a way as to make the chemical bond-breaking and bond-forming processes take place more readily. It is important to remember that enzymes don't change the ΔG of a reaction. In other words, they don't change whether a reaction is exergonic (spontaneous) or endergonic. This is because they don't change the free energy of the reactants or products. They only reduce the activation energy required to reach the transition state ([**Figure 1**]).

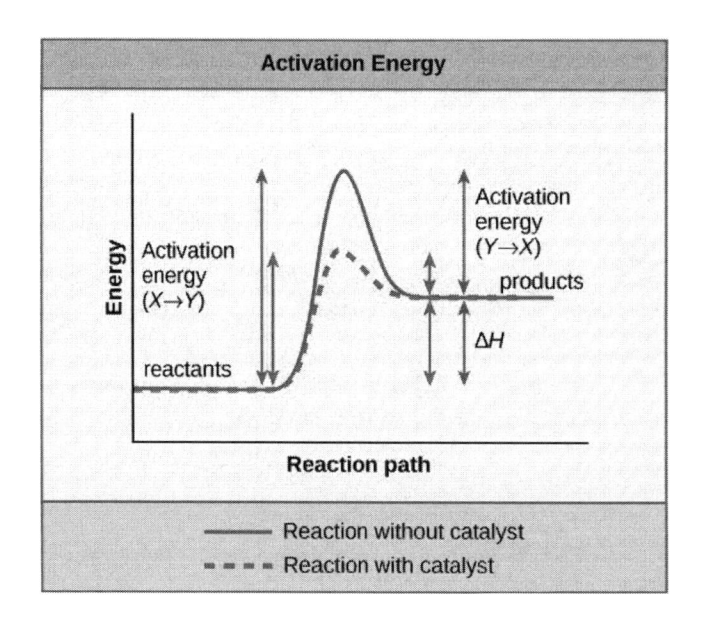

*Figure 1: Enzymes lower the activation energy of the reaction but do not change the free energy of the reaction.*

## ENZYME ACTIVE SITE AND SUBSTRATE SPECIFICITY

The chemical reactants to which an enzyme binds are the enzyme's substrates. There may be one or more substrates, depending on the particular chemical reaction. In some reactions, a single-reactant substrate is broken down into multiple products. In others, two substrates may come together to create one larger molecule. Two reactants might also enter a reaction, both become modified, and leave the reaction as two products. The location within the enzyme where the substrate binds is called the enzyme's active site. The active site is where the "action" happens, so to speak. Since enzymes are proteins, there is a unique combination of amino acid residues (also called side chains, or R groups) within the active site. Each residue is characterized by different properties. Residues can be large or small, weakly acidic or basic, hydrophilic or hydrophobic, positively or negatively charged, or neutral. The unique combination of amino acid residues, their positions, sequences, structures, and properties, creates a very specific chemical environment within the active site. This specific environment is suited to bind, albeit briefly, to a specific chemical substrate (or substrates). Due to this jigsaw puzzle-like match between an enzyme and its substrates (which adapts to find the best fit between the transition state and the active site), enzymes are known for their specificity. The "best fit" results from the shape and the amino acid functional group's attraction to the substrate. There is a specifically matched enzyme for each substrate and, thus, for each chemical reaction; however, there is flexibility as well.

The fact that active sites are so perfectly suited to provide specific environmental conditions also means that they are subject to influences by the local environment. It is true that increasing the environmental temperature generally increases reaction rates, enzyme-catalyzed or otherwise. However, increasing or decreasing the temperature outside of an optimal range can affect chemical bonds within the active site in such a way that they are less well suited to bind substrates. High temperatures will eventually cause enzymes, like other biological molecules, to denature, a process that changes the natural properties of a substance. Likewise, the pH of the local environment can also affect enzyme function. Active site amino acid residues have their own acidic or basic properties that are optimal for catalysis. These residues are sensitive to changes in pH that can impair the way substrate molecules bind. Enzymes are suited to function best within a certain pH range, and, as with temperature, extreme pH values (acidic or basic) of the environment can cause enzymes to denature.

### Induced Fit and Enzyme Function

For many years, scientists thought that enzyme-substrate binding took place in a simple "lock-and-key" fashion. This model asserted that the enzyme and substrate fit together perfectly in one instantaneous step. However, current research supports a more refined view called induced fit ([Figure 2]). The induced-fit model expands upon the lock-and-key model by describing a more dynamic interaction between enzyme and substrate. As the enzyme and substrate come together, their interaction causes a mild shift in the enzyme's structure that confirms an ideal binding arrangement between the enzyme and the transition state of the substrate. This ideal binding maximizes the enzyme's ability to catalyze its reaction.

View an animation of induced fit at this website: https://www.youtube.com/watch?v=ueup2PTkFW8.

When an enzyme binds its substrate, an enzyme-substrate complex is formed. This complex lowers the activation energy of the reaction and promotes its rapid progression in one of many ways. On a basic level, enzymes promote chemical reactions that involve more than one substrate by bringing the substrates together in an optimal orientation. The appropriate region (atoms and bonds) of one molecule is juxtaposed to the appropriate region of the other molecule with which it must react. Another way in which enzymes promote the reaction of their substrates is by creating an optimal environment within the active site for the reaction to occur. Certain chemical reactions might proceed best in a slightly acidic or non-polar environment. The chemical properties that emerge from the particular arrangement of amino acid residues within an active site create the perfect environment for an enzyme's specific substrates to react.

You've learned that the activation energy required for many reactions includes the energy involved in manipulating or slightly contorting chemical bonds so that they can easily break and allow others to reform. Enzymatic action can aid this process. The enzyme-substrate complex can lower the activation energy by contorting substrate molecules in such a way as to facilitate bond-breaking, helping to reach the transition state. Finally, enzymes can also lower activation energies by taking part in the chemical reaction itself. The amino acid residues can provide certain ions or chemical groups that actually form covalent bonds with substrate molecules as a necessary step of the reaction process. In these cases, it is important to remember that the enzyme will always return to its original state at the completion of the reaction. One of the hallmark properties of enzymes is that they remain ultimately unchanged by the reactions they catalyze. After an enzyme is done catalyzing a reaction, it releases its product(s).

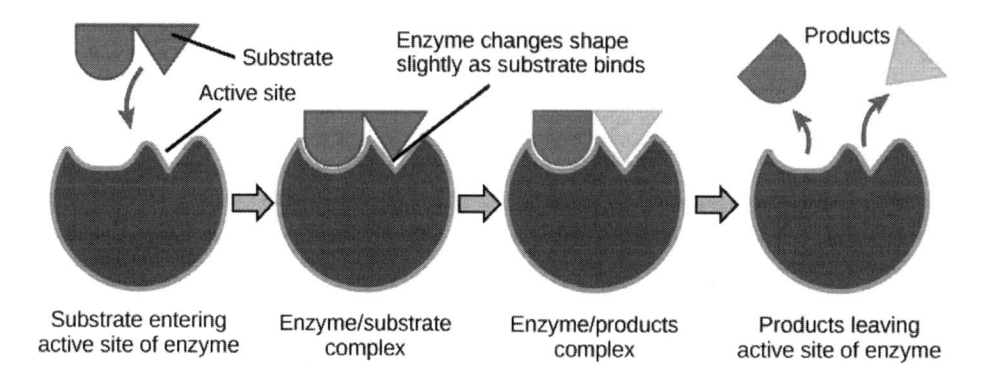

*Figure 2: According to the induced-fit model, both enzyme and substrate undergo dynamic conformational changes upon binding. The enzyme contorts the substrate into its transition state, thereby increasing the rate of the reaction.*

## CONTROL OF METABOLISM THROUGH ENZYME REGULATION

It would seem ideal to have a scenario in which all of the enzymes encoded in an organism's genome existed in abundant supply and functioned optimally under all cellular conditions, in all cells, at all times. In reality, this is far from the case. A variety of mechanisms ensure that this does not happen. Cellular needs and conditions vary from cell to cell, and change within individual cells over time. The required enzymes and energetic demands of stomach cells are different from those of fat storage cells, skin cells, blood cells, and nerve cells. Furthermore, a digestive cell works much harder to process and break down nutrients during the time that closely follows a meal compared with many hours after a meal. As these cellular demands and conditions vary, so do the amounts and functionality of different enzymes.

Since the rates of biochemical reactions are controlled by activation energy, and enzymes lower and determine activation energies for chemical reactions, the relative amounts and functioning of the variety of enzymes within a cell ultimately determine which reactions will proceed and at which rates. This determination is tightly controlled. In certain cellular environments, enzyme activity is partly controlled by environmental factors, like pH and temperature. There are other mechanisms through which cells control the activity of enzymes and determine the rates at which various biochemical reactions will occur.

## Regulation of Enzymes by Molecules

Enzymes can be regulated in ways that either promote or reduce their activity. There are many different kinds of molecules that inhibit or promote enzyme function, and various mechanisms exist for doing so. In some cases of enzyme inhibition, for example, an inhibitor molecule is similar enough to a substrate that it can bind to the active site and simply block the substrate from binding. When this happens, the enzyme is inhibited through competitive inhibition, because an inhibitor molecule competes with the substrate for active site binding ([Figure 3]). On the other hand, in noncompetitive inhibition, an inhibitor molecule binds to the enzyme in a location other than an allosteric site and still manages to block substrate binding to the active site.

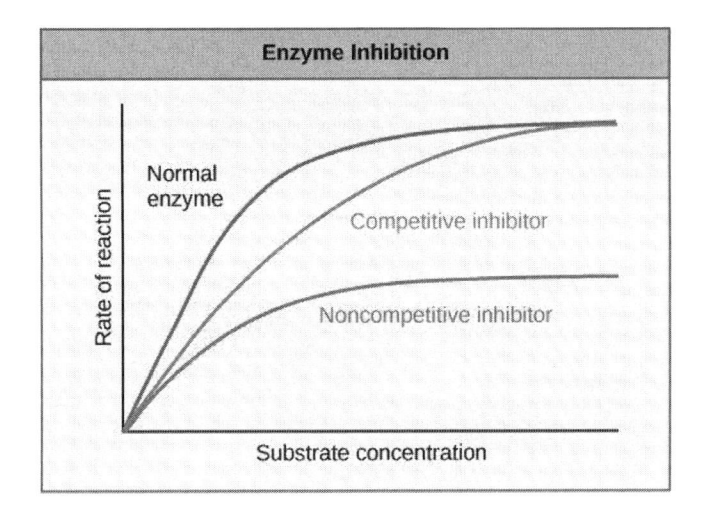

*Figure 3: Competitive and noncompetitive inhibition affect the rate of reaction differently. Competitive inhibitors affect the initial rate but do not affect the maximal rate, whereas noncompetitive inhibitors affect the maximal rate.*

Some inhibitor molecules bind to enzymes in a location where their binding induces a conformational change that reduces the affinity of the enzyme for its substrate. This type of inhibition is called allosteric inhibition ([Figure 4]). Most allosterically regulated enzymes are made up of more than one polypeptide, meaning that they have more than one protein subunit. When an allosteric inhibitor binds to an enzyme, all active sites on the protein subunits are changed slightly such that they bind their substrates with less efficiency. There are allosteric activators as

## 254   Biological Macromolecules

well as inhibitors. Allosteric activators bind to locations on an enzyme away from the active site, inducing a conformational change that increases the affinity of the enzyme's active site(s) for its substrate(s).

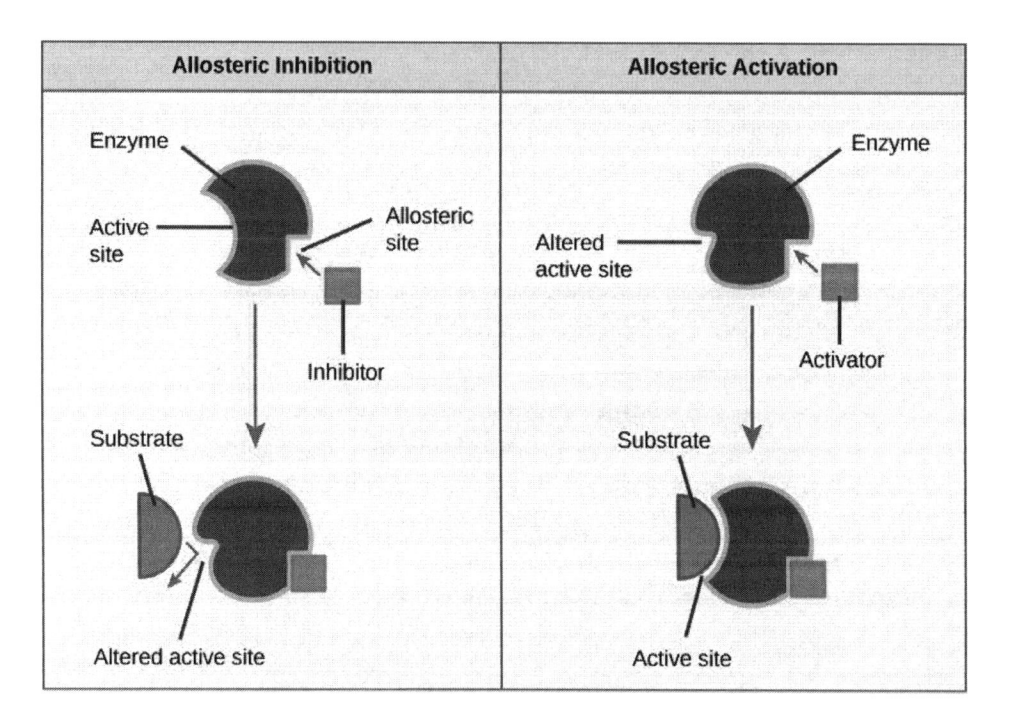

*Figure 4: Allosteric inhibitors modify the active site of the enzyme so that substrate binding is reduced or prevented. In contrast, allosteric activators modify the active site of the enzyme so that the affinity for the substrate increases.*

Everyday Connection

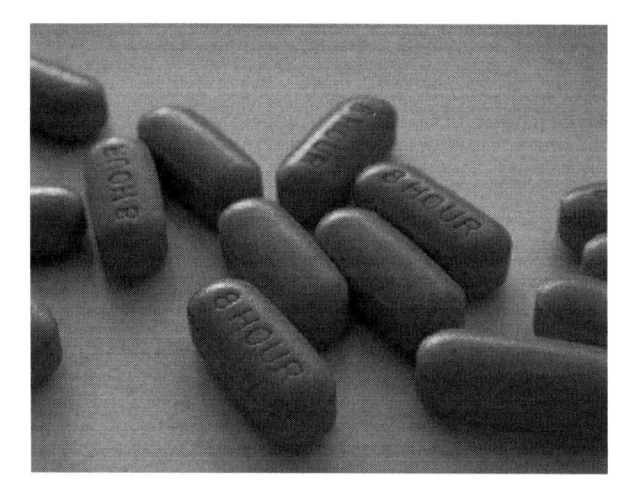

*Figure 5: Have you ever wondered how pharmaceutical drugs are developed? (credit: Deborah Austin)*

Drug Discovery by Looking for Inhibitors of Key Enzymes in Specific PathwaysEnzymes are key components of metabolic pathways. Understanding how enzymes work and how they can be regulated is a key principle behind the development of many of the pharmaceutical drugs ([Figure 5]) on the market today. Biologists working in this field collaborate with other scientists, usually chemists, to design drugs.

Consider statins for example—which is the name given to the class of drugs that reduces cholesterol levels. These compounds are essentially inhibitors of the enzyme HMG-CoA reductase. HMG-CoA reductase is the enzyme that synthesizes cholesterol from lipids in the body. By inhibiting this enzyme, the levels of cholesterol synthesized in the body can be reduced. Similarly, acetaminophen, popularly marketed under the brand name Tylenol, is an inhibitor of the enzyme cyclooxygenase. While it is effective in providing relief from fever and inflammation (pain), its mechanism of action is still not completely understood.

How are drugs developed? One of the first challenges in drug development is identifying the specific molecule that the drug is intended to target. In the case of statins, HMG-CoA reductase is the drug target. Drug targets are identified through painstaking research in the laboratory. Identifying the target alone is not sufficient; scientists also need to know how the target acts inside the cell and which reactions go awry in the case of disease. Once the target and the pathway are identified, then the actual process of drug design begins. During this stage, chemists and biologists work together to design and synthesize molecules that can either block or activate a particular reaction. However, this is only the beginning: both if and when a drug prototype is successful in performing its function, then it must undergo many tests from in vitro experiments to clinical trials before it can get FDA approval to be on the market.

Many enzymes don't work optimally, or even at all, unless bound to other specific non-protein helper molecules, either temporarily through ionic or hydrogen bonds or permanently through stronger covalent bonds. Two types of helper molecules are cofactors and coenzymes. Binding to these molecules promotes optimal conformation and function for their respective enzymes. Cofactors are inorganic ions such as iron ($Fe^{++}$) and magnesium ($Mg^{++}$). One example of an enzyme that requires a metal ion as a cofactor is the enzyme that builds DNA molecules, DNA polymerase, which requires bound zinc ion ($Zn^{++}$) to function. Coenzymes are organic helper molecules, with a basic atomic structure made up of carbon and hydrogen, which are required for enzyme action. The most common sources of coenzymes are dietary vitamins ([Figure 6]). Some vitamins are precursors to coenzymes and others act directly as coenzymes. Vitamin C is a coenzyme for multiple enzymes that take part in building the important connective tissue component, collagen. An important step in the breakdown of glucose to yield energy is catalysis by a multi-enzyme complex called pyruvate dehydrogenase. Pyruvate dehydrogenase is a complex of several enzymes that actually requires one cofactor (a magnesium ion) and five different organic coenzymes to catalyze its specific chemical reaction. Therefore, enzyme function is, in part, regulated by an abundance of various cofactors and coenzymes, which are supplied primarily by the diets of most organisms.

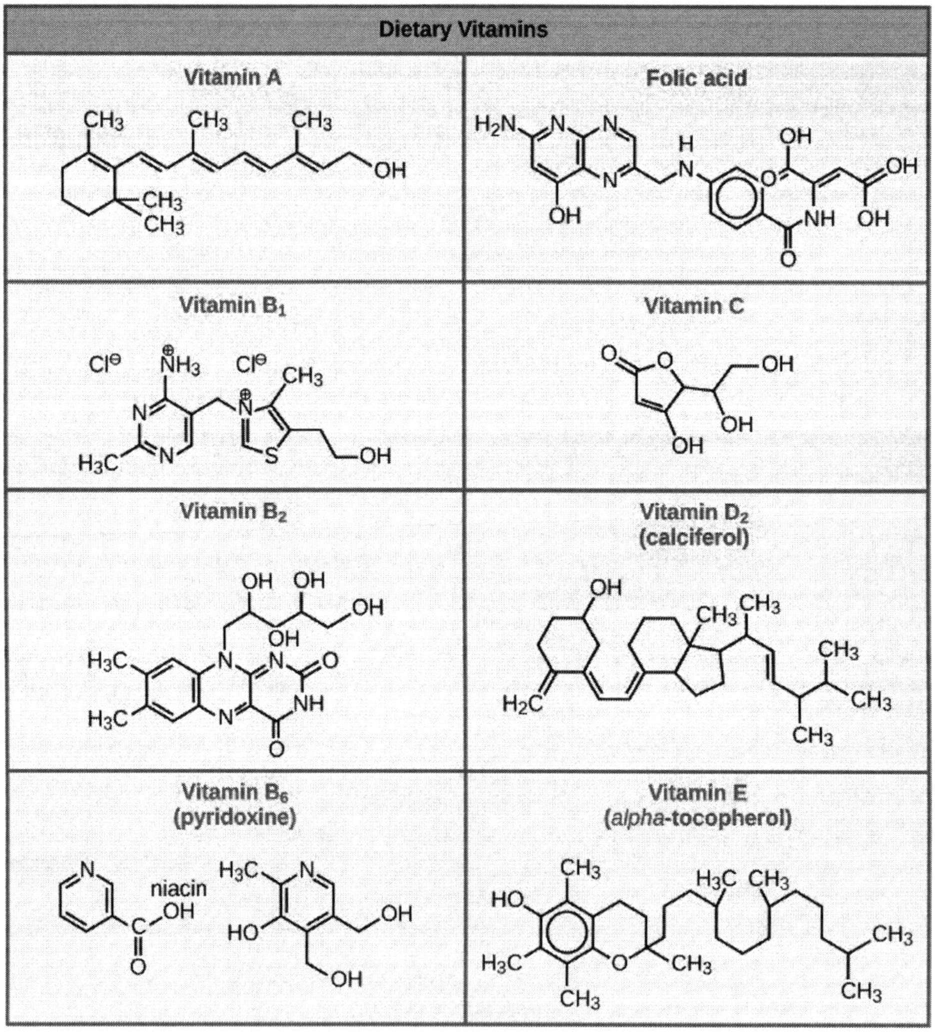

*Figure 6: Vitamins are important coenzymes or precursors of coenzymes, and are required for enzymes to function properly. Multivitamin capsules usually contain mixtures of all the vitamins at different percentages.*

## Enzyme Compartmentalization

In eukaryotic cells, molecules such as enzymes are usually compartmentalized into different organelles. This allows for yet another level of regulation of enzyme activity. Enzymes required only for certain cellular processes can be housed separately along with their substrates, allowing for more efficient chemical reactions. Examples of this sort of enzyme regulation based on location and proximity include the enzymes involved in the latter stages of cellular respiration, which take place exclusively in the mitochondria, and the enzymes involved in the digestion of cellular debris and foreign materials, located within lysosomes.

## Feedback Inhibition in Metabolic Pathways

Molecules can regulate enzyme function in many ways. A major question remains, however: What are these molecules and where do they come from? Some are cofactors and coenzymes, ions, and organic molecules, as you've learned. What other molecules in the cell provide enzymatic regulation, such as allosteric modulation, and competitive and noncompetitive inhibition? The answer is that a wide variety of molecules can perform these roles. Some of these molecules include pharmaceutical and non-pharmaceutical drugs, toxins, and poisons from the environment. Perhaps the most relevant sources of enzyme regulatory molecules, with respect to cellular metabolism, are the products of the cellular metabolic reactions themselves. In a most efficient and elegant way, cells have evolved to use the products of their own reactions for feedback inhibition of enzyme activity. Feedback inhibition involves the use of a reaction product to regulate its own further production ([**Figure 7**]). The cell responds to the abundance of specific products by slowing down production during anabolic or catabolic reactions. Such reaction products may inhibit the enzymes that catalyzed their production through the mechanisms described above.

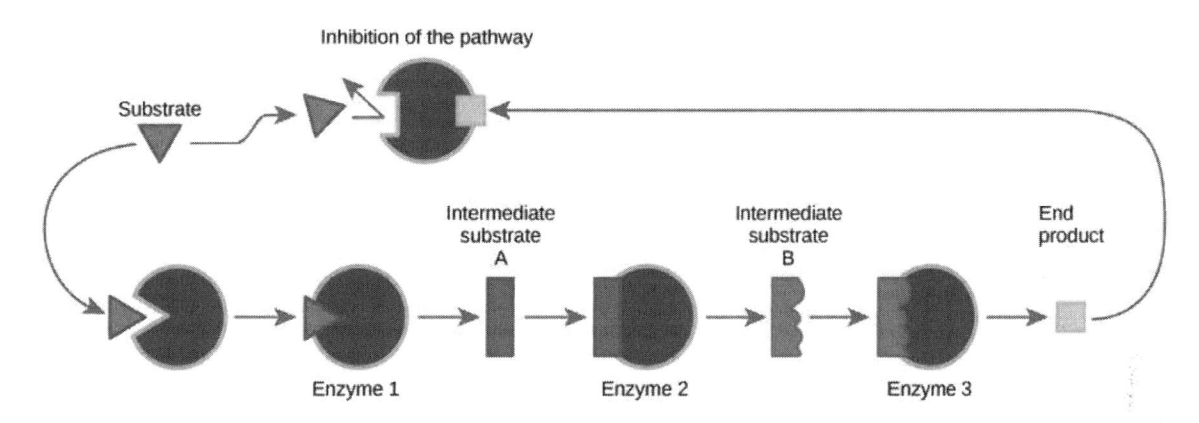

*Figure 7: Metabolic pathways are a series of reactions catalyzed by multiple enzymes. Feedback inhibition, where the end product of the pathway inhibits an upstream step, is an important regulatory mechanism in cells.*

The production of both amino acids and nucleotides is controlled through feedback inhibition. Additionally, ATP is an allosteric regulator of some of the enzymes involved in the catabolic breakdown of sugar, the process that produces ATP. In this way, when ATP is abundant, the cell can prevent its further production. Remember that ATP is an unstable molecule that can spontaneously dissociate into ADP. If too much ATP were present in a cell, much of it would go to waste. On the other hand, ADP serves as a positive allosteric regulator (an allosteric activator) for some of the same enzymes that are inhibited by ATP. Thus, when relative levels of ADP are high compared to ATP, the cell is triggered to produce more ATP through the catabolism of sugar.

## SECTION SUMMARY

Enzymes are chemical catalysts that accelerate chemical reactions at physiological temperatures by lowering their activation energy. Enzymes are usually proteins consisting of one or more polypeptide chains. Enzymes have an active site that provides a unique chemical environment, made up of certain amino acid R groups (residues). This unique environment is perfectly suited to convert particular chemical reactants for that enzyme, called substrates, into unstable intermediates called transition states. Enzymes and substrates are thought to bind with an induced fit, which means that enzymes undergo slight conformational adjustments upon substrate contact, leading to full, optimal binding. Enzymes bind to substrates and catalyze reactions in four different ways: bringing substrates together in an optimal orientation, compromising the bond structures of substrates so that bonds can be more easily broken, providing optimal environmental conditions for a reaction to occur, or participating directly in their chemical reaction by forming transient covalent bonds with the substrates.

Enzyme action must be regulated so that in a given cell at a given time, the desired reactions are being catalyzed and the undesired reactions are not. Enzymes are regulated by cellular conditions, such as temperature and pH. They are also regulated through their location within a cell, sometimes being compartmentalized so that they can only catalyze reactions under certain circumstances. Inhibition and activation of enzymes via other molecules are other important ways that enzymes are regulated. Inhibitors can act competitively, noncompetitively, or allosterically; noncompetitive inhibitors are usually allosteric. Activators can also enhance the function of enzymes allosterically. The most common method by which cells regulate the enzymes in metabolic pathways is through feedback inhibition. During feedback inhibition, the products of a metabolic pathway serve as inhibitors (usually allosteric) of one or more of the enzymes (usually the first committed enzyme of the pathway) involved in the pathway that produces them.

## Glossary

**active site:** specific region of the enzyme to which the substrate binds

**allosteric inhibition:** inhibition by a binding event at a site different from the active site, which induces a conformational change and reduces the affinity of the enzyme for its substrate

**coenzyme:** small organic molecule, such as a vitamin or its derivative, which is required to enhance the activity of an enzyme

**cofactor:** inorganic ion, such as iron and magnesium ions, required for optimal regulation of enzyme activity

**competitive inhibition:** type of inhibition in which the inhibitor competes with the substrate molecule by binding to the active site of the enzyme

**denature:** process that changes the natural properties of a substance

**feedback inhibition:** effect of a product of a reaction sequence to decrease its further production by inhibiting the activity of the first enzyme in the pathway that produces it

**induced fit:** dynamic fit between the enzyme and its substrate, in which both components modify their structures to allow for ideal binding

**substrate:** molecule on which the enzyme acts

# Nucleic Acids

### Learning Objectives

By the end of this section, you will be able to:

- Describe the structure of nucleic acids and define the two types of nucleic acids
- Explain the structure and role of DNA
- Explain the structure and roles of RNA

Nucleic acids are the most important macromolecules for the continuity of life. They carry the genetic blueprint of a cell and carry instructions for the functioning of the cell.

## DNA AND RNA

The two main types of nucleic acids are deoxyribonucleic acid (DNA) and ribonucleic acid (RNA). DNA is the genetic material found in all living organisms, ranging from single-celled bacteria to multicellular mammals. It is found in the nucleus of eukaryotes and in the organelles, chloroplasts, and mitochondria. In prokaryotes, the DNA is not enclosed in a membranous envelope.

The entire genetic content of a cell is known as its genome, and the study of genomes is genomics. In eukaryotic cells but not in prokaryotes, DNA forms a complex with histone proteins to form chromatin, the substance of eukaryotic chromosomes. A chromosome may contain tens of thousands of genes. Many genes contain the information to make protein products; other genes code for RNA products. DNA controls all of the cellular activities by turning the genes "on" or "off."

The other type of nucleic acid, RNA, is mostly involved in protein synthesis. The DNA molecules never leave the nucleus but instead use an intermediary to communicate with the rest of the cell. This intermediary is the messenger RNA (mRNA). Other types of RNA—like rRNA, tRNA, and microRNA—are involved in protein synthesis and its regulation.

DNA and RNA are made up of monomers known as nucleotides. The nucleotides combine with each other to form a polynucleotide, DNA or RNA. Each nucleotide is made up of three components: a nitrogenous base, a pentose (five-carbon) sugar, and a phosphate group ([Figure 1]). Each nitrogenous base in a nucleotide is attached to a sugar molecule, which is attached to one or more phosphate groups.

260  Biological Macromolecules

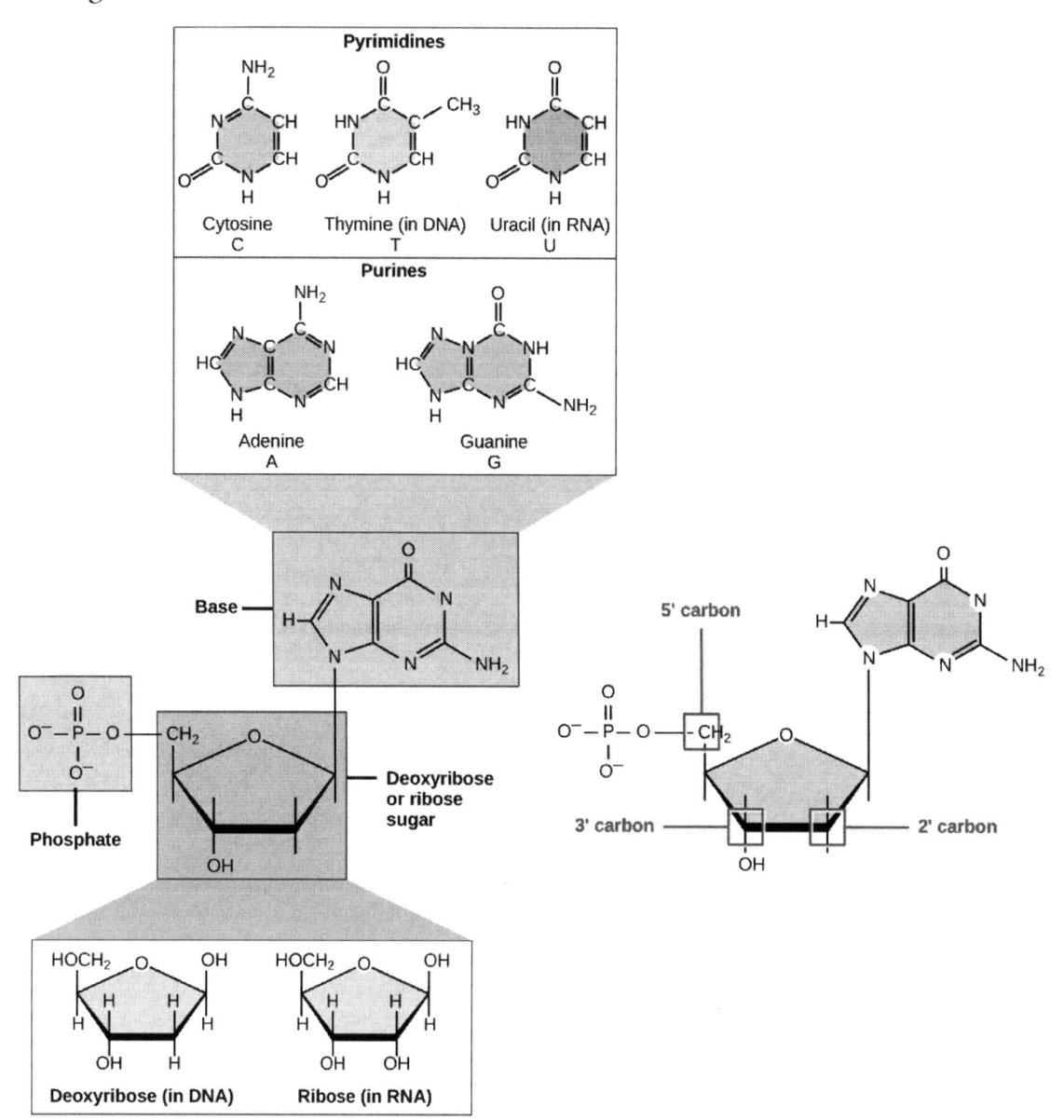

*Figure 1: A nucleotide is made up of three components: a nitrogenous base, a pentose sugar, and one or more phosphate groups. Carbon residues in the pentose are numbered 1' through 5' (the prime distinguishes these residues from those in the base, which are numbered without using a prime notation). The base is attached to the 1' position of the ribose, and the phosphate is attached to the 5' position. When a polynucleotide is formed, the 5' phosphate of the incoming nucleotide attaches to the 3' hydroxyl group at the end of the growing chain. Two types of pentose are found in nucleotides, deoxyribose (found in DNA) and ribose (found in RNA). Deoxyribose is similar in structure to ribose, but it has an H instead of an OH at the 2' position. Bases can be divided into two categories: purines and pyrimidines. Purines have a double ring structure, and pyrimidines have a single ring.*

The nitrogenous bases, important components of nucleotides, are organic molecules and are so named because they contain carbon and nitrogen. They are bases because they contain an amino group that has the potential of binding an extra hydrogen, and thus, decreases the hydrogen ion concentration in its environment, making it more basic. Each nucleotide in DNA contains one of four possible nitrogenous bases: adenine (A), guanine (G) cytosine (C), and thymine (T).

Adenine and guanine are classified as purines. The primary structure of a purine is two carbon-nitrogen rings. Cytosine, thymine, and uracil are classified as pyrimidines which have a single carbon-nitrogen ring as their primary structure ([Figure 1]). Each of these basic carbon-nitrogen rings has different functional groups attached to it. In molecular biology shorthand, the nitrogenous bases are simply known by their symbols A, T, G, C, and U. DNA contains A, T, G, and C whereas RNA contains A, U, G, and C.

The pentose sugar in DNA is deoxyribose, and in RNA, the sugar is ribose ([Figure 1]). The difference between the sugars is the presence of the hydroxyl group on the second carbon of the ribose and hydrogen on the second carbon of the deoxyribose. The carbon atoms of the sugar molecule are numbered as 1′, 2′, 3′, 4′, and 5′ (1′ is read as "one prime"). The phosphate residue is attached to the hydroxyl group of the 5′ carbon of one sugar and the hydroxyl group of the 3′ carbon of the sugar of the next nucleotide, which forms a 5′–3′ phosphodiester linkage. The phosphodiester linkage is not formed by simple dehydration reaction like the other linkages connecting monomers in macromolecules: its formation involves the removal of two phosphate groups. A polynucleotide may have thousands of such phosphodiester linkages.

## DNA DOUBLE-HELIX STRUCTURE

DNA has a double-helix structure ([Figure 2]). The sugar and phosphate lie on the outside of the helix, forming the backbone of the DNA. The nitrogenous bases are stacked in the interior, like the steps of a staircase, in pairs; the pairs are bound to each other by hydrogen bonds. Every base pair in the double helivx is separated from the next base pair by 0.34 nm. The two strands of the helix run in opposite directions, meaning that the 5′ carbon end of one strand will face the 3′ carbon end of its matching strand. (This is referred to as antiparallel orientation and is important to DNA replication and in many nucleic acid interactions.)

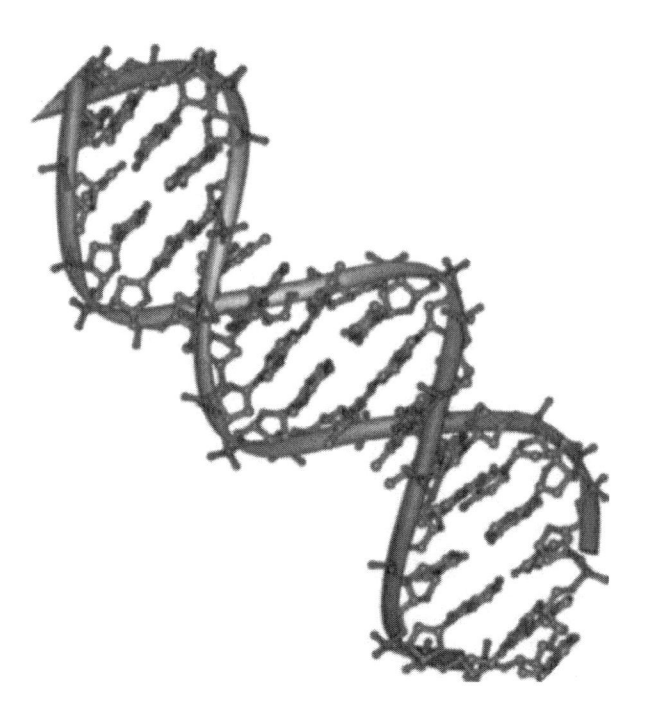

*Figure 2: Native DNA is an antiparallel double helix. The phosphate backbone (indicated by the curvy lines) is on the outside, and the bases are on the inside. Each base from one strand interacts via hydrogen bonding with a base from the opposing strand. (credit: Jerome Walker/Dennis Myts)*

## 262 Biological Macromolecules

Only certain types of base pairing are allowed. For example, a certain purine can only pair with a certain pyrimidine. This means A can pair with T, and G can pair with C, as shown in **[Figure 3]**. This is known as the base complementary rule. In other words, the DNA strands are complementary to each other. If the sequence of one strand is AATTGGCC, the complementary strand would have the sequence TTAACCGG. During DNA replication, each strand is copied, resulting in a daughter DNA double helix containing one parental DNA strand and a newly synthesized strand.

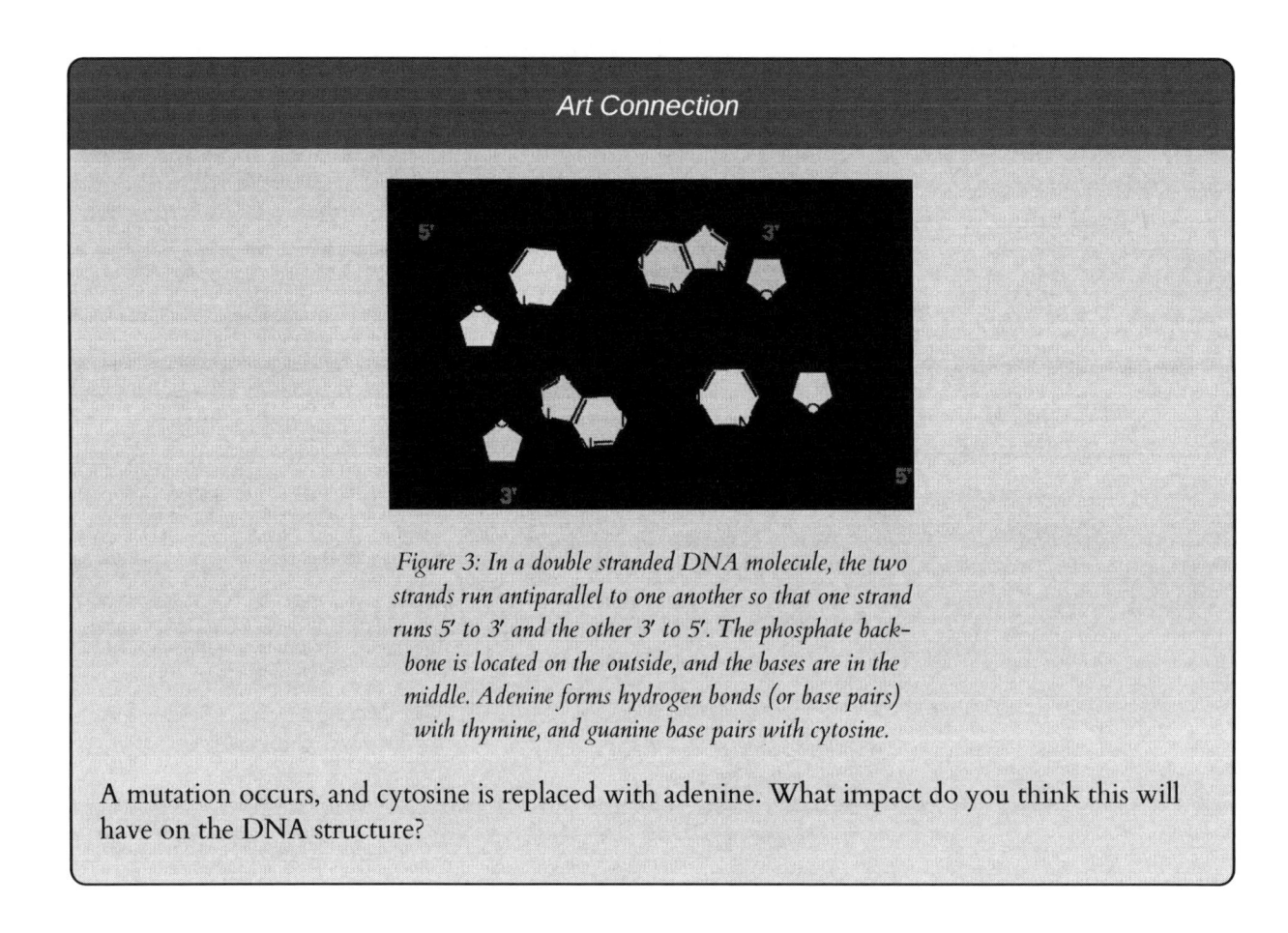

### Art Connection

*Figure 3: In a double stranded DNA molecule, the two strands run antiparallel to one another so that one strand runs 5' to 3' and the other 3' to 5'. The phosphate backbone is located on the outside, and the bases are in the middle. Adenine forms hydrogen bonds (or base pairs) with thymine, and guanine base pairs with cytosine.*

A mutation occurs, and cytosine is replaced with adenine. What impact do you think this will have on the DNA structure?

### RNA

Ribonucleic acid, or RNA, is mainly involved in the process of protein synthesis under the direction of DNA. RNA is usually single-stranded and is made of ribonucleotides that are linked by phosphodiester bonds. A ribonucleotide in the RNA chain contains ribose (the pentose sugar), one of the four nitrogenous bases (A, U, G, and C), and the phosphate group.

There are four major types of RNA: messenger RNA (mRNA), ribosomal RNA (rRNA), transfer RNA (tRNA), and microRNA (miRNA). The first, mRNA, carries the message from DNA, which controls all of the cellular activities in a cell. If a cell requires a certain protein to be synthesized, the gene for this product is turned "on" and the messenger RNA is synthesized in the nucleus. The RNA base sequence is complementary to the coding sequence of the DNA from which it has been copied. However, in RNA, the base T is absent and U is present instead. If the DNA strand has a sequence AATTGCGC, the sequence of the complementary RNA is UUAACGCG. In the cytoplasm, the mRNA interacts with ribosomes and other cellular machinery (**[Figure 4]**).

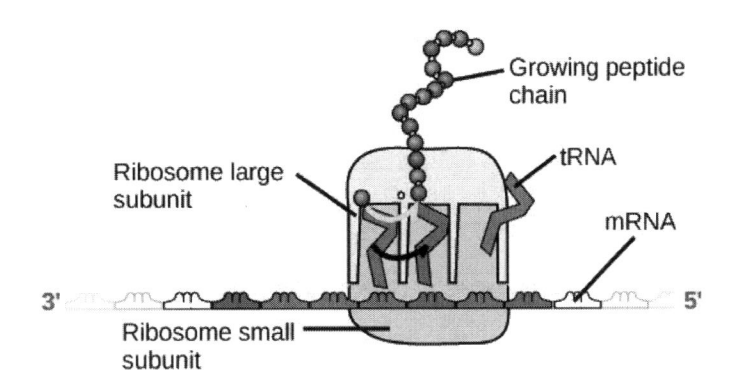

*Figure 4: A ribosome has two parts: a large subunit and a small subunit. The mRNA sits in between the two subunits. A tRNA molecule recognizes a codon on the mRNA, binds to it by complementary base pairing, and adds the correct amino acid to the growing peptide chain.*

The mRNA is read in sets of three bases known as codons. Each codon codes for a single amino acid. In this way, the mRNA is read and the protein product is made. Ribosomal RNA (rRNA) is a major constituent of ribosomes on which the mRNA binds. The rRNA ensures the proper alignment of the mRNA and the ribosomes; the rRNA of the ribosome also has an enzymatic activity (peptidyl transferase) and catalyzes the formation of the peptide bonds between two aligned amino acids. Transfer RNA (tRNA) is one of the smallest of the four types of RNA, usually 70–90 nucleotides long. It carries the correct amino acid to the site of protein synthesis. It is the base pairing between the tRNA and mRNA that allows for the correct amino acid to be inserted in the polypeptide chain. microRNAs are the smallest RNA molecules and their role involves the regulation of gene expression by interfering with the expression of certain mRNA messages. [**Figure 1**] summarizes features of DNA and RNA.

Features of DNA and RNA

|  | DNA | RNA |
| --- | --- | --- |
| Function | Carries genetic information | Involved in protein synthesis |
| Location | Remains in the nucleus | Leaves the nucleus |
| Structure | Double helix | Usually single-stranded |
| Sugar | Deoxyribose | Ribose |
| Pyrimidines | Cytosine, thymine | Cytosine, uracil |
| Purines | Adenine, guanine | Adenine, guanine |

Even though the RNA is single stranded, most RNA types show extensive intramolecular base pairing between complementary sequences, creating a predictable three-dimensional structure essential for their function.

As you have learned, information flow in an organism takes place from DNA to RNA to protein. DNA dictates the structure of mRNA in a process known as transcription, and RNA dictates the structure of protein in a process known as translation. This is known as the Central Dogma of Life, which holds true for all organisms; however, exceptions to the rule occur in connection with viral infections.

To learn more about DNA, explore the **Howard Hughes Medical Institute BioInteractive animations** on the topic of DNA.

## SECTION SUMMARY

Nucleic acids are molecules made up of nucleotides that direct cellular activities such as cell division and protein synthesis. Each nucleotide is made up of a pentose sugar, a nitrogenous base, and a phosphate group. There are two types of nucleic acids: DNA and RNA. DNA carries the genetic blueprint of the cell and is passed on from parents to offspring (in the form of chromosomes). It has a double-helical structure with the two strands running in opposite directions, connected by hydrogen bonds, and complementary to each other. RNA is single-stranded and is made of a pentose sugar (ribose), a nitrogenous base, and a phosphate group. RNA is involved in protein synthesis and its regulation. Messenger RNA (mRNA) is copied from the DNA, is exported from the nucleus to the cytoplasm, and contains information for the construction of proteins. Ribosomal RNA (rRNA) is a part of the ribosomes at the site of protein synthesis, whereas transfer RNA (tRNA) carries the amino acid to the site of protein synthesis. microRNA regulates the use of mRNA for protein synthesis.

### Glossary

**deoxyribonucleic acid (DNA):** double-helical molecule that carries the hereditary information of the cell

**messenger RNA (mRNA):** RNA that carries information from DNA to ribosomes during protein synthesis

**nucleic acid:** biological macromolecule that carries the genetic blueprint of a cell and carries instructions for the functioning of the cell

**nucleotide:** monomer of nucleic acids; contains a pentose sugar, one or more phosphate groups, and a nitrogenous base

**phosphodiester:** linkage covalent chemical bond that holds together the polynucleotide chains with a phosphate group linking two pentose sugars of neighboring nucleotides

**polynucleotide:** long chain of nucleotides

**purine:** type of nitrogenous base in DNA and RNA; adenine and guanine are purines

**pyrimidine:** type of nitrogenous base in DNA and RNA; cytosine, thymine, and uracil are pyrimidines

**ribonucleic acid (RNA):** single-stranded, often internally base paired, molecule that is involved in protein synthesis

**ribosomal RNA (rRNA):** RNA that ensures the proper alignment of the mRNA and the ribosomes during protein synthesis and catalyzes the formation of the peptide linkage

**transcription:** process through which messenger RNA forms on a template of DNA

**transfer RNA (tRNA):** RNA that carries activated amino acids to the site of protein synthesis on the ribosome

**translation:** process through which RNA directs the formation of protein

# Exercises

### PART 1

1. A nucleotide of DNA may contain _____.
   1. ribose, uracil, and a phosphate group
   2. deoxyribose, uracil, and a phosphate group
   3. deoxyribose, thymine, and a phosphate group
   4. ribose, thymine, and a phosphate group
2. The building blocks of nucleic acids are _____.
   1. sugars
   2. nitrogenous bases
   3. peptides
   4. nucleotides
3. What are the structural differences between RNA and DNA?
4. What are the four types of RNA and how do they function?

### PART 2

1. Which of the following is not true about enzymes:
   1. They increase ΔG of reactions
   2. They are usually made of amino acids
   3. They lower the activation energy of chemical reactions
   4. Each one is specific to the particular substrate(s) to which it binds
2. An allosteric inhibitor does which of the following?
   1. Binds to an enzyme away from the active site and changes the conformation of the active site, increasing its affinity for substrate binding
   2. Binds to the active site and blocks it from binding substrate
   3. Binds to an enzyme away from the active site and changes the conformation of the active site, decreasing its affinity for the substrate
   4. Binds directly to the active site and mimics the substrate
3. Which of the following analogies best describe the induced-fit model of enzyme-substrate binding?
   1. A hug between two people
   2. A key fitting into a lock

3. A square peg fitting through the square hole and a round peg fitting through the round hole of a children's toy

4. The fitting together of two jigsaw puzzle pieces.

4. With regard to enzymes, why are vitamins necessary for good health? Give examples.

5. Explain in your own words how enzyme feedback inhibition benefits a cell.

## PART 3

1. The monomers that make up proteins are called _____.

    1. nucleotides
    2. disaccharides
    3. amino acids
    4. chaperones

2. The α helix and the β-pleated sheet are part of which protein structure?

    1. primary
    2. secondary
    3. tertiary
    4. quaternary

3. Explain what happens if even one amino acid is substituted for another in a polypeptide chain. Provide a specific example.

4. Describe the differences in the four protein structures.

## PART 4

1. Saturated fats have all of the following characteristics except:

    1. they are solid at room temperature
    2. they have single bonds within the carbon chain
    3. they are usually obtained from animal sources
    4. they tend to dissolve in water easily

2. Phospholipids are important components of _____.

    1. the plasma membrane of animal cells
    2. the ring structure of steroids
    3. the waxy covering on leaves
    4. the double bond in hydrocarbon chains

3. Explain at least three functions that lipids serve in plants and/or animals.

4. Why have trans fats been banned from some restaurants? How are they created?

## PART 5

1. An example of a monosaccharide is _____.

    1. fructose
    2. glucose
    3. galactose
    4. all of the above

2. Cellulose and starch are examples of:

    1. monosaccharides
    2. disaccharides
    3. lipids
    4. polysaccharides

3. Plant cell walls contain which of the following in abundance?

    1. starch
    2. cellulose
    3. glycogen
    4. lactose

4. Lactose is a disaccharide formed by the formation of a _____ bond between glucose and _____.

    1. glycosidic; lactose
    2. glycosidic; galactose
    3. hydrogen; sucrose
    4. hydrogen; fructose

5. Describe the similarities and differences between glycogen and starch.

6. Why is it impossible for humans to digest food that contains cellulose?

# Check Your Knowledge: Self-Test

1. $C_6H_{12}O_6$ is the chemical formula for a _____.
   1. polymer of carbohydrate
   2. pentose monosaccharide
   3. hexose monosaccharide
   4. all of the above
2. What organic compound do brain cells primarily rely on for fuel?
   1. glucose
   2. glycogen
   3. galactose
   4. glycerol
3. Which of the following is a functional group that is part of a building block of proteins?
   1. phosphate
   2. adenine
   3. amino
   4. ribose
4. A pentose sugar is a part of the monomer used to build which type of macromolecule?
   1. polysaccharides
   2. nucleic acids
   3. phosphorylated glucose
   4. glycogen
5. A phospholipid _____.
   1. has both polar and nonpolar regions
   2. is made up of a triglyceride bonded to a phosphate group
   3. is a building block of ATP
   4. can donate both cations and anions in solution
6. In DNA, nucleotide bonding forms a compound with a characteristic shape known as a(n) _____.
   1. beta chain
   2. pleated sheet
   3. alpha helix
   4. double helix

7. Uracil _____.
   1. contains nitrogen
   2. is a pyrimidine
   3. is found in RNA
   4. all of the above

8. The ability of an enzyme's active sites to bind only substrates of compatible shape and charge is known as _____.
   1. selectivity
   2. specificity
   3. subjectivity
   4. specialty

9. If the disaccharide maltose is formed from two glucose monosaccharides, which are hexose sugars, how many atoms of carbon, hydrogen, and oxygen does maltose contain and why?

10. Why are biological macromolecules considered organic?

11. What role do electrons play in dehydration synthesis and hydrolysis?

12. Amino acids have the generic structure seen below, where R represents different carbon-based side chains.

    Describe how the structure of amino acids allows them to be linked into long peptide chains to form proteins.

13. Describe the similarities and differences between glycogen and starch.

14. Why is it impossible for humans to digest food that contains cellulose?

15. Explain at least three functions that lipids serve in plants and/or animals.

16. Why are fatty acids better than glycogen for storing large amounts of chemical energy?

17. What are the structural differences between RNA and DNA?

18. What are the four types of RNA and how do they function?

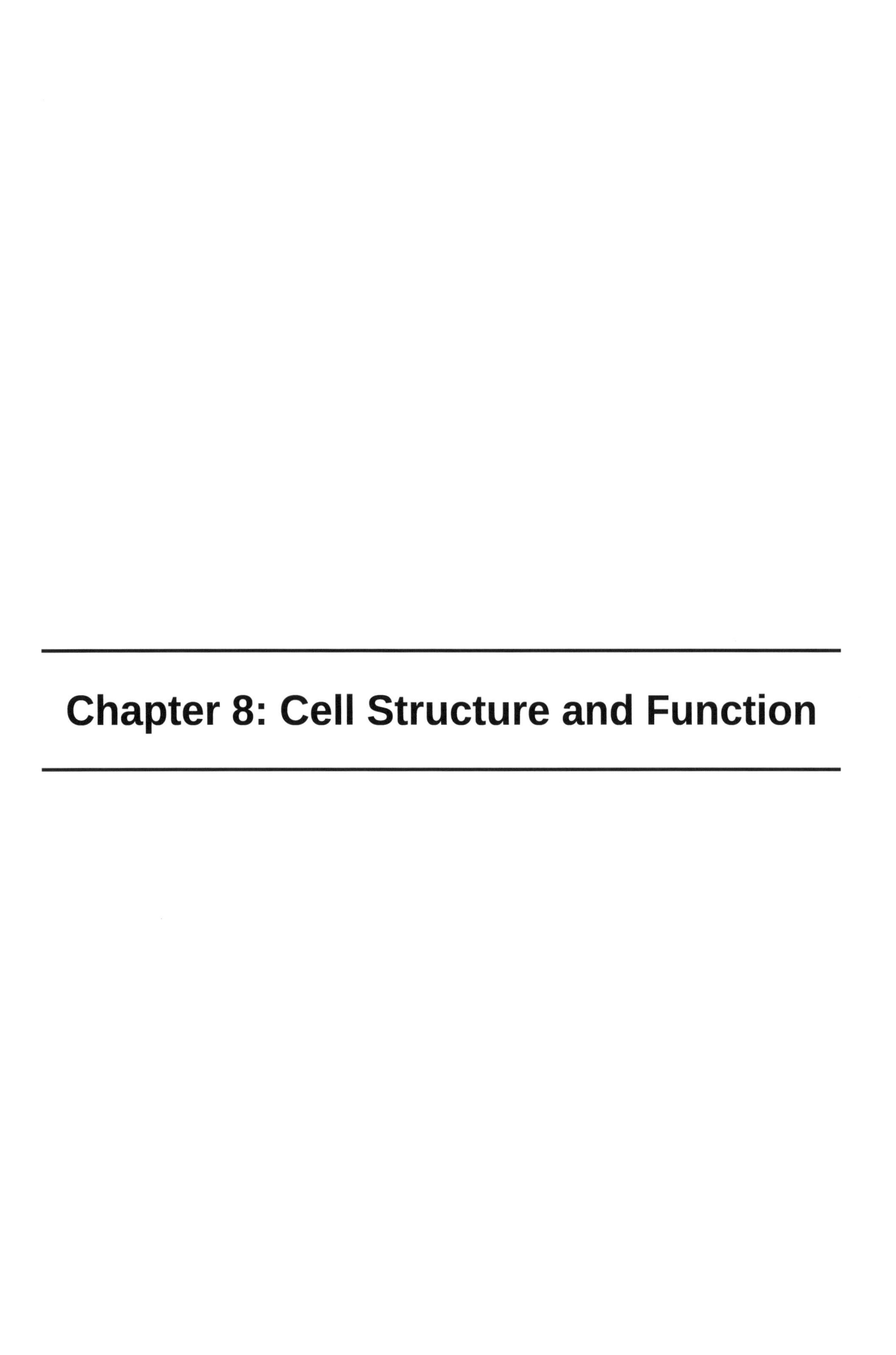

# Chapter 8: Cell Structure and Function

# Introduction

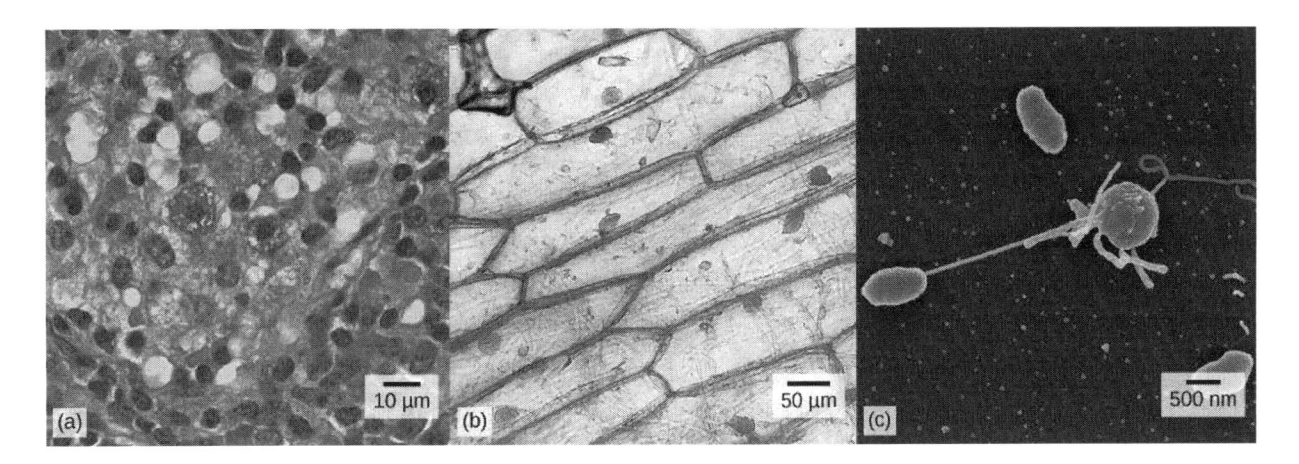

*(a) Nasal sinus cells (viewed with a light microscope), (b) onion cells (viewed with a light microscope), and (c) Vibrio tasmaniensis bacterial cells (viewed using a scanning electron microscope) are from very different organisms, yet all share certain characteristics of basic cell structure. (credit a: modification of work by Ed Uthman, MD; credit b: modification of work by Umberto Salvagnin; credit c: modification of work by Anthony D'Onofrio; scale-bar data from Matt Russell)*

Close your eyes and picture a brick wall. What is the basic building block of that wall? It is a single brick, of course. Like a brick wall, your body is composed of basic building blocks, and the building blocks of your body are cells.

Your body has many kinds of cells, each specialized for a specific purpose. Just as a home is made from a variety of building materials, the human body is constructed from many cell types. For example, epithelial cells protect the surface of the body and cover the organs and body cavities within. Bone cells help to support and protect the body. Cells of the immune system fight invading bacteria. Additionally, red blood cells carry oxygen throughout the body. Each of these cell types plays a vital role during the growth, development, and day-to-day maintenance of the body. In spite of their enormous variety, however, all cells share certain fundamental characteristics.

# Studying Cells

**Learning Objectives**

By the end of this section, you will be able to:

- Describe the role of cells in organisms
- Compare and contrast light microscopy and electron microscopy
- Summarize cell theory

A cell is the smallest unit of a living thing. A living thing, whether made of one cell (like bacteria) or many cells (like a human), is called an organism. Thus, cells are the basic building blocks of all organisms.

Several cells of one kind that interconnect with each other and perform a shared function form tissues, several tissues combine to form an organ (your stomach, heart, or brain), and several organs make up an organ system (such as the digestive system, circulatory system, or nervous system). Several systems that function together form an organism (like a human being). Here, we will examine the structure and function of cells.

There are many types of cells, all grouped into one of two broad categories: prokaryotic and eukaryotic. For example, both animal and plant cells are classified as eukaryotic cells, whereas bacterial cells are classified as prokaryotic. Before discussing the criteria for determining whether a cell is prokaryotic or eukaryotic, let's first examine how biologists study cells.

## MICROSCOPY

Cells vary in size. With few exceptions, individual cells cannot be seen with the naked eye, so scientists use microscopes (micro- = "small"; -scope = "to look at") to study them. A microscope is an instrument that magnifies an object. Most photographs of cells are taken with a microscope, and these images can also be called micrographs.

The optics of a microscope's lenses change the orientation of the image that the user sees. A specimen that is right-side up and facing right on the microscope slide will appear upside-down and facing left when viewed through a microscope, and vice versa. Similarly, if the slide is moved left while looking through the microscope, it will appear to move right, and if moved down, it will seem to move up. This occurs because microscopes use two sets of lenses to magnify the image. Because of the manner by which light travels through the lenses, this system of two lenses produces an inverted image (binocular, or dissecting microscopes, work in a similar manner, but include an additional magnification system that makes the final image appear to be upright).

### Light Microscopes

To give you a sense of cell size, a typical human red blood cell is about eight millionths of a meter or eight micrometers (abbreviated as eight μm) in diameter; the head of a pin of is about two thousandths of a meter (two mm) in diameter. That means about 250 red blood cells could fit on the head of a pin.

Most student microscopes are classified as light microscopes ([Figure 1]a). Visible light passes and is bent through the lens system to enable the user to see the specimen. Light microscopes are advantageous for viewing living organisms, but since individual cells are generally transparent, their components are not distinguishable unless they are colored with special stains. Staining, however, usually kills the cells.

Light microscopes commonly used in the undergraduate college laboratory magnify up to approximately 400 times. Two parameters that are important in microscopy are magnification and resolving power. Magnification is the process of enlarging an object in appearance. Resolving power is the ability of a microscope to distinguish two adjacent structures as separate: the higher the resolution, the better the clarity and detail of the image. When oil immersion lenses are used for the study of small objects, magnification is usually increased to 1,000 times. In order to gain a better understanding of cellular structure and function, scientists typically use electron microscopes.

*Figure 1: (a) Most light microscopes used in a college biology lab can magnify cells up to approximately 400 times and have a resolution of about 200 nanometers. (b) Electron microscopes provide a much higher magnification, 100,000x, and a have a resolution of 50 picometers. (credit a: modification of work by "GcG"/Wikimedia Commons; credit b: modification of work by Evan Bench)*

### Electron Microscopes

In contrast to light microscopes, electron microscopes ([Figure 1]b) use a beam of electrons instead of a beam of light. Not only does this allow for higher magnification and, thus, more detail ([Figure 1]), it also provides higher resolving power. The method used to prepare the specimen for viewing with an electron microscope kills the specimen. Electrons have short wavelengths (shorter than photons) that move best in a vacuum, so living cells cannot be viewed with an electron microscope.

In a scanning electron microscope, a beam of electrons moves back and forth across a cell's surface, creating details of cell surface characteristics. In a transmission electron microscope, the electron beam penetrates the cell and provides details of a cell's internal structures. As you might imagine, electron microscopes are significantly more bulky and expensive than light microscopes.

276 Stoichiometry and The Mole

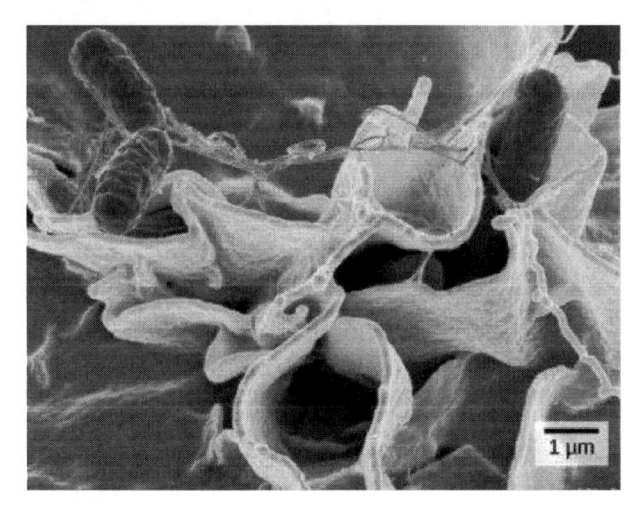

*Figure 2: (b) This scanning electron microscope micrograph shows Salmonella bacteria (in red) invading human cells (yellow). Even though subfigure (b) shows a different Salmonella specimen than subfigure*

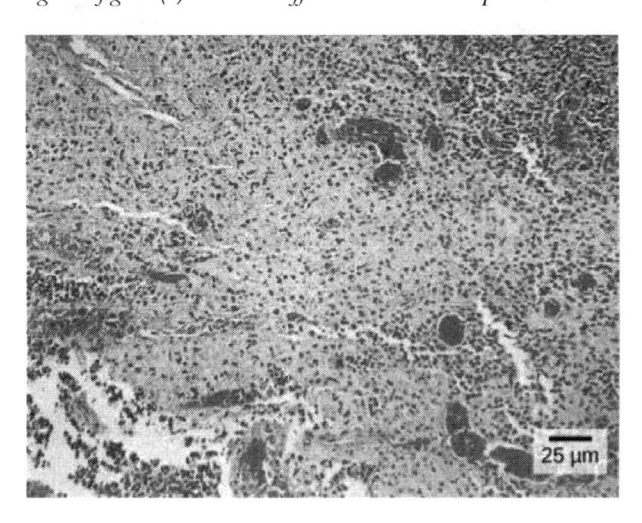

*Figure 2: (a) These Salmonella bacteria appear as tiny purple dots when viewed with a light microscope.(a), you can still observe the comparative increase in magnification and detail. (credit a: modification of work by CDC/Armed Forces Institute of Pathology, Charles N. Farmer, Rocky Mountain Laboratories; credit b: modification of work by NIAID, NIH; scale-bar data from Matt Russell)*

For another perspective on cell size, try the HowBig interactive at this site: http://openstaxcollege.org/l/cell_sizes.

## CELL THEORY

The microscopes we use today are far more complex than those used in the 1600s by Antony van Leeuwenhoek, a Dutch shopkeeper who had great skill in crafting lenses. Despite the limitations of his now-ancient lenses, van Leeuwenhoek observed the movements of protista (a type of single-celled organism) and sperm, which he collectively termed "animalcules."

In a 1665 publication called *Micrographia*, experimental scientist Robert Hooke coined the term "cell" for the box-like structures he observed when viewing cork tissue through a lens. In the 1670s, van Leeuwenhoek discovered bacteria and protozoa. Later advances in lenses, microscope construction, and staining techniques enabled other scientists to see some components inside cells.

By the late 1830s, botanist Matthias Schleiden and zoologist Theodor Schwann were studying tissues and proposed the unified cell theory, which states that all living things are composed of one or more cells, the cell is the basic unit of life, and new cells arise from existing cells. Rudolf Virchow later made important contributions to this theory.

## CAREER CONNECTION

Cytotechnology: have you ever heard of a medical test called a Pap smear ([**Figure 3**])? In this test, a doctor takes a small sample of cells from the uterine cervix of a patient and sends it to a medical lab where a cytotechnologist stains the cells and examines them for any changes that could indicate cervical cancer or a microbial infection.

Cytotechnologists (cyto- = "cell") are professionals who study cells via microscopic examinations and other laboratory tests. They are trained to determine which cellular changes are within normal limits and which are abnormal. Their focus is not limited to cervical cells; they study cellular specimens that come from all organs. When they notice abnormalities, they consult a pathologist, who is a medical doctor who can make a clinical diagnosis.

Cytotechnologists play a vital role in saving people's lives. When abnormalities are discovered early, a patient's treatment can begin sooner, which usually increases the chances of a successful outcome.

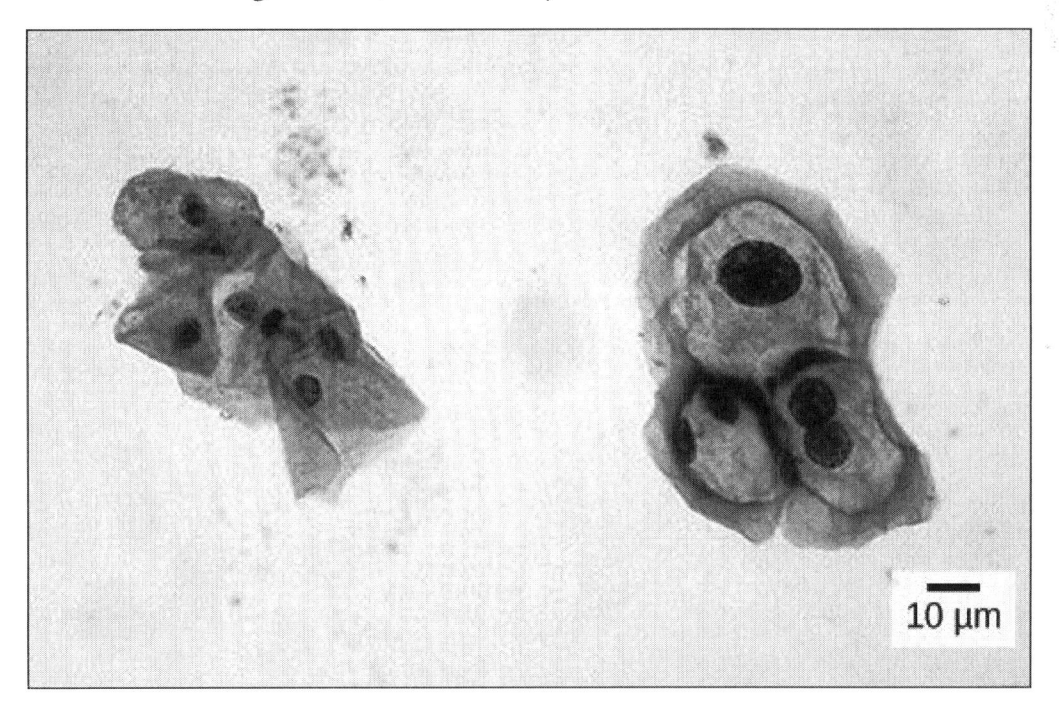

*Figure 3: These uterine cervix cells, viewed through a light microscope, were obtained from a Pap smear. Normal cells are on the left. The cells on the right are infected with human papillomavirus (HPV). Notice that the infected cells are larger; also, two of these cells each have two nuclei instead of one, the normal number. (credit: modification of work by Ed Uthman, MD; scale-bar data from Matt Russell)*

## SECTION SUMMARY

A cell is the smallest unit of life. Most cells are so tiny that they cannot be seen with the naked eye. Therefore, scientists use microscopes to study cells. Electron microscopes provide higher magnification, higher resolution, and more detail than light microscopes. The unified cell theory states that all organisms are composed of one or more cells, the cell is the basic unit of life, and new cells arise from existing cells

## Glossary

**cell theory:** see unified cell theory

**electron microscope:** an instrument that magnifies an object using a beam of electrons passed and bent through a lens system to visualize a specimen

**light microscope:** an instrument that magnifies an object using a beam visible light passed and bent through a lens system to visualize a specimen

**microscope:** an instrument that magnifies an object

**unified cell theory:** a biological concept that states that all organisms are composed of one or more cells; the cell is the basic unit of life; and new cells arise from existing cells

# Prokaryotic Cells

## LEARNING OBJECTIVES

By the end of this section, you will be able to:

- Name examples of prokaryotic and eukaryotic organisms
- Compare and contrast prokaryotic cells and eukaryotic cells
- Describe the relative sizes of different kinds of cells
- Explain why cells must be small

Cells fall into one of two broad categories: prokaryotic and eukaryotic. Only the predominantly single-celled organisms of the domains Bacteria and Archaea are classified as prokaryotes (pro- = "before"; -kary- = "nucleus"). Cells of animals, plants, fungi, and protists are all eukaryotes (ceu- = "true") and are made up of eukaryotic cells.

## COMPONENTS OF PROKARYOTIC CELLS

All cells share four common components: 1) a plasma membrane, an outer covering that separates the cell's interior from its surrounding environment; 2) cytoplasm, consisting of a jelly-like cytosol within the cell in which other cellular components are found; 3) DNA, the genetic material of the cell; and 4) ribosomes, which synthesize proteins. However, prokaryotes differ from eukaryotic cells in several ways.

A prokaryote is a simple, mostly single-celled (unicellular) organism that lacks a nucleus, or any other membrane-bound organelle. We will shortly come to see that this is significantly different in eukaryotes. Prokaryotic DNA is found in a central part of the cell: the nucleoid ([Figure 1]).

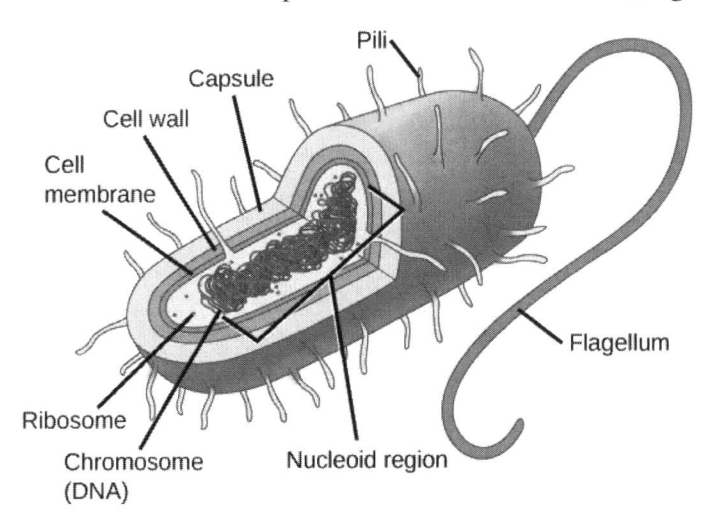

*Figure 1: This figure shows the generalized structure of a prokaryotic cell. All prokaryotes have chromosomal DNA localized in a nucleoid, ribosomes, a cell membrane, and a cell wall. The other structures shown are present in some, but not all, bacteria.*

# 280 Stoichiometry and The Mole

Most prokaryotes have a peptidoglycan cell wall and many have a polysaccharide capsule ([**Figure 1**]). The cell wall acts as an extra layer of protection, helps the cell maintain its shape, and prevents dehydration. The capsule enables the cell to attach to surfaces in its environment. Some prokaryotes have flagella, pili, or fimbriae. Flagella are used for locomotion. Pili are used to exchange genetic material during a type of reproduction called conjugation. Fimbriae are used by bacteria to attach to a host cell.

*Career Connection*

MicrobiologistThe most effective action anyone can take to prevent the spread of contagious illnesses is to wash his or her hands. Why? Because microbes (organisms so tiny that they can only be seen with microscopes) are ubiquitous. They live on doorknobs, money, your hands, and many other surfaces. If someone sneezes into his hand and touches a doorknob, and afterwards you touch that same doorknob, the microbes from the sneezer's mucus are now on your hands. If you touch your hands to your mouth, nose, or eyes, those microbes can enter your body and could make you sick.

However, not all microbes (also called microorganisms) cause disease; most are actually beneficial. You have microbes in your gut that make vitamin K. Other microorganisms are used to ferment beer and wine.

Microbiologists are scientists who study microbes. Microbiologists can pursue a number of careers. Not only do they work in the food industry, they are also employed in the veterinary and medical fields. They can work in the pharmaceutical sector, serving key roles in research and development by identifying new sources of antibiotics that could be used to treat bacterial infections.

Environmental microbiologists may look for new ways to use specially selected or genetically engineered microbes for the removal of pollutants from soil or groundwater, as well as hazardous elements from contaminated sites. These uses of microbes are called bioremediation technologies. Microbiologists can also work in the field of bioinformatics, providing specialized knowledge and insight for the design, development, and specificity of computer models of, for example, bacterial epidemics.

*Cell Size*

At 0.1 to 5.0 µm in diameter, prokaryotic cells are significantly smaller than eukaryotic cells, which have diameters ranging from 10 to 100 µm ([**Figure 2**]). The small size of prokaryotes allows ions and organic molecules that enter them to quickly diffuse to other parts of the cell. Similarly, any wastes produced within a prokaryotic cell can quickly diffuse out. This is not the case in eukaryotic cells, which have developed different structural adaptations to enhance intracellular transport.

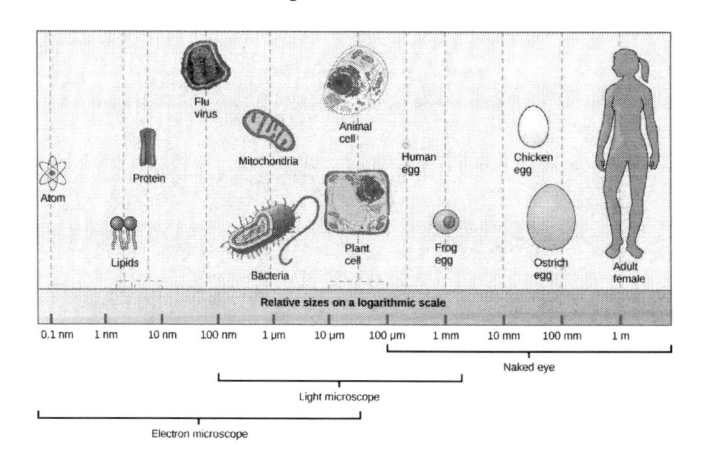

*Figure 2: This figure shows relative sizes of microbes on a logarithmic scale (recall that each unit of increase in a logarithmic scale represents a 10-fold increase in the quantity being measured).*

Small size, in general, is necessary for all cells, whether prokaryotic or eukaryotic. Let's examine why that is so. First, we'll consider the area and volume of a typical cell. Not all cells are spherical in shape, but most tend to approximate a sphere. You may remember from your high school geometry course that the formula for the surface area of a sphere is $4\pi r^2$, while the formula for its volume is $4\pi r^3/3$. Thus, as the radius of a cell increases, its surface area increases as the square of its radius, but its volume increases as the cube of its radius (much more rapidly). Therefore, as a cell increases in size, its surface area-to-volume ratio decreases. This same principle would apply if the cell had the shape of a cube (**[Figure 3]**). If the cell grows too large, the plasma membrane will not have sufficient surface area to support the rate of diffusion required for the increased volume. In other words, as a cell grows, it becomes less efficient. One way to become more efficient is to divide; another way is to develop organelles that perform specific tasks. These adaptations lead to the development of more sophisticated cells called eukaryotic cells.

### Art Connection

A piece of rebar is weighed and then submerged in a graduated cylinder partially filled with water, with results as shown.

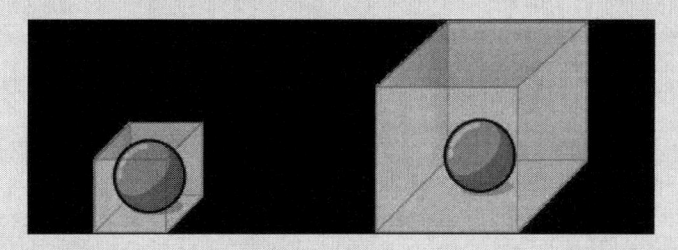

*Figure 3: Notice that as a cell increases in size, its surface area-to-volume ratio decreases. When there is insufficient surface area to support a cell's increasing volume, a cell will either divide or die. The cell on the left has a volume of 1 mm3 and a surface area of 6 mm2, with a surface area-to-volume ratio of 6 to 1, whereas the cell on the right has a volume of 8 mm3 and a surface area of 24 mm2, with a surface area-to-volume ratio of 3 to 1.*

Prokaryotic cells are much smaller than eukaryotic cells. What advantages might small cell size confer on a cell? What advantages might large cell size have?

### Section Summary

Prokaryotes are predominantly single-celled organisms of the domains Bacteria and Archaea. All prokaryotes have plasma membranes, cytoplasm, ribosomes, and DNA that is not membrane-bound. Most have peptidoglycan cell walls and many have polysaccharide capsules. Prokaryotic cells range in diameter from 0.1 to 5.0 μm.

As a cell increases in size, its surface area-to-volume ratio decreases. If the cell grows too large, the plasma membrane will not have sufficient surface area to support the rate of diffusion required for the increased volume.

### Glossary

**nucleoid:** central part of a prokaryotic cell in which the chromosome is found

**prokaryote:** unicellular organism that lacks a nucleus or any other membrane-bound organelle

# Eukaryotic Cells

### LEARNING OBJECTIVES

By the end of this section, you will be able to:

- Describe the structure of eukaryotic cells
- Compare animal cells with plant cells
- State the role of the plasma membrane
- Summarize the functions of the major cell organelles

Have you ever heard the phrase "form follows function?" It's a philosophy practiced in many industries. In architecture, this means that buildings should be constructed to support the activities that will be carried out inside them. For example, a skyscraper should be built with several elevator banks; a hospital should be built so that its emergency room is easily accessible.

Our natural world also utilizes the principle of form following function, especially in cell biology, and this will become clear as we explore eukaryotic cells ([**Figure 1**]). Unlike prokaryotic cells, eukaryotic cells have: 1) a membrane-bound nucleus; 2) numerous membrane-bound organelles such as the endoplasmic reticulum, Golgi apparatus, chloroplasts, mitochondria, and others; and 3) several, rod-shaped chromosomes. Because a eukaryotic cell's nucleus is surrounded by a membrane, it is often said to have a "true nucleus." The word "organelle" means "little organ," and, as already mentioned, organelles have specialized cellular functions, just as the organs of your body have specialized functions.

At this point, it should be clear to you that eukaryotic cells have a more complex structure than prokaryotic cells. Organelles allow different functions to be compartmentalized in different areas of the cell. Before turning to organelles, let's first examine two important components of the cell: the plasma membrane and the cytoplasm.

## Art Connection

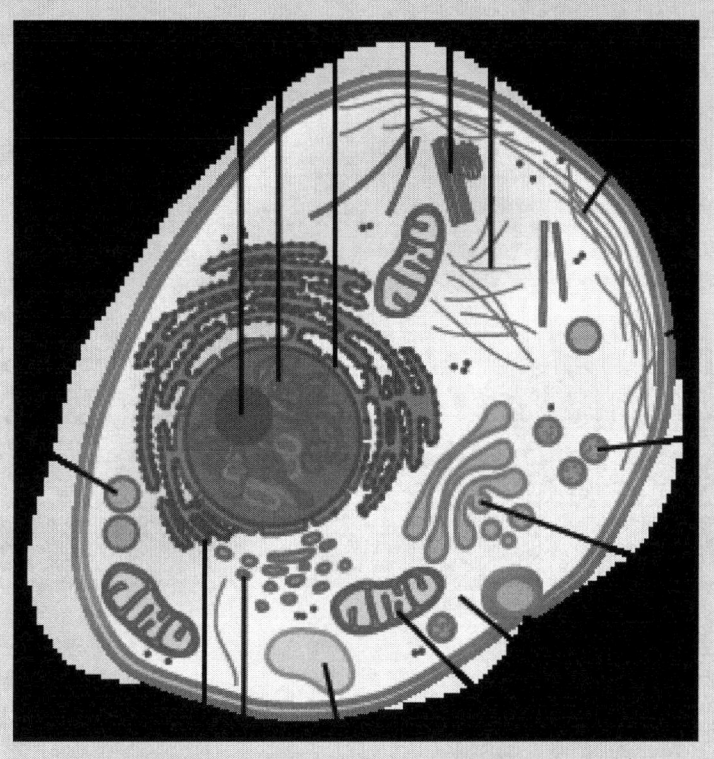

*Figure 1: These figures show the major organelles and other cell components of (a) a typical animal cell. Plant cells do not have lysosomes or centrosomes.*

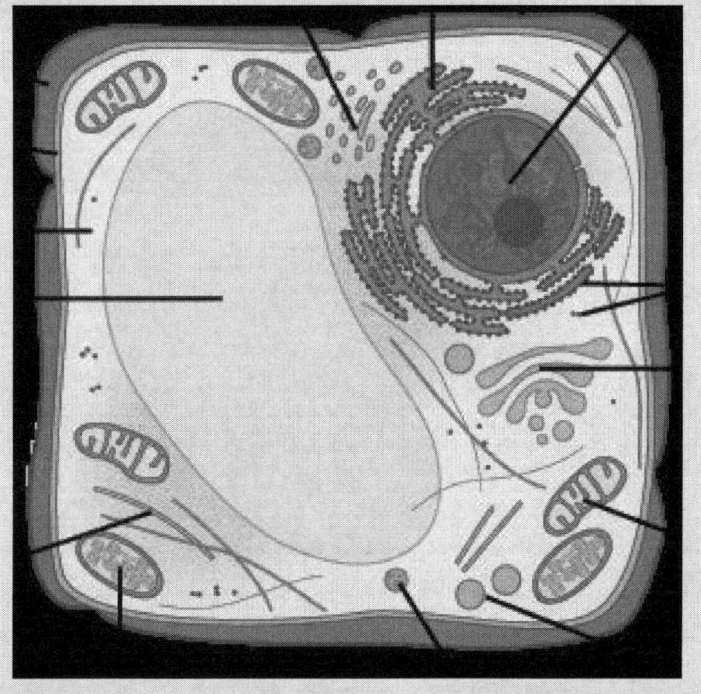

*Figure 1: (b) a typical eukaryotic plant cell. The plant cell has a cell wall, chloroplasts, plastids, and a central vacuole—structures not found in animal cells. Plant cells do not have lysosomes or centrosomes.*

If the nucleolus were not able to carry out its function, what other cellular organelles would be affected?

## The Plasma Membrane

Like prokaryotes, eukaryotic cells have a plasma membrane ([Figure 2]), a phospholipid bilayer with embedded proteins that separates the internal contents of the cell from its surrounding environment. A phospholipid is a lipid molecule with two fatty acid chains and a phosphate-containing group. The plasma membrane controls the passage of organic molecules, ions, water, and oxygen into and out of the cell. Wastes (such as carbon dioxide and ammonia) also leave the cell by passing through the plasma membrane.

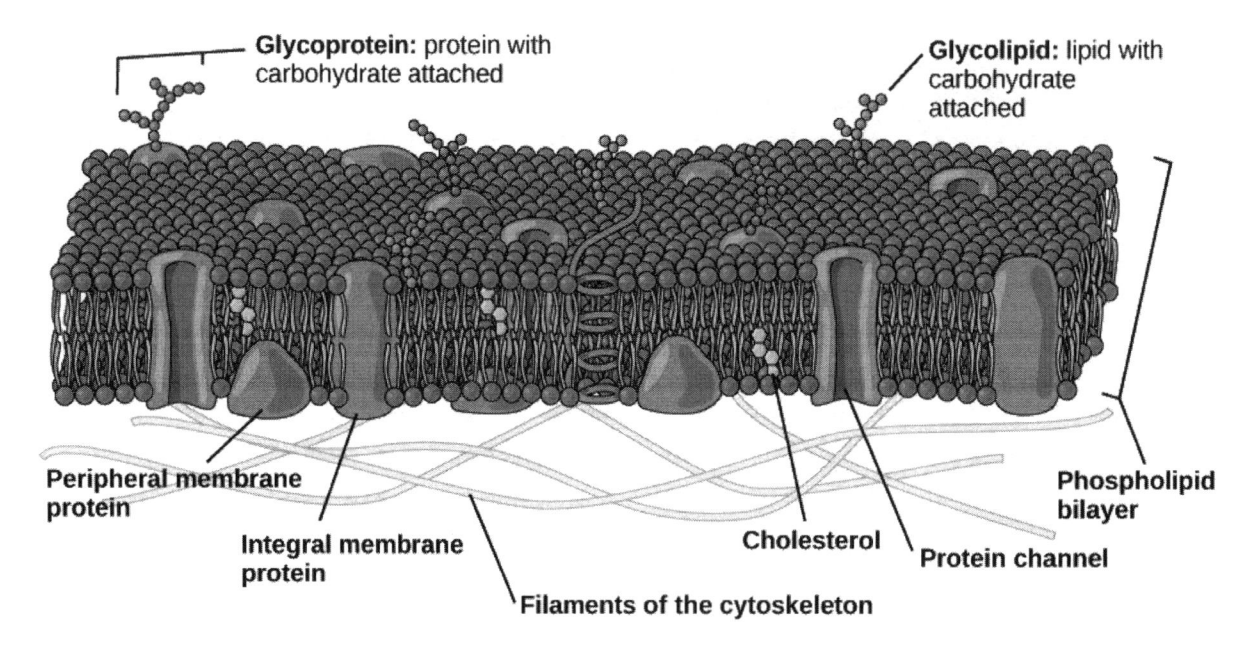

*Figure 2: The eukaryotic plasma membrane is a phospholipid bilayer with proteins and cholesterol embedded in it.*

The plasma membranes of cells that specialize in absorption are folded into fingerlike projections called microvilli (singular = microvillus); ([Figure 3]). Such cells are typically found lining the small intestine, the organ that absorbs nutrients from digested food. This is an excellent example of form following function. People with celiac disease have an immune response to gluten, which is a protein found in wheat, barley, and rye. The immune response damages microvilli, and thus, afflicted individuals cannot absorb nutrients. This leads to malnutrition, cramping, and diarrhea. Patients suffering from celiac disease must follow a gluten-free diet.

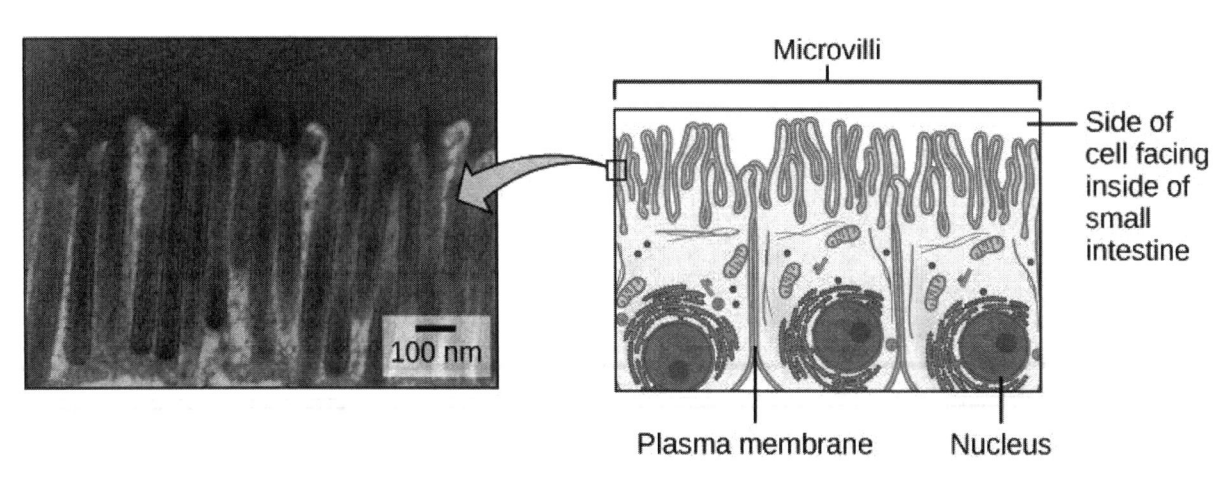

*Figure 3: Microvilli, shown here as they appear on cells lining the small intestine, increase the surface area available for absorption. These microvilli are only found on the area of the plasma membrane that faces the cavity from which substances will be absorbed. (credit "micrograph": modification of work by Louisa Howard)*

## THE CYTOPLASM

The cytoplasm is the entire region of a cell between the plasma membrane and the nuclear envelope (a structure to be discussed shortly). It is made up of organelles suspended in the gel-like cytosol, the cytoskeleton, and various chemicals ([**Figure 1**]). Even though the cytoplasm consists of 70 to 80 percent water, it has a semi-solid consistency, which comes from the proteins within it. However, proteins are not the only organic molecules found in the cytoplasm. Glucose and other simple sugars, polysaccharides, amino acids, nucleic acids, fatty acids, and derivatives of glycerol are found there, too. Ions of sodium, potassium, calcium, and many other elements are also dissolved in the cytoplasm. Many metabolic reactions, including protein synthesis, take place in the cytoplasm.

## THE NUCLEUS

Typically, the nucleus is the most prominent organelle in a cell ([**Figure 1**]). The nucleus (plural = nuclei) houses the cell's DNA and directs the synthesis of ribosomes and proteins. Let's look at it in more detail ([**Figure 4**]).

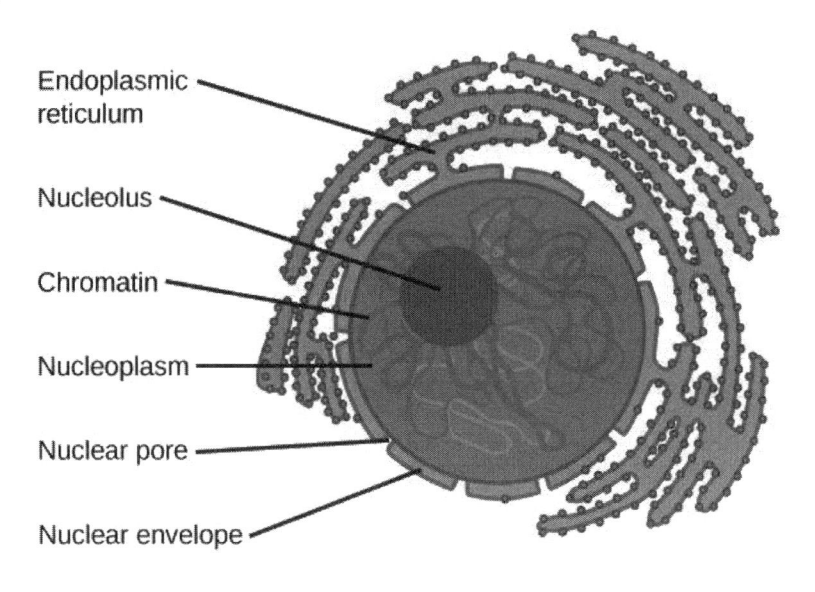

*Figure 4: The nucleus stores chromatin (DNA plus proteins) in a gel-like substance called the nucleoplasm. The nucleolus is a condensed region of chromatin where ribosome synthesis occurs. The boundary of the nucleus is called the nuclear envelope. It consists of two phospholipid bilayers: an outer membrane and an inner membrane. The nuclear membrane is continuous with the endoplasmic reticulum. Nuclear pores allow substances to enter and exit the nucleus.*

### The Nuclear Envelope

The nuclear envelope is a double-membrane structure that constitutes the outermost portion of the nucleus ([**Figure 4**]). Both the inner and outer membranes of the nuclear envelope are phospholipid bilayers.

The nuclear envelope is punctuated with pores that control the passage of ions, molecules, and RNA between the nucleoplasm and cytoplasm. The nucleoplasm is the semi-solid fluid inside the nucleus, where we find the chromatin and the nucleolus.

### Chromatin and Chromosomes

To understand chromatin, it is helpful to first consider chromosomes. Chromosomes are structures within the nucleus that are made up of DNA, the hereditary material. You may remember that in prokaryotes, DNA is organized into a single circular chromosome. In eukaryotes, chromosomes are linear structures. Every eukaryotic species has a specific number of chromosomes in the nuclei of its

286  Stoichiometry and The Mole

body's cells. For example, in humans, the chromosome number is 46, while in fruit flies, it is eight. Chromosomes are only visible and distinguishable from one another when the cell is getting ready to divide. When the cell is in the growth and maintenance phases of its life cycle, proteins are attached to chromosomes, and they resemble an unwound, jumbled bunch of threads. These unwound protein-chromosome complexes are called chromatin ([**Figure 5**]); chromatin describes the material that makes up the chromosomes both when condensed and decondensed.

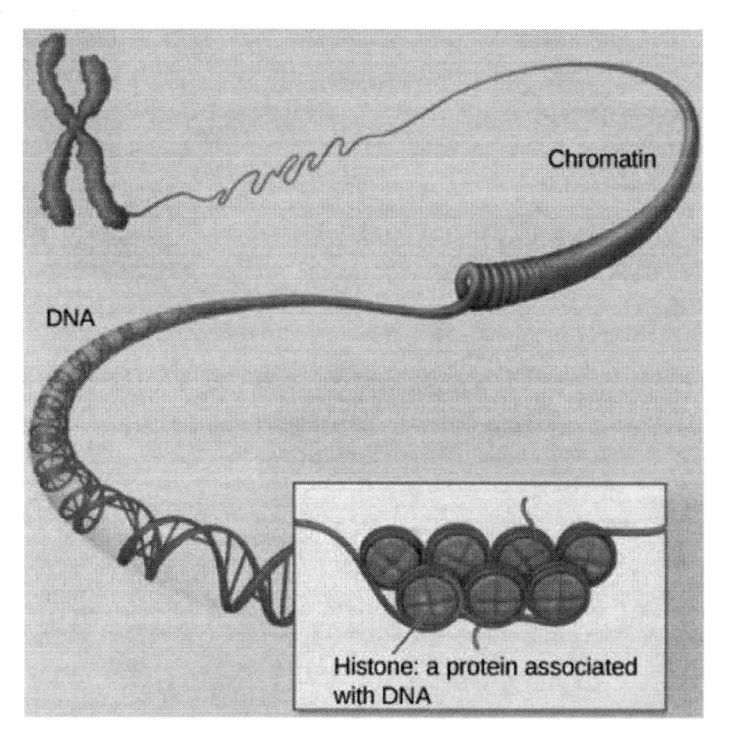

*Figure 5: (a) This image shows various levels of the organization of chromatin (DNA and protein).*

*Figure 5 (b) This image shows paired chromosomes. (credit b: modification of work by NIH; scale-bar data from Matt Russell)*

### The Nucleolus

We already know that the nucleus directs the synthesis of ribosomes, but how does it do this? Some chromosomes have sections of DNA that encode ribosomal RNA. A darkly staining area within the nucleus called the nucleolus (plural = nucleoli) aggregates the ribosomal RNA with associated proteins to assemble the ribosomal subunits that are then transported out through the pores in the nuclear envelope to the cytoplasm.

## RIBOSOMES

Ribosomes are the cellular structures responsible for protein synthesis. When viewed through an electron microscope, ribosomes appear either as clusters (polyribosomes) or single, tiny dots that float freely in the cytoplasm. They may be attached to the cytoplasmic side of the plasma membrane or the cytoplasmic side of the endoplasmic reticulum and the outer membrane of the nuclear envelope ([Figure 1]). Electron microscopy has shown us that ribosomes, which are large complexes of protein and RNA, consist of two subunits, aptly called large and small ([Figure 6]). Ribosomes receive their "orders" for protein synthesis from the nucleus where the DNA is transcribed into messenger RNA (mRNA). The mRNA travels to the ribosomes, which translate the code provided by the sequence of the nitrogenous bases in the mRNA into a specific order of amino acids in a protein. Amino acids are the building blocks of proteins.

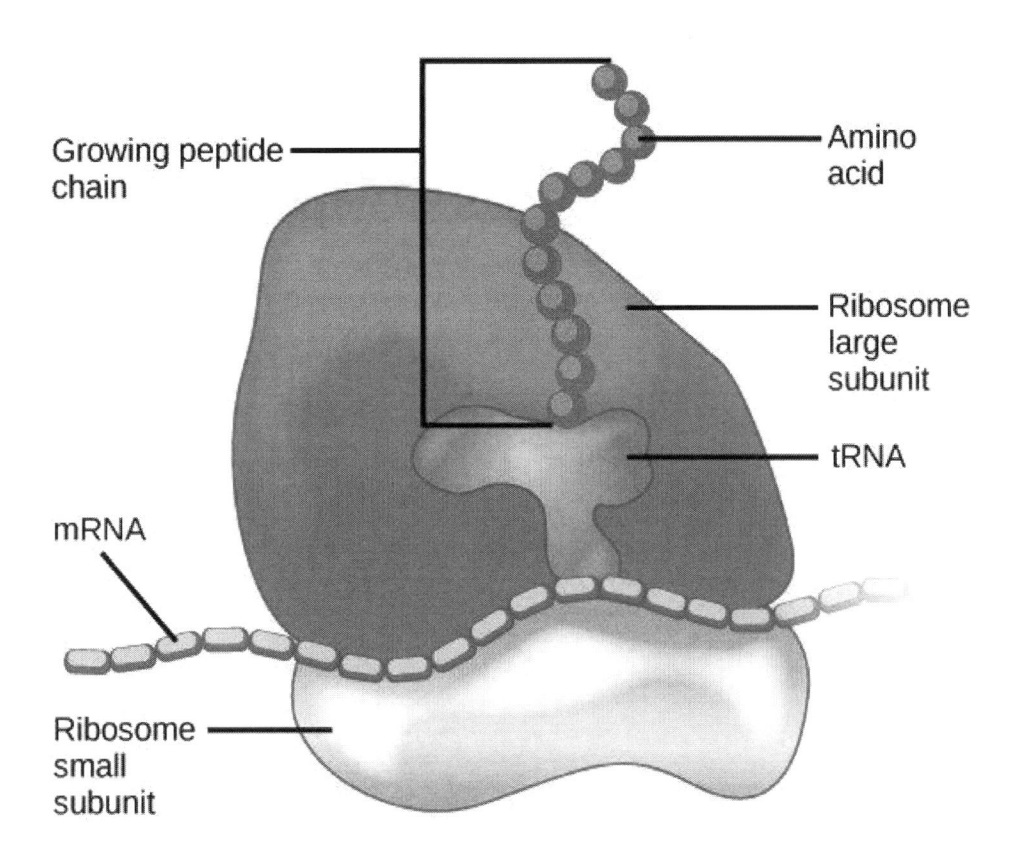

*Figure 6: Ribosomes are made up of a large subunit (top) and a small subunit (bottom). During protein synthesis, ribosomes assemble amino acids into proteins.*

Because proteins synthesis is an essential function of all cells (including enzymes, hormones, antibodies, pigments, structural components, and surface receptors), ribosomes are found in practically every cell. Ribosomes are particularly abundant in cells that synthesize large amounts of protein. For example, the pancreas is responsible for creating several digestive enzymes and the cells that produce these enzymes contain many ribosomes. Thus, we see another example of form following function.

## MITOCHONDRIA

Mitochondria (singular = mitochondrion) are often called the "powerhouses" or "energy factories" of a cell because they are responsible for making adenosine triphosphate (ATP), the cell's main energy-carrying molecule. ATP represents the short-term stored energy of the cell. Cellular respiration is the process of making ATP using the chemical energy found in glucose and other nutrients. In mitochondria, this process uses oxygen and produces carbon dioxide as a waste product. In fact, the carbon dioxide that you exhale with every breath comes from the cellular reactions that produce carbon dioxide as a byproduct.

In keeping with our theme of form following function, it is important to point out that muscle cells have a very high concentration of mitochondria that produce ATP. Your muscle cells need a lot of energy to keep your body moving. When your cells don't get enough oxygen, they do not make a lot of ATP. Instead, the small amount of ATP they make in the absence of oxygen is accompanied by the production of lactic acid.

Mitochondria are oval-shaped, double membrane organelles ([**Figure 7**]) that have their own ribosomes and DNA. Each membrane is a phospholipid bilayer embedded with proteins. The inner layer has folds called cristae. The area surrounded by the folds is called the mitochondrial matrix. The cristae and the matrix have different roles in cellular respiration.

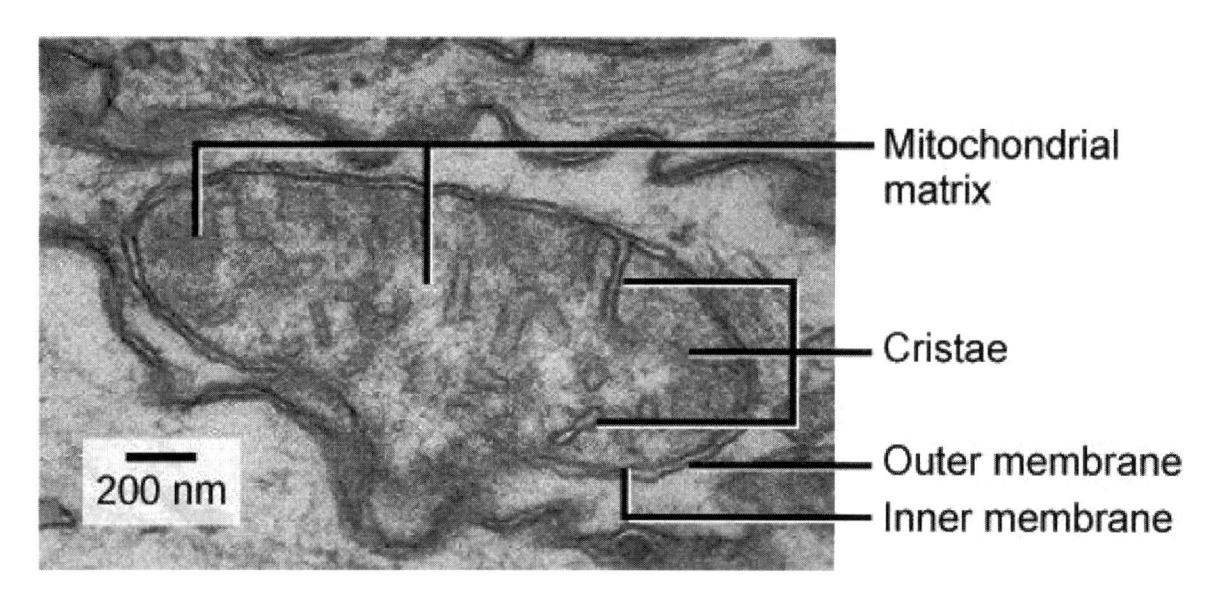

*Figure 7: This electron micrograph shows a mitochondrion as viewed with a transmission electron microscope. This organelle has an outer membrane and an inner membrane. The inner membrane contains folds, called cristae, which increase its surface area. The space between the two membranes is called the intermembrane space, and the space inside the inner membrane is called the mitochondrial matrix. ATP synthesis takes place on the inner membrane. (credit: modification of work by Matthew Britton; scale-bar data from Matt Russell)*

## PEROXISOMES

Peroxisomes are small, round organelles enclosed by single membranes. They carry out oxidation reactions that break down fatty acids and amino acids. They also detoxify many poisons that may enter the body. (Many of these oxidation reactions release hydrogen peroxide, $H_2O_2$, which would be damaging to cells; however, when these reactions are confined to peroxisomes, enzymes safely break down the $H_2O_2$ into oxygen and water.) For example, alcohol is detoxified by peroxisomes in liver cells. Glyoxysomes, which are specialized peroxisomes in plants, are responsible for converting stored fats into sugars.

## VESICLES AND VACUOLES

Vesicles and vacuoles are membrane-bound sacs that function in storage and transport. Other than the fact that vacuoles are somewhat larger than vesicles, there is a very subtle distinction between them: The membranes of vesicles can fuse with either the plasma membrane or other membrane systems within the cell. Additionally, some agents such as enzymes within plant vacuoles break down macromolecules. The membrane of a vacuole does not fuse with the membranes of other cellular components.

## ANIMAL CELLS VERSUS PLANT CELLS

At this point, you know that each eukaryotic cell has a plasma membrane, cytoplasm, a nucleus, ribosomes, mitochondria, peroxisomes, and in some, vacuoles, but there are some striking differences between animal and plant cells. While both animal and plant cells have microtubule organizing centers (MTOCs), animal cells also have centrioles associated with the MTOC: a complex called the centrosome. Animal cells each have a centrosome and lysosomes, whereas plant cells do not. Plant cells have a cell wall, chloroplasts and other specialized plastids, and a large central vacuole, whereas animal cells do not.

### The Centrosome

The centrosome is a microtubule-organizing center found near the nuclei of animal cells. It contains a pair of centrioles, two structures that lie perpendicular to each other ([Figure 8]). Each centriole is a cylinder of nine triplets of microtubules.

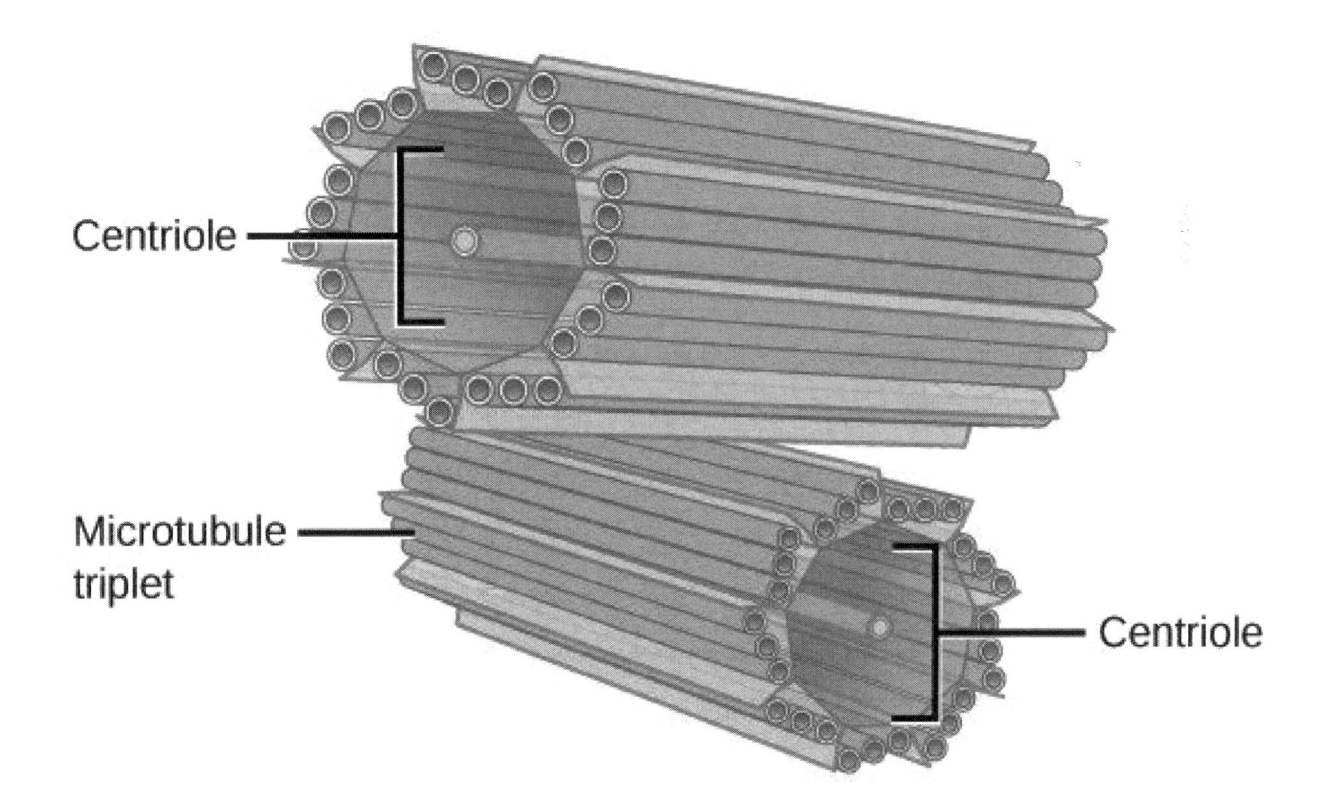

*Figure 8: The centrosome consists of two centrioles that lie at right angles to each other. Each centriole is a cylinder made up of nine triplets of microtubules. Nontubulin proteins (indicated by the green lines) hold the microtubule triplets together.*

The centrosome (the organelle where all microtubules originate) replicates itself before a cell divides, and the centrioles appear to have some role in pulling the duplicated chromosomes to opposite ends of the dividing cell. However, the exact function of the centrioles in cell division isn't clear, because cells that have had the centrosome removed can still divide, and plant cells, which lack centrosomes, are capable of cell division.

### Lysosomes

Animal cells have another set of organelles not found in plant cells: lysosomes. The lysosomes are the cell's "garbage disposal." In plant cells, the digestive processes take place in vacuoles. Enzymes within the lysosomes aid the breakdown of proteins, polysaccharides, lipids, nucleic acids, and even worn-out organelles. These enzymes are active at a much lower pH than that of the cytoplasm. Therefore, the pH within lysosomes is more acidic than the pH of the cytoplasm. Many reactions that take place in the cytoplasm could not occur at a low pH, so again, the advantage of compartmentalizing the eukaryotic cell into organelles is apparent.

### The Cell Wall

If you examine [Figure 1]b, the diagram of a plant cell, you will see a structure external to the plasma membrane called the cell wall. The cell wall is a rigid covering that protects the cell, provides structural support, and gives shape to the cell. Fungal and protistan cells also have cell walls. While the chief component of prokaryotic cell walls is peptidoglycan, the major organic molecule in the plant cell wall is cellulose ([Figure 9]), a polysaccharide made up of glucose units. Have you ever noticed that when you bite into a raw vegetable, like celery, it crunches? That's because you are tearing the rigid cell walls of the celery cells with your teeth.

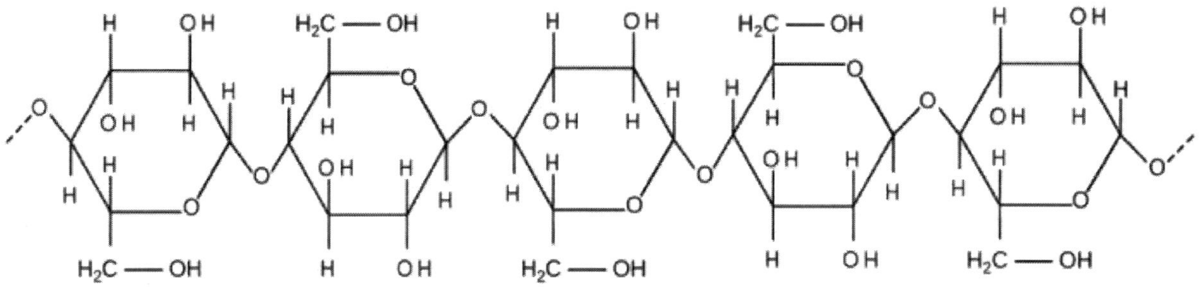

*Figure 9: Cellulose is a long chain of β-glucose molecules connected by a 1-4 linkage. The dashed lines at each end of the figure indicate a series of many more glucose units. The size of the page makes it impossible to portray an entire cellulose molecule.*

### Chloroplasts

Like the mitochondria, chloroplasts have their own DNA and ribosomes, but chloroplasts have an entirely different function. Chloroplasts are plant cell organelles that carry out photosynthesis. Photosynthesis is the series of reactions that use carbon dioxide, water, and light energy to make glucose and oxygen. This is a major difference between plants and animals; plants (autotrophs) are able to make their own food, like sugars, while animals (heterotrophs) must ingest their food.

Like mitochondria, chloroplasts have outer and inner membranes, but within the space enclosed by a chloroplast's inner membrane is a set of interconnected and stacked fluid-filled membrane sacs called thylakoids ([Figure 10]). Each stack of thylakoids is called a granum (plural = grana). The fluid enclosed by the inner membrane that surrounds the grana is called the stroma.

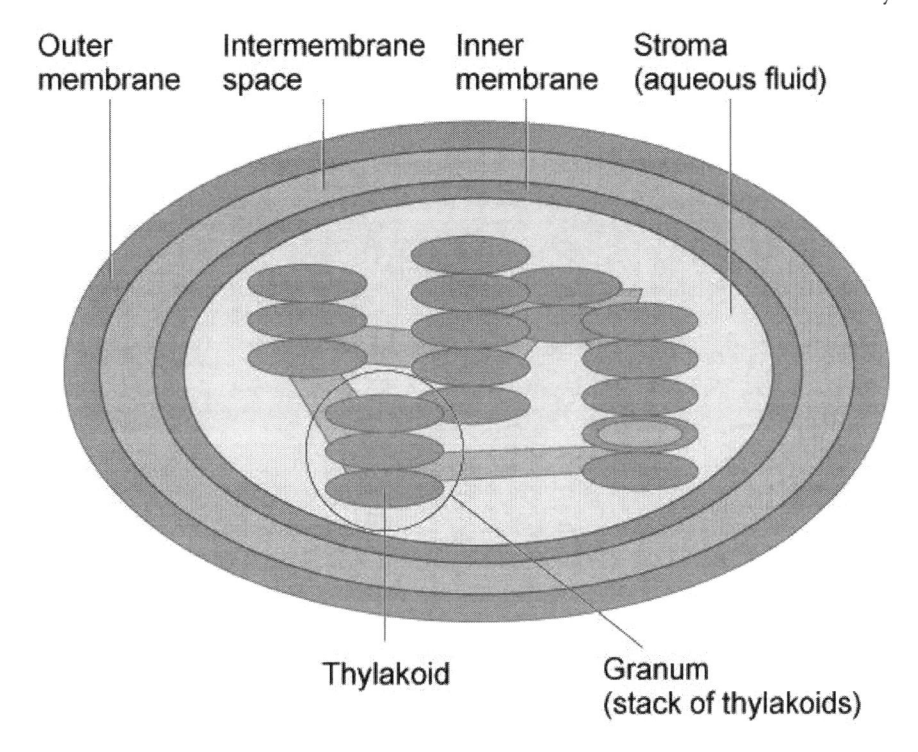

*Figure 10: The chloroplast has an outer membrane, an inner membrane, and membrane structures called thylakoids that are stacked into grana. The space inside the thylakoid membranes is called the thylakoid space. The light harvesting reactions take place in the thylakoid membranes, and the synthesis of sugar takes place in the fluid inside the inner membrane, which is called the stroma. Chloroplasts also have their own genome, which is contained on a single circular chromosome.*

The chloroplasts contain a green pigment called chlorophyll, which captures the light energy that drives the reactions of photosynthesis. Like plant cells, photosynthetic protists also have chloroplasts. Some bacteria perform photosynthesis, but their chlorophyll is not relegated to an organelle.

Evolution Connection

Endosymbiosis

We have mentioned that both mitochondria and chloroplasts contain DNA and ribosomes. Have you wondered why? Strong evidence points to endosymbiosis as the explanation.

Symbiosis is a relationship in which organisms from two separate species depend on each other for their survival. Endosymbiosis (endo- = "within") is a mutually beneficial relationship in which one organism lives inside the other. Endosymbiotic relationships abound in nature. We have already mentioned that microbes that produce vitamin K live inside the human gut. This relationship is beneficial for us because we are unable to synthesize vitamin K. It is also beneficial for the microbes because they are protected from other organisms and from drying out, and they receive abundant food from the environment of the large intestine.

Scientists have long noticed that bacteria, mitochondria, and chloroplasts are similar in size. We also know that bacteria have DNA and ribosomes, just as mitochondria and chloroplasts do. Scientists believe that host cells and bacteria formed an endosymbiotic relationship when the host cells ingested both aerobic and autotrophic bacteria (cyanobacteria) but did not destroy them. Through many

### The Central Vacuole

Previously, we mentioned vacuoles as essential components of plant cells. If you look at **[Figure 1]b**, you will see that plant cells each have a large central vacuole that occupies most of the area of the cell. The central vacuole plays a key role in regulating the cell's concentration of water in changing environmental conditions. Have you ever noticed that if you forget to water a plant for a few days, it wilts? That's because as the water concentration in the soil becomes lower than the water concentration in the plant, water moves out of the central vacuoles and cytoplasm. As the central vacuole shrinks, it leaves the cell wall unsupported. This loss of support to the cell walls of plant cells results in the wilted appearance of the plant.

The central vacuole also supports the expansion of the cell. When the central vacuole holds more water, the cell gets larger without having to invest a lot of energy in synthesizing new cytoplasm.

## SECTION SUMMARY

Like a prokaryotic cell, a eukaryotic cell has a plasma membrane, cytoplasm, and ribosomes, but a eukaryotic cell is typically larger than a prokaryotic cell, has a true nucleus (meaning its DNA is surrounded by a membrane), and has other membrane-bound organelles that allow for compartmentalization of functions. The plasma membrane is a phospholipid bilayer embedded with proteins. The nucleus's nucleolus is the site of ribosome assembly. Ribosomes are either found in the cytoplasm or attached to the cytoplasmic side of the plasma membrane or endoplasmic reticulum. They perform protein synthesis. Mitochondria participate in cellular respiration; they are responsible for the majority of ATP produced in the cell. Peroxisomes hydrolyze fatty acids, amino acids, and some toxins. Vesicles and vacuoles are storage and transport compartments. In plant cells, vacuoles also help break down macromolecules.

Animal cells also have a centrosome and lysosomes. The centrosome has two bodies perpendicular to each other, the centrioles, and has an unknown purpose in cell division. Lysosomes are the digestive organelles of animal cells.

Plant cells and plant-like cells each have a cell wall, chloroplasts, and a central vacuole. The plant cell wall, whose primary component is cellulose, protects the cell, provides structural support, and gives shape to the cell. Photosynthesis takes place in chloroplasts. The central vacuole can expand without having to produce more cytoplasm.

### Glossary

**cell wall:** rigid cell covering made of various molecules that protects the cell, provides structural support, and gives shape to the cell

**central vacuole:** large plant cell organelle that regulates the cell's storage compartment, holds water, and plays a significant role in cell growth as the site of macromolecule degradation

**centrosome:** region in animal cells made of two centrioles

**chlorophyll:** green pigment that captures the light energy that drives the light reactions of photosynthesis

**chloroplast:** plant cell organelle that carries out photosynthesis

**chromatin:** protein-DNA complex that serves as the building material of chromosomes

**chromosome:** structure within the nucleus that is made up of chromatin that contains DNA, the hereditary material

**cytoplasm:** entire region between the plasma membrane and the nuclear envelope, consisting of organelles suspended in the gel-like cytosol, the cytoskeleton, and various chemicals

**cytosol:** gel-like material of the cytoplasm in which cell structures are suspended

**eukaryotic cell:** cell that has a membrane-bound nucleus and several other membrane-bound compartments or sacs

**lysosome:** organelle in an animal cell that functions as the cell's digestive component; it breaks down proteins, polysaccharides, lipids, nucleic acids, and even worn-out organelles

**mitochondria:** (singular = mitochondrion) cellular organelles responsible for carrying out cellular respiration, resulting in the production of ATP, the cell's main energy-carrying molecule

**nuclear envelope:** double-membrane structure that constitutes the outermost portion of the nucleus

**nucleolus:** darkly staining body within the nucleus that is responsible for assembling the subunits of the ribosomes

**nucleoplasm:** semi-solid fluid inside the nucleus that contains the chromatin and nucleolus

**nucleus:** cell organelle that houses the cell's DNA and directs the synthesis of ribosomes and proteins

**organelle:** compartment or sac within a cell

**peroxisome:** small, round organelle that contains hydrogen peroxide, oxidizes fatty acids and amino acids, and detoxifies many poisons

**plasma membrane:** phospholipid bilayer with embedded (integral) or attached (peripheral) proteins, and separates the internal content of the cell from its surrounding environment

**ribosome:** cellular structure that carries out protein synthesis

**vacuole:** membrane-bound sac, somewhat larger than a vesicle, which functions in cellular storage and transport

**vesicle:** small, membrane-bound sac that functions in cellular storage and transport; its membrane is capable of fusing with the plasma membrane and the membranes of the endoplasmic reticulum and Golgi apparatus

# The Endomembrane System and Proteins

### LEARNING OBJECTIVES

By the end of this section, you will be able to:

- List the components of the endomembrane system
- Recognize the relationship between the endomembrane system and its functions

The endomembrane system (endo = "within") is a group of membranes and organelles ([Figure 1]) in eukaryotic cells that works together to modify, package, and transport lipids and proteins. It includes the nuclear envelope, lysosomes, and vesicles, which we've already mentioned, and the endoplasmic reticulum and Golgi apparatus, which we will cover shortly. Although not technically *within* the cell, the plasma membrane is included in the endomembrane system because, as you will see, it interacts with the other endomembranous organelles. The endomembrane system does not include the membranes of either mitochondria or chloroplasts.

### Art Connection

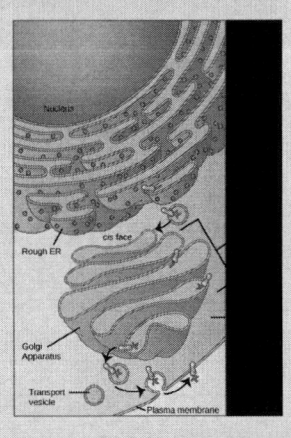

*Figure 1: "Membrane and secretory proteins are synthesized in the rough endoplasmic reticulum (RER). The RER also sometimes modifies proteins. In this illustration, a (green) integral membrane protein in the ER is modified by attachment of a (purple) carbohydrate. Vesicles with the integral protein bud from the ER and fuse with the cis face of the Golgi apparatus. As the protein passes along the Golgi's cisternae, it is further modified by the addition of more carbohydrates. After its synthesis is complete, it exits as integral membrane protein of the vesicle that bud from the Golgi's trans face and when the vesicle fuses with the cell membrane the protein becomes integral portion of that cell membrane. (credit: modification of work by Magnus Manske)*

If a peripheral membrane protein were synthesized in the lumen (inside) of the ER, would it end up on the inside or outside of the plasma membrane?

## THE ENDOPLASMIC RETICULUM

The endoplasmic reticulum (ER) ([Figure 1]) is a series of interconnected membranous sacs and tubules that collectively modifies proteins and synthesizes lipids. However, these two functions are performed in separate areas of the ER: the rough ER and the smooth ER, respectively.

The hollow portion of the ER tubules is called the lumen or cisternal space. The membrane of the ER, which is a phospholipid bilayer embedded with proteins, is continuous with the nuclear envelope.

### Rough ER

The rough endoplasmic reticulum (RER) is so named because the ribosomes attached to its cytoplasmic surface give it a studded appearance when viewed through an electron microscope ([Figure 2]).

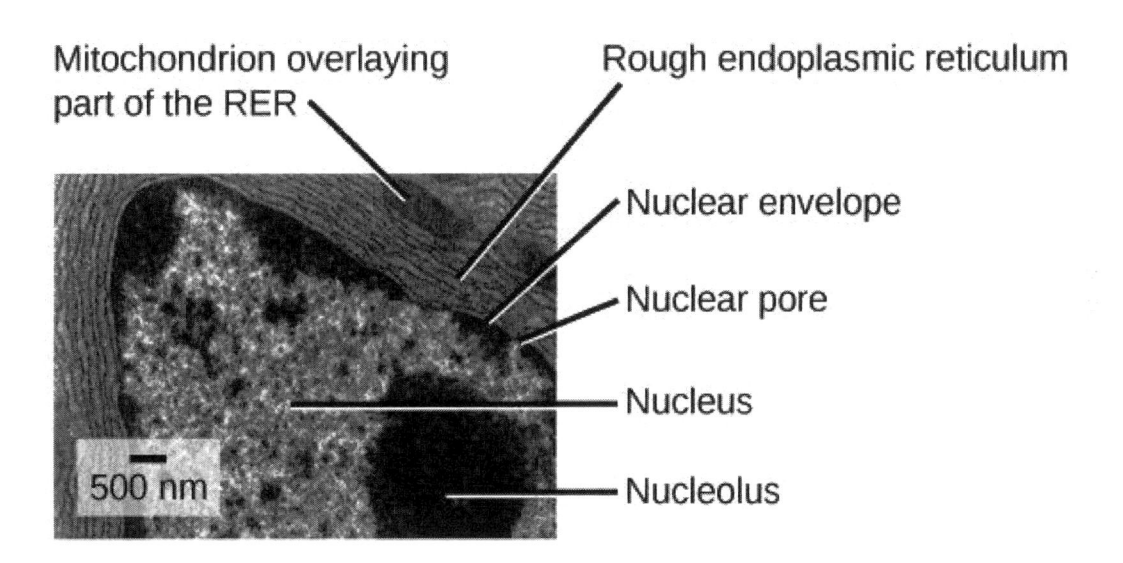

*Figure 2: This transmission electron micrograph shows the rough endoplasmic reticulum and other organelles in a pancreatic cell. (credit: modification of work by Louisa Howard)*

Ribosomes transfer their newly synthesized proteins into the lumen of the RER where they undergo structural modifications, such as folding or the acquisition of side chains. These modified proteins will be incorporated into cellular membranes—the membrane of the ER or those of other organelles—or secreted from the cell (such as protein hormones, enzymes). The RER also makes phospholipids for cellular membranes.

If the phospholipids or modified proteins are not destined to stay in the RER, they will reach their destinations via transport vesicles that bud from the RER's membrane ([Figure 1]).

Since the RER is engaged in modifying proteins (such as enzymes, for example) that will be secreted from the cell, you would be correct in assuming that the RER is abundant in cells that secrete proteins. This is the case with cells of the liver, for example.

### Smooth ER

The smooth endoplasmic reticulum (SER) is continuous with the RER but has few or no ribosomes on its cytoplasmic surface ([link]). Functions of the SER include synthesis of carbohydrates, lipids, and steroid hormones; detoxification of medications and poisons; and storage of calcium ions.

In muscle cells, a specialized SER called the sarcoplasmic reticulum is responsible for storage of the calcium ions that are needed to trigger the coordinated contractions of the muscle cells.

> You can watch an excellent animation of the endomembrane system here: https://www.youtube.com/watch?v=Fcxc8Gv7NiU. At the end of the animation, there is a short self-assessment.

Cardiology: heart disease is the leading cause of death in the United States. This is primarily due to our sedentary lifestyle and our high trans-fat diets.

Heart failure is just one of many disabling heart conditions. Heart failure does not mean that the heart has stopped working. Rather, it means that the heart can't pump with sufficient force to transport oxygenated blood to all the vital organs. Left untreated, heart failure can lead to kidney failure and failure of other organs.

The wall of the heart is composed of cardiac muscle tissue. Heart failure occurs when the endoplasmic reticula of cardiac muscle cells do not function properly. As a result, an insufficient number of calcium ions are available to trigger a sufficient contractile force.

Cardiologists (cardi- = "heart"; -ologist = "one who studies") are doctors who specialize in treating heart diseases, including heart failure. Cardiologists can make a diagnosis of heart failure via physical examination, results from an electrocardiogram (ECG, a test that measures the electrical activity of the heart), a chest X-ray to see whether the heart is enlarged, and other tests. If heart failure is diagnosed, the cardiologist will typically prescribe appropriate medications and recommend a reduction in table salt intake and a supervised exercise program.

## THE GOLGI APPARATUS

We have already mentioned that vesicles can bud from the ER and transport their contents elsewhere, but where do the vesicles go? Before reaching their final destination, the lipids or proteins within the transport vesicles still need to be sorted, packaged, and tagged so that they wind up in the right place. Sorting, tagging, packaging, and distribution of lipids and proteins takes place in the Golgi apparatus (also called the Golgi body), a series of flattened membranes ([[Figure 3]]).

The receiving side of the Golgi apparatus is called the *cis* face. The opposite side is called the *trans* face. The transport vesicles that formed from the ER travel to the *cis* face, fuse with it, and empty their contents into the lumen of the Golgi apparatus. As the proteins and lipids travel through the Golgi, they undergo further modifications that allow them to be sorted. The most frequent modification is the addition of short chains of sugar molecules. These newly modified proteins and lipids are then tagged with phosphate groups or other small molecules so that they can be routed to their proper destinations.

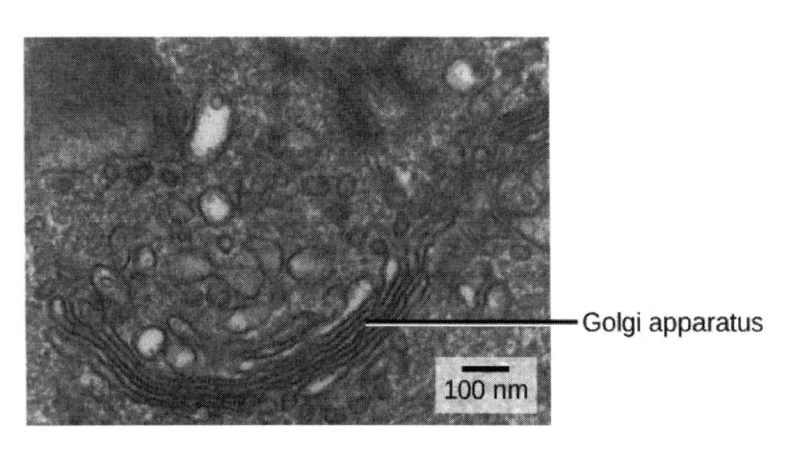

*Figure 3: The Golgi apparatus in this white blood cell is visible as a stack of semicircular, flattened rings in the lower portion of the image. Several vesicles can be seen near the Golgi apparatus. (credit: modification of work by Louisa Howard)*

Finally, the modified and tagged proteins are packaged into secretory vesicles that bud from the *trans* face of the Golgi. While some of these vesicles deposit their contents into other parts of the cell where they will be used, other secretory vesicles fuse with the plasma membrane and release their contents outside the cell.

In another example of form following function, cells that engage in a great deal of secretory activity (such as cells of the salivary glands that secrete digestive enzymes or cells of the immune system that secrete antibodies) have an abundance of Golgi.

In plant cells, the Golgi apparatus has the additional role of synthesizing polysaccharides, some of which are incorporated into the cell wall and some of which are used in other parts of the cell.

Career Connection

Genetics: Many diseases arise from genetic mutations that prevent the synthesis of critical proteins. One such disease is Lowe disease (also called oculocerebrorenal syndrome, because it affects the eyes, brain, and kidneys). In Lowe disease, there is a deficiency in an enzyme localized to the Golgi apparatus. Children with Lowe disease are born with cataracts, typically develop kidney disease after the first year of life, and may have impaired mental abilities.

Lowe disease is a genetic disease caused by a mutation on the X chromosome. The X chromosome is one of the two human sex chromosome, as these chromosomes determine a person's sex. Females possess two X chromosomes while males possess one X and one Y chromosome. In females, the genes on only one of the two X chromosomes are expressed. Therefore, females who carry the Lowe disease gene on one of their X chromosomes have a 50/50 chance of having the disease. However, males only have one X chromosome and the genes on this chromosome are always expressed. Therefore, males will always have Lowe disease if their X chromosome carries the Lowe disease gene. The location of the mutated gene, as well as the locations of many other mutations that cause genetic diseases, has now been identified. Through prenatal testing, a woman can find out if the fetus she is carrying may be afflicted with one of several genetic diseases.

Geneticists analyze the results of prenatal genetic tests and may counsel pregnant women on available options. They may also conduct genetic research that leads to new drugs or foods, or perform DNA analyses that are used in forensic investigations.

## LYSOSOMES

In addition to their role as the digestive component and organelle-recycling facility of animal cells, lysosomes are considered to be parts of the endomembrane system. Lysosomes also use their hydrolytic enzymes to destroy pathogens (disease-causing organisms) that might enter the cell. A good example of this occurs in a group of white blood cells called macrophages, which are part of your body's immune system. In a process known as phagocytosis or endocytosis, a section of the

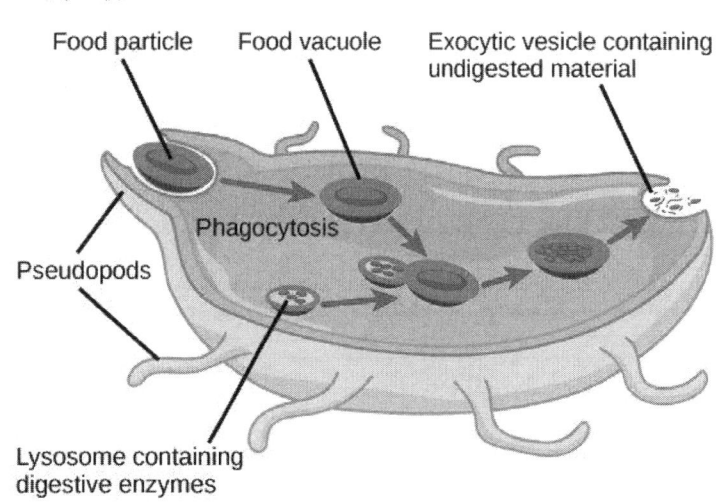

*Figure 4: A macrophage has engulfed (phagocytized) a potentially pathogenic bacterium and then fuses with a lysosomes within the cell to destroy the pathogen. Other organelles are present in the cell but for simplicity are not shown.*

plasma membrane of the macrophage invaginates (folds in) and engulfs a pathogen. The invaginated section, with the pathogen inside, then pinches itself off from the plasma membrane and becomes a vesicle. The vesicle fuses with a lysosome. The lysosome's hydrolytic enzymes then destroy the pathogen ([Figure 4]).

## SECTION SUMMARY

The endomembrane system includes the nuclear envelope, lysosomes, vesicles, the ER, and Golgi apparatus, as well as the plasma membrane. These cellular components work together to modify, package, tag, and transport proteins and lipids that form the membranes.

The RER modifies proteins and synthesizes phospholipids used in cell membranes. The SER synthesizes carbohydrates, lipids, and steroid hormones; engages in the detoxification of medications and poisons; and stores calcium ions. Sorting, tagging, packaging, and distribution of lipids and proteins take place in the Golgi apparatus. Lysosomes are created by the budding of the membranes of the RER and Golgi. Lysosomes digest macromolecules, recycle worn-out organelles, and destroy pathogens.

### Glossary

**endomembrane system:** group of organelles and membranes in eukaryotic cells that work together modifying, packaging, and transporting lipids and proteins

**endoplasmic reticulum (ER):** series of interconnected membranous structures within eukaryotic cells that collectively modify proteins and synthesize lipids

**Golgi apparatus:** eukaryotic organelle made up of a series of stacked membranes that sorts, tags, and packages lipids and proteins for distribution

**rough endoplasmic reticulum (RER):** region of the endoplasmic reticulum that is studded with ribosomes and engages in protein modification and phospholipid synthesis

**smooth endoplasmic reticulum (SER):** region of the endoplasmic reticulum that has few or no ribosomes on its cytoplasmic surface and synthesizes carbohydrates, lipids, and steroid hormones; detoxifies certain chemicals (like pesticides, preservatives, medications, and environmental pollutants), and stores calcium ions

# The Cytoskeleton

## LEARNING OBJECTIVES

By the end of this section, you will be able to:

- Describe the cytoskeleton
- Compare the roles of microfilaments, intermediate filaments, and microtubules
- Compare and contrast cilia and flagella
- Summarize the differences among the components of prokaryotic cells, animal cells, and plant cells

If you were to remove all the organelles from a cell, would the plasma membrane and the cytoplasm be the only components left? No. Within the cytoplasm, there would still be ions and organic molecules, plus a network of protein fibers that help maintain the shape of the cell, secure some organelles in specific positions, allow cytoplasm and vesicles to move within the cell, and enable cells within multicellular organisms to move. Collectively, this network of protein fibers is known as the cytoskeleton. There are three types of fibers within the cytoskeleton: microfilaments, intermediate filaments, and microtubules ([**Figure 1**]). Here, we will examine each.

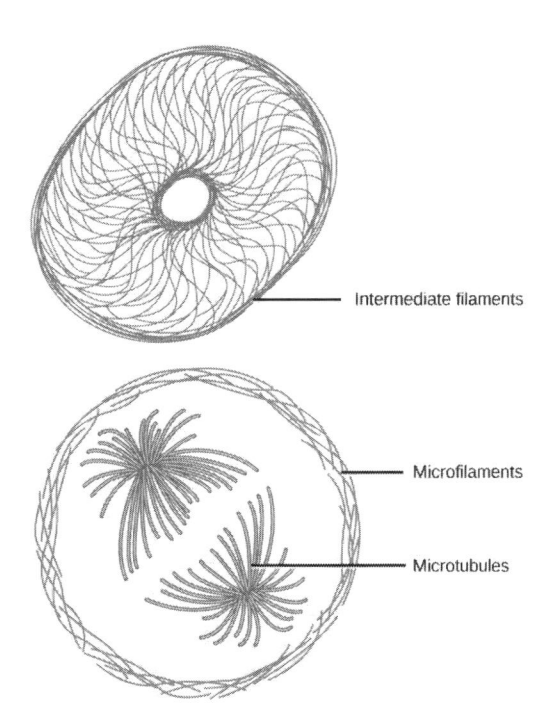

*Figure 1: Microfilaments thicken the cortex around the inner edge of a cell; like rubber bands, they resist tension. Microtubules are found in the interior of the cell where they maintain cell shape by resisting compressive forces. Intermediate filaments are found throughout the cell and hold organelles in place.*

## MICROFILAMENTS

Of the three types of protein fibers in the cytoskeleton, microfilaments are the narrowest. They function in cellular movement, have a diameter of about 7 nm, and are made of two intertwined strands of a globular protein called actin ([**Figure 2**]). For this reason, microfilaments are also known as actin filaments.

Actin is powered by ATP to assemble its filamentous form, which serves as a track for the movement of a motor protein called myosin. This enables actin to engage in cellular events requiring motion, such as cell division in animal cells and cytoplasmic streaming, which is the circular movement of the cell cytoplasm in plant cells. Actin and myosin are plentiful in muscle cells. When your actin and myosin filaments slide past each other, your muscles contract.

Microfilaments also provide some rigidity and shape to the cell. They can depolymerize (disassemble) and reform quickly, thus enabling a cell to change its shape and move. White blood cells (your body's infection-fighting cells) make good use of this ability. They can move to the site of an infection and phagocytize the pathogen.

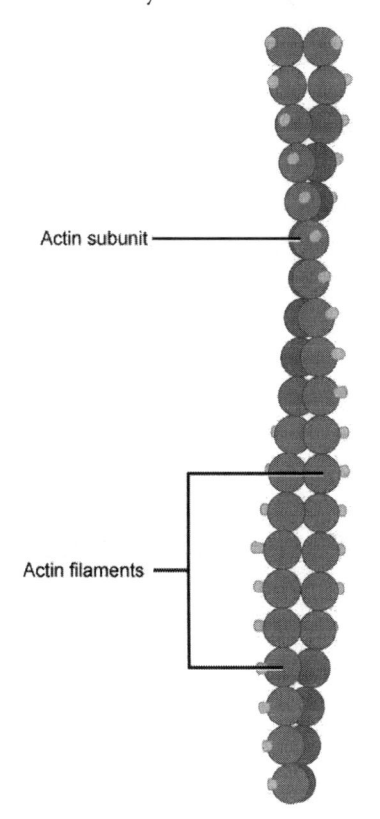

*Figure 2: Microfilaments are made of two intertwined strands of actin.*

To see an example of a white blood cell in action, watch a short time-lapse video of the cell capturing two bacteria. It engulfs one and then moves on to the other.

## INTERMEDIATE FILAMENTS

Intermediate filaments are made of several strands of fibrous proteins that are wound together ([**Figure 3**]). These elements of the cytoskeleton get their name from the fact that their diameter, 8 to 10 nm, is between those of microfilaments and microtubules.

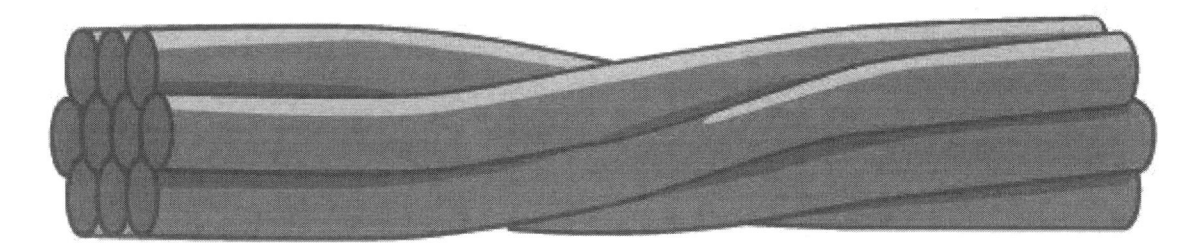

*Figure 3: Intermediate filaments consist of several intertwined strands of fibrous proteins.*

Intermediate filaments have no role in cell movement. Their function is purely structural. They bear tension, thus maintaining the shape of the cell, and anchor the nucleus and other organelles in place. [**Figure 1**] shows how intermediate filaments create a supportive scaffolding inside the cell.

The intermediate filaments are the most diverse group of cytoskeletal elements. Several types of fibrous proteins are found in the intermediate filaments. You are probably most familiar with keratin, the fibrous protein that strengthens your hair, nails, and the epidermis of the skin.

## MICROTUBULES

As their name implies, microtubules are small hollow tubes. The walls of the microtubule are made of polymerized dimers of α-tubulin and β-tubulin, two globular proteins ([Figure 4]). With a diameter of about 25 nm, microtubules are the widest components of the cytoskeleton. They help the cell resist compression, provide a track along which vesicles move through the cell, and pull replicated chromosomes to opposite ends of a dividing cell. Like microfilaments, microtubules can dissolve and reform quickly.

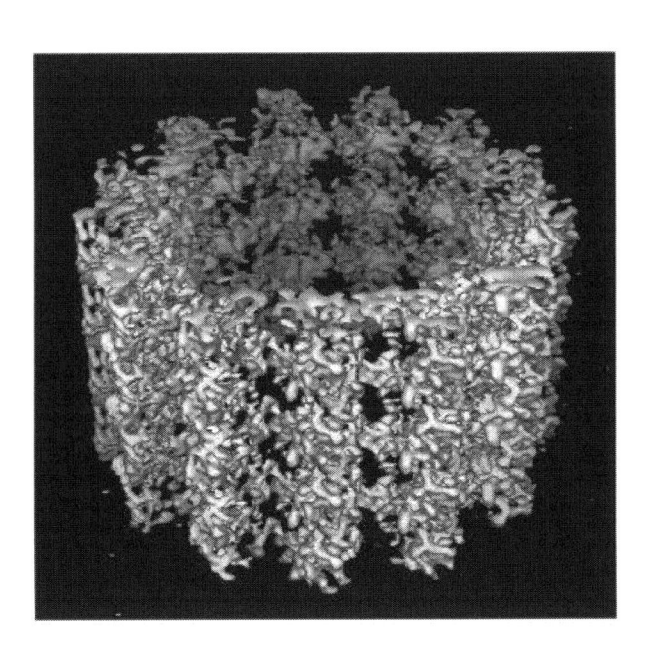

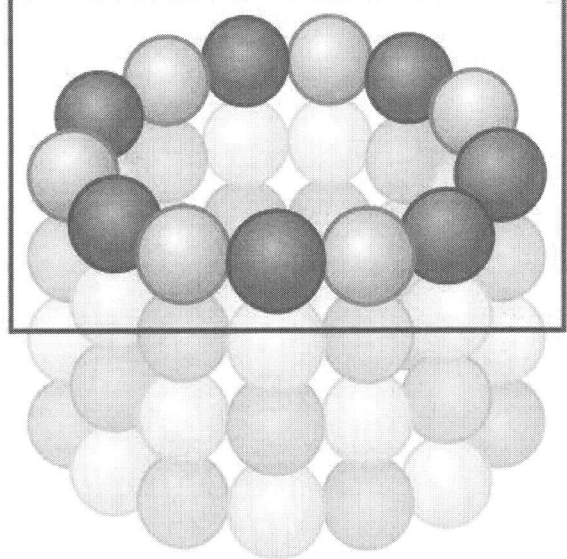

*Figure 4: Microtubules are hollow. Their walls consist of 13 polymerized dimers of α-tubulin and β-tubulin (right image). The left image shows the molecular structure of the tube.*

Microtubules are also the structural elements of flagella, cilia, and centrioles (the latter are the two perpendicular bodies of the centrosome). In fact, in animal cells, the centrosome is the microtubule-organizing center. In eukaryotic cells, flagella and cilia are quite different structurally from their counterparts in prokaryotes, as discussed below.

### Flagella and Cilia

To refresh your memory, flagella (singular = flagellum) are long, hair-like structures that extend from the plasma membrane and are used to move an entire cell (for example, sperm, *Euglena*). When present, the cell has just one flagellum or a few flagella. When cilia (singular = cilium) are present, however, many of them extend along the entire surface of the plasma membrane. They are short, hair-like structures that are used to move entire cells (such as paramecia) or substances along the outer surface of the cell (for example, the cilia of cells lining the Fallopian tubes that move the ovum toward the uterus, or cilia lining the cells of the respiratory tract that trap particulate matter and move it toward your nostrils.)

Despite their differences in length and number, flagella and cilia share a common structural arrangement of microtubules called a "9 + 2 array." This is an appropriate name because a single flagellum or cilium is made of a ring of nine microtubule doublets, surrounding a single microtubule doublet in the center ([Figure 5]).

302  Stoichiometry and The Mole

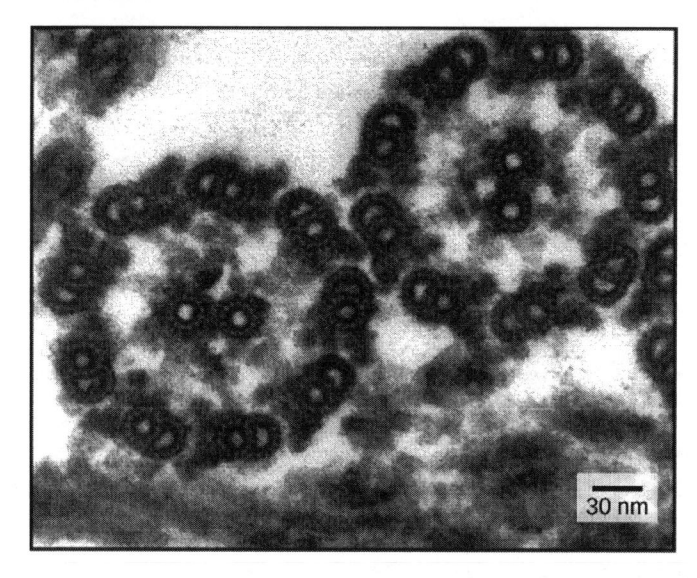

*Figure 5: This transmission electron micrograph of two flagella shows the 9 + 2 array of microtubules: nine microtubule doublets surround a single microtubule doublet. (credit: modification of work by Dartmouth Electron Microscope Facility, Dartmouth College; scale-bar data from Matt Russell)*

You have now completed a broad survey of the components of prokaryotic and eukaryotic cells.

| Components of Prokaryotic and Eukaryotic Cells ||||||
|---|---|---|---|---|
| **Cell Component** | **Function** | **Present in Prokaryotes?** | **Present in Animal Cells?** | **Present in Plant Cells?** |
| Plasma membrane | Separates cell from external environment; controls passage of organic molecules, ions, water, oxygen, and wastes into and out of cell | Yes | Yes | Yes |
| Cytoplasm | Provides turgor pressure to plant cells as fluid inside the central vacuole; site of many metabolic reactions; medium in which organelles are found | Yes | Yes | Yes |
| Nucleolus | Darkened area within the nucleus where ribosomal subunits are synthesized. | No | Yes | Yes |
| Nucleus | Cell organelle that houses DNA and directs synthesis of ribosomes and proteins | No | Yes | Yes |
| Ribosomes | Protein synthesis | Yes | Yes | Yes |
| Mitochondria | ATP production/cellular respiration | No | Yes | Yes |
| Peroxisomes | Oxidizes and thus breaks down fatty acids and amino acids, and detoxifies poisons | No | Yes | Yes |
| Vesicles and vacuoles | Storage and transport; digestive function in plant cells | No | Yes | Yes |
| Centrosome | Unspecified role in cell division in animal cells; source of microtubules in animal cells | No | Yes | No |
| Lysosomes | Digestion of macromolecules; recycling of worn-out organelles | No | Yes | No |
| Cell wall | Protection, structural support and maintenance of cell shape | Yes, primarily peptidoglycan | No | Yes, primarily cellulose |
| Chloroplasts | Photosynthesis | No | No | Yes |
| Endoplasmic reticulum | Modifies proteins and synthesizes lipids | No | Yes | Yes |

| Components of Prokaryotic and Eukaryotic Cells | | | | |
|---|---|---|---|---|
| Cell Component | Function | Present in Prokaryotes? | Present in Animal Cells? | Present in Plant Cells? |
| Golgi apparatus | Modifies, sorts, tags, packages, and distributes lipids and proteins | No | Yes | Yes |
| Cytoskeleton | Maintains cell's shape, secures organelles in specific positions, allows cytoplasm and vesicles to move within cell, and enables unicellular organisms to move independently | Yes | Yes | Yes |
| Flagella | Cellular locomotion | Some | Some | No, except for some plant sperm cells. |
| Cilia | Cellular locomotion, movement of particles along extracellular surface of plasma membrane, and filtration | Some | Some | No |

## SECTION SUMMARY

The cytoskeleton has three different types of protein elements. From narrowest to widest, they are the microfilaments (actin filaments), intermediate filaments, and microtubules. Microfilaments are often associated with myosin. They provide rigidity and shape to the cell and facilitate cellular movements. Intermediate filaments bear tension and anchor the nucleus and other organelles in place. Microtubules help the cell resist compression, serve as tracks for motor proteins that move vesicles through the cell, and pull replicated chromosomes to opposite ends of a dividing cell. They are also the structural element of centrioles, flagella, and cilia.

### Glossary

**cilium:** (plural = cilia) short, hair-like structure that extends from the plasma membrane in large numbers and is used to move an entire cell or move substances along the outer surface of the cell

**cytoskeleton:** network of protein fibers that collectively maintain the shape of the cell, secure some organelles in specific positions, allow cytoplasm and vesicles to move within the cell, and enable unicellular organisms to move independently

**flagellum:** (plural = flagella) long, hair-like structure that extends from the plasma membrane and is used to move the cell

**intermediate filament:** cytoskeletal component, composed of several intertwined strands of fibrous protein, that bears tension, supports cell-cell junctions, and anchors cells to extracellular structures

**microfilament:** narrowest element of the cytoskeleton system; it provides rigidity and shape to the cell and enables cellular movements

**microtubule:** widest element of the cytoskeleton system; it helps the cell resist compression, provides a track along which vesicles move through the cell, pulls replicated chromosomes to opposite ends of a dividing cell, and is the structural element of centrioles, flagella, and cilia

# Exercises

## PART I

1. In your everyday life, you have probably noticed that certain instruments are ideal for certain situations. For example, you would use a spoon rather than a fork to eat soup because a spoon is shaped for scooping, while soup would slip between the tines of a fork. The use of ideal instruments also applies in science. In what situation(s) would the use of a light microscope be ideal, and why?

2. In what situation(s) would the use of a scanning electron microscope be ideal, and why?

3. In what situation(s) would a transmission electron microscope be ideal, and why?

4. What are the advantages and disadvantages of each of these types of microscopes?

## PART II

1. Antibiotics are medicines that are used to fight bacterial infections. These medicines kill prokaryotic cells without harming human cells. What part or parts of the bacterial cell do you think antibiotics target? Why?

2. Explain why not all microbes are harmful:

## PART III

1. You already know that ribosomes are abundant in red blood cells. In what other cells of the body would you find them in great abundance? Why?

2. What are the structural and functional similarities and differences between mitochondria and chloroplasts?

3. List the organelles of an eukaryotic cell and describe their functions:

## PART IV

1. In the context of cell biology, what do we mean by form follows function? What are at least two examples of this concept?

2. In your opinion, is the nuclear membrane part of the endomembrane system? Why or why not? Defend your answer:

## PART V

1. What are the similarities and differences between the structures of centrioles and flagella?

2. How do cilia and flagella differ?

# Check Your Knowledge: Self-Test

1. When viewing a specimen through a light microscope, scientists use _____ to distinguish the individual components of cells.

    a) beam of electrons

    b) radioactive isotopes

    c) special stains

    d) high temperatures

2. The _____ is the basic unit of life.

    a) organism

    b) cell

    c) tissue

    d) organ

3. Prokaryotes depend on _____ to obtain some materials and to get rid of wastes.

    a) ribosomes

    b) flagella

    c) cell division

    d) diffusion

4. Bacteria that lack fimbriae are less likely to _____.

    a) adhere to cell surfaces

    b) swim through bodily fluids

    c) synthesize proteins

    d) retain the ability to divide

5. Cell that lack a nucleus and have no membrane bound organelles are classified as:

    a) eukaryotic

    b) prokaryotic

6. Human cells are eukaryotic, therefore…

    a) they have no nucleus

    b) they have a cell wall

    c) they have membrane-bound organelles

7. Which of the following is surrounded by two phospholipid bilayers?

    a) the ribosomes

    b) the vesicles

c) the cytoplasm

d) the nucleoplasm

8. Peroxisomes got their name because hydrogen peroxide is:

   a) used in their detoxification reactions

   b) produced during their oxidation reactions

   c) incorporated into their membranes

   d) a cofactor for the organelles' enzymes

9. In plant cells, the function of the lysosomes is carried out by _____.

   a) vacuoles

   b) peroxisomes

   c) ribosomes

   d) nuclei

10. Which of the following is found both in eukaryotic and prokaryotic cells?

    a) nucleus

    b) mitochondrion

    c) vacuole

    d) ribosomes

11. Name the organelle in which photosynthesis takes place in plant cells:

    a) mitochondria

    b) chloroplast

    c) lysosomes

    d) nucleus

12. Which of the following can only be found in animal cells?

    a) lysosomes

    b) mitochondria

    c) nucleus

    d) plasma membrane

13. Which of the following structures store and transport molecules in cells?

    a) peroxisomes

    b) nucleus

    c) vesicles

    d) mitochondria

14. Imagine a cell in which oxidation reactions occur frequently. Based on this information, which of the following would be abundant in this cell?

    a) mitochondria

    b) vesicles

c) peroxisomes

d) rough endoplasmic reticulum

15. Imagine a cell that requires a lot of energy to perform its function. Based on this information, which of the following would be abundant in this cell?

    a) mitochondria

    b) vesicles

    c) peroxisomes

    d) rough endoplasmic reticulum

16. Which of the following is not a component of the endomembrane system?

    a) mitochondrion

    b) Golgi apparatus

    c) endoplasmic reticulum

    d) lysosome

17. The process by which a cell engulfs a foreign particle – such as a bacterium – is known as:

    a) endosymbiosis

    b) phagocytosis

    c) hydrolysis

    d) membrane synthesis

18. Which of the following is most likely to have the greatest concentration of smooth endoplasmic reticulum?

    a) a cell that secretes enzymes

    b) a cell that destroys pathogens

    c) a cell that makes steroid hormones

    d) a cell that engages in photosynthesis

19. Which of the following is most likely to have the greatest concentration of rough endoplasmic reticulum?

    a) a cell that secretes proteins

    b) a cell that engages in photosynthesis

    c) a cell that engages in detoxification

    d) a cell that stores lipids

20. Which of the following sequences correctly lists in order the steps involved in the incorporation of a proteinaceous molecule into the structure of the plasma membrane?

    a) synthesis of the protein on the ribosome; modification in the Golgi apparatus; packaging in the endoplasmic reticulum; tagging in the vesicle

    b) synthesis of the protein on the lysosome; tagging in the Golgi; packaging in the vesicle; distribution in the endoplasmic reticulum

c) synthesis of the protein on the ribosome; modification in the endoplasmic reticulum; tagging in the Golgi; distribution via the vesicle

d) synthesis of the protein on the lysosome; packaging in the vesicle; distribution via the Golgi; tagging in the endoplasmic reticulum

21. Which of the following have the ability to disassemble and reform quickly?

    a) microfilaments and intermediate filaments

    b) microfilaments and microtubules

    c) intermediate filaments and microtubules

    d) only intermediate filaments

22. Which of the following do not play a role in intracellular movement?

    a) microfilaments and intermediate filaments

    b) microfilaments and microtubules

    c) intermediate filaments and microtubules

    d) only intermediate filaments

# Chapter 9: Cell Transport

# Introduction

*Despite its seeming hustle and bustle, Grand Central Station functions with a high level of organization: People and objects move from one location to another, they cross or are contained within certain boundaries, and they provide a constant flow as part of larger activity. Analogously, a plasma membrane's functions involve movement within the cell and across boundaries in the process of intracellular and intercellular activities. (credit: modification of work by Randy Le'Moine)*

The plasma membrane, which is also called the cell membrane, has many functions, but the most basic one is to define the borders of the cell and keep the cell functional. The plasma membrane is selectively permeable. This means that the membrane allows some materials to freely enter or leave the cell, while other materials cannot move freely, but require the use of a specialized structure, and occasionally, even energy investment for crossing.

# Components and Structure of the Membrane

> **Learning Objectives**
>
> By the end of this section, you will be able to:
> - Understand the fluid mosaic model of cell membranes
> - Describe the functions of phospholipids, proteins, and carbohydrates in membranes
> - Discuss membrane fluidity

A cell's plasma membrane defines the cell, outlines its borders, and determines the nature of its interaction with its environment (see **[Figure 1]** for a summary). Cells exclude some substances, take in others, and excrete still others, all in controlled quantities. The plasma membrane must be very flexible to allow certain cells, such as red blood cells and white blood cells, to change shape as they pass through narrow capillaries. These are the more obvious functions of a plasma membrane. In addition, the surface of the plasma membrane carries markers that allow cells to recognize one another, which is vital for tissue and organ formation during early development, and which later plays a role in the "self" versus "non-self" distinction of the immune response.

Among the most sophisticated functions of the plasma membrane is the ability to transmit signals by means of complex, integral proteins known as receptors. These proteins act both as receivers of extracellular inputs and as activators of intracellular processes. These membrane receptors provide extracellular attachment sites for effectors like hormones and growth factors, and they activate intracellular response cascades when their effectors are bound. Occasionally, receptors are hijacked by viruses (HIV, human immunodeficiency virus, is one example) that use them to gain entry into cells, and at times, the genes encoding receptors become mutated, causing the process of signal transduction to malfunction with disastrous consequences.

## *FLUID MOSAIC MODEL*

The existence of the plasma membrane was identified in the 1890s, and its chemical components were identified in 1915. The principal components identified at that time were lipids and proteins. The first widely accepted model of the plasma membrane's structure was proposed in 1935 by Hugh Davson and James Danielli; it was based on the "railroad track" appearance of the plasma membrane in early electron micrographs. They theorized that the structure of the plasma membrane resembles a sandwich, with protein being analogous to the bread, and lipids being analogous to the filling. In the 1950s, advances in microscopy, notably transmission electron microscopy (TEM), allowed researchers to see that the core of the plasma membrane consisted of a double, rather than a single, layer. A new model that better explains both the microscopic observations and the function of that plasma membrane was proposed by S. J. Singer and Garth L. Nicolson in 1972.

The explanation proposed by Singer and Nicolson is called the fluid mosaic model. The model has evolved somewhat over time, but it still best accounts for the structure and functions of the plasma membrane as we now understand them. The fluid mosaic model describes the structure of the plasma membrane as a mosaic of components—including phospholipids, cholesterol, proteins, and carbohydrates—that gives the membrane a fluid character. Plasma membranes range from 5 to 10 nm in thickness. For comparison, human red blood cells, visible via light microscopy, are approximately 8 μm wide, or approximately 1,000 times wider than a plasma membrane. The membrane does look a bit like a sandwich ([Figure 1]).

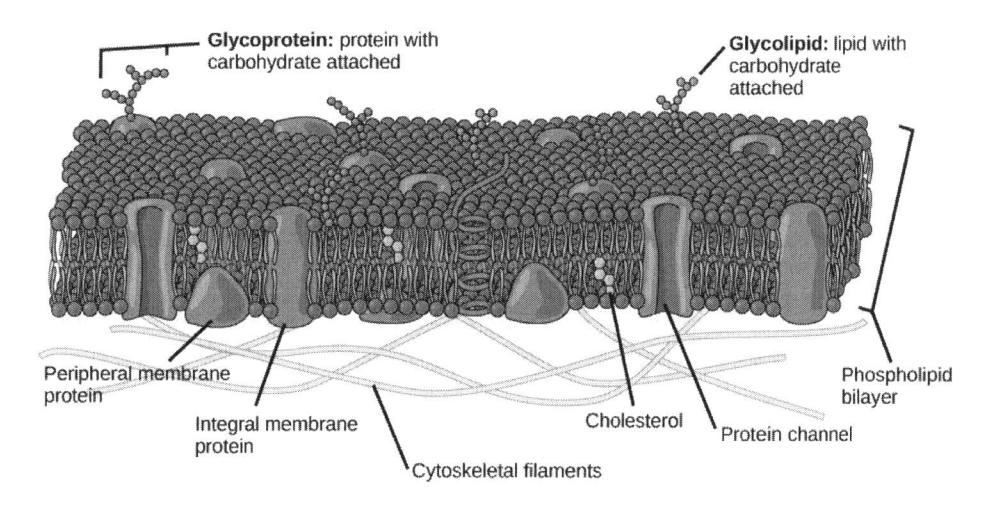

*Figure 1: The fluid mosaic model of the plasma membrane describes the plasma membrane as a fluid combination of phospholipids, cholesterol, and proteins. Carbohydrates attached to lipids (glycolipids) and to proteins (glycoproteins) extend from the outward-facing surface of the membrane.*

The principal components of a plasma membrane are lipids (phospholipids and cholesterol), proteins, and carbohydrates attached to some of the lipids and some of the proteins. A phospholipid is a molecule consisting of glycerol, two fatty acids, and a phosphate-linked head group. Cholesterol, another lipid composed of four fused carbon rings, is found alongside the phospholipids in the core of the membrane. The proportions of proteins, lipids, and carbohydrates in the plasma membrane vary with cell type, but for a typical human cell, protein accounts for about 50 percent of the composition by mass, lipids (of all types) account for about 40 percent of the composition by mass, with the remaining 10 percent of the composition by mass being carbohydrates. However, the concentration of proteins and lipids varies with different cell membranes. For example, myelin, an outgrowth of the membrane of specialized cells that insulates the axons of the peripheral nerves, contains only 18 percent protein and 76 percent lipid. The mitochondrial inner membrane contains 76 percent protein and only 24 percent lipid. The plasma membrane of human red blood cells is 30 percent lipid. Carbohydrates are present only on the exterior surface of the plasma membrane and are attached to proteins, forming glycoproteins, or attached to lipids, forming glycolipids.

## Phospholipids

The main fabric of the membrane is composed of amphiphilic, phospholipid molecules. The hydrophilic or "water-loving" areas of these molecules (which look like a collection of balls in an artist's rendition of the model) ([Figure 1]) are in contact with the aqueous fluid both inside and outside the cell. Hydrophobic, or water-hating molecules, tend to be non-polar. They interact with other non-polar molecules in chemical reactions, but generally do not interact with polar molecules. When

# 314  Cell Transport

placed in water, hydrophobic molecules tend to form a ball or cluster. The hydrophilic regions of the phospholipids tend to form hydrogen bonds with water and other polar molecules on both the exterior and interior of the cell. Thus, the membrane surfaces that face the interior and exterior of the cell are hydrophilic. In contrast, the interior of the cell membrane is hydrophobic and will not interact with water. Therefore, phospholipids form an excellent two-layer cell membrane that separates fluid within the cell from the fluid outside of the cell.

A phospholipid molecule ([Figure 2]) consists of a three-carbon glycerol backbone with two fatty acid molecules attached to carbons 1 and 2, and a phosphate-containing group attached to the third carbon. This arrangement gives the overall molecule an area described as its head (the phosphate-containing group), which has a polar character or negative charge, and an area called the tail (the fatty acids), which has no charge. The head can form hydrogen bonds, but the tail cannot. A molecule with this arrangement of a positively or negatively charged area and an uncharged, or non-polar, area is referred to as amphiphilic or "dual-loving."

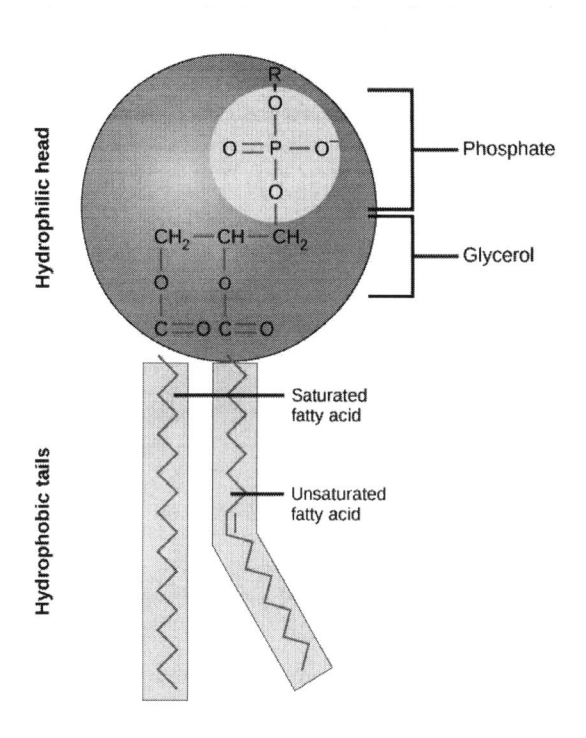

*Figure 2: This phospholipid molecule is composed of a hydrophilic head and two hydrophobic tails. The hydrophilic head group consists of a phosphate-containing group attached to a glycerol molecule. The hydrophobic tails, each containing either a saturated or an unsaturated fatty acid, are long hydrocarbon chains.*

This characteristic is vital to the structure of a plasma membrane because, in water, phospholipids tend to become arranged with their hydrophobic tails facing each other and their hydrophilic heads facing out. In this way, they form a lipid bilayer—a barrier composed of a double layer of phospholipids that separates the water and other materials on one side of the barrier from the water and other materials on the other side. In fact, phospholipids heated in an aqueous solution tend to spontaneously form small spheres or droplets (called micelles or liposomes), with their hydrophilic heads forming the exterior and their hydrophobic tails on the inside ([Figure 3]).

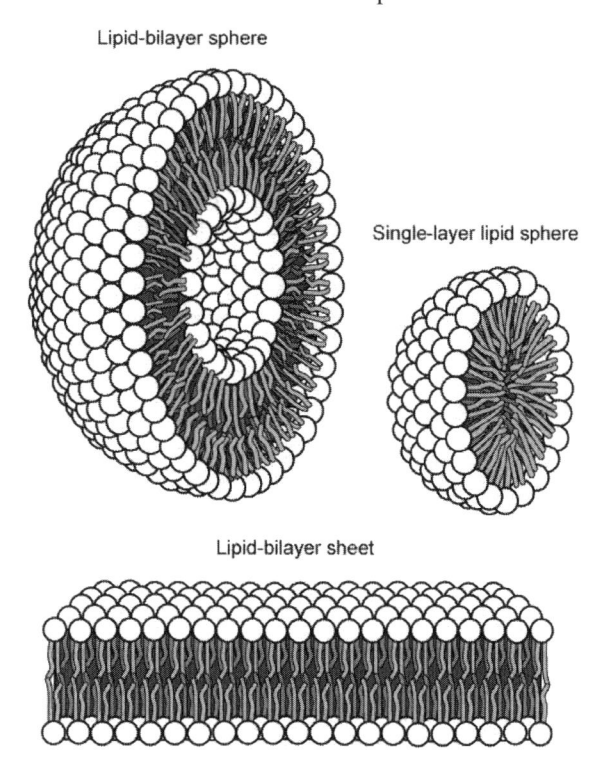

*Figure 3: In an aqueous solution, phospholipids tend to arrange themselves with their polar heads facing outward and their hydrophobic tails facing inward. (credit: modification of work by Mariana Ruiz Villareal)*

## Proteins

Proteins make up the second major component of plasma membranes. Integral proteins (some specialized types are called integrins) are, as their name suggests, integrated completely into the membrane structure, and their hydrophobic membrane-spanning regions interact with the hydrophobic region of the the phospholipid bilayer (**[Figure 1]**). Single-pass integral membrane proteins usually have a hydrophobic transmembrane segment that consists of 20–25 amino acids. Some span only part of the membrane—associating with a single layer—while others stretch from one side of the membrane to the other, and are exposed on either side. Some complex proteins are composed of up to 12 segments of a single protein, which are extensively folded and embedded in the membrane (**[Figure 1]**). This type of protein has a hydrophilic region or regions, and one or several mildly hydrophobic regions. This arrangement of regions of the protein tends to orient the protein alongside the phospholipids, with the hydrophobic region of the protein adjacent to the tails of the phospholipids and the hydrophilic region or regions of the protein protruding from the membrane and in contact with the cytosol or extracellular fluid.

Peripheral proteins are found on the exterior and interior surfaces of membranes, attached either to integral proteins or to phospholipids. Peripheral proteins, along with integral proteins, may serve as enzymes, as structural attachments for the fibers of the cytoskeleton, or as part of the cell's recognition

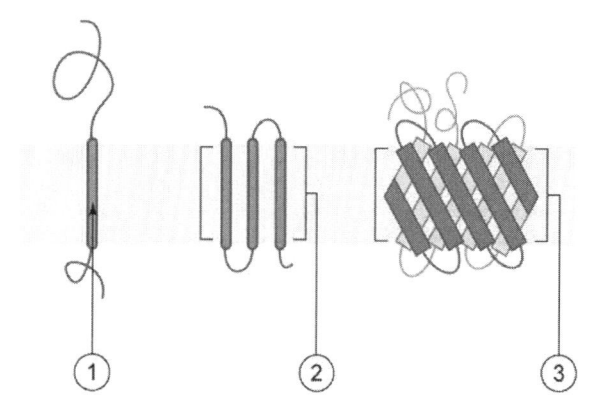

*Figure 4: Integral membranes proteins may have one or more alpha-helices that span the membrane (examples 1 and 2), or they may have beta-sheets that span the membrane (example 3). (credit: "Foobar"/Wikimedia Commons)*

sites. These are sometimes referred to as "cell-specific" proteins. The body recognizes its own proteins and attacks foreign proteins associated with invasive pathogens.

## Carbohydrates

Carbohydrates are the third major component of plasma membranes. They are always found on the exterior surface of cells and are bound either to proteins (forming glycoproteins) or to lipids (forming glycolipids) ([Figure 1]). These carbohydrate chains may consist of 2–60 monosaccharide units and can be either straight or branched. Along with peripheral proteins, carbohydrates form specialized sites on the cell surface that allow cells to recognize each other. These sites have unique patterns that allow the cell to be recognized, much the way that the facial features unique to each person allow him or her to be recognized. This recognition function is very important to cells, as it allows the immune system to differentiate between body cells (called "self") and foreign cells or tissues (called "non-self"). Similar types of glycoproteins and glycolipids are found on the surfaces of viruses and may change frequently, preventing immune cells from recognizing and attacking them.

These carbohydrates on the exterior surface of the cell—the carbohydrate components of both glycoproteins and glycolipids—are collectively referred to as the glycocalyx (meaning "sugar coating"). The glycocalyx is highly hydrophilic and attracts large amounts of water to the surface of the cell. This aids in the interaction of the cell with its watery environment and in the cell's ability to obtain substances dissolved in the water. As discussed above, the glycocalyx is also important for cell identification, self/non-self determination, and embryonic development, and is used in cell-cell attachments to form tissues.

## Evolution Connection

How Viruses Infect Specific OrgansGlycoprotein and glycolipid patterns on the surfaces of cells give many viruses an opportunity for infection. HIV and hepatitis viruses infect only specific organs or cells in the human body. HIV is able to penetrate the plasma membranes of a subtype of lymphocytes called T-helper cells, as well as some monocytes and central nervous system cells. The hepatitis virus attacks liver cells.

These viruses are able to invade these cells, because the cells have binding sites on their surfaces that are specific to and compatible with certain viruses ([Figure 5]). Other recognition sites on the virus's surface interact with the human immune system, prompting the body to produce antibodies. Antibodies are made in response to the antigens or proteins associated with invasive pathogens, or in response to foreign cells, such as might occur with an organ transplant. These same sites serve as places for antibodies to attach and either destroy or inhibit the activity of the virus. Unfortunately, these recognition sites on HIV change at a rapid rate because of mutations, making the production of an effective vaccine against the virus very difficult, as the virus evolves and adapts. A person infected with HIV will quickly develop different populations, or variants, of the virus that are distinguished by differences in these recognition sites. This rapid change of surface

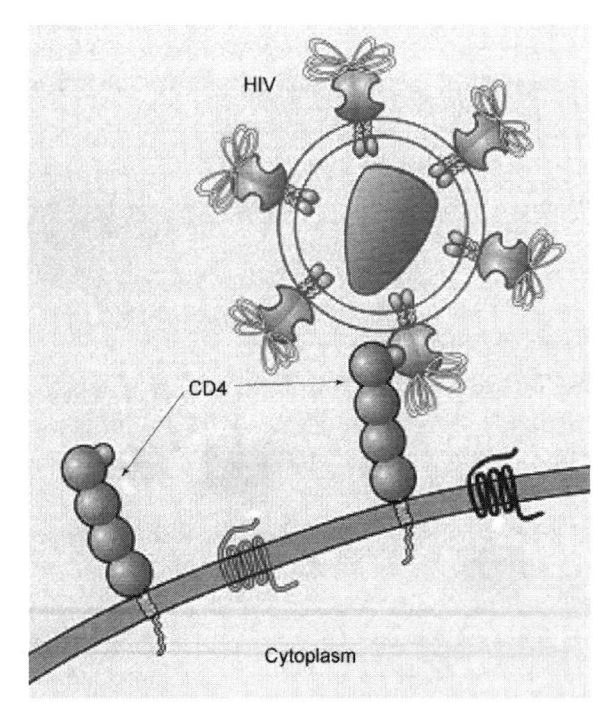

*Figure 5: HIV binds to the CD4 receptor, a glycoprotein on the surfaces of T cells. (credit: modification of work by NIH, NIAID)*

markers decreases the effectiveness of the person's immune system in attacking the virus, because the antibodies will not recognize the new variations of the surface patterns. In the case of HIV, the problem is compounded by the fact that the virus specifically infects and destroys cells involved in the immune response, further incapacitating the host.

## MEMBRANE FLUIDITY

The mosaic characteristic of the membrane, described in the fluid mosaic model, helps to illustrate its nature. The integral proteins and lipids exist in the membrane as separate but loosely attached molecules. These resemble the separate, multicolored tiles of a mosaic picture, and they float, moving somewhat with respect to one another. The membrane is not like a balloon, however, that can expand and contract; rather, it is fairly rigid and can burst if penetrated or if a cell takes in too much water. However, because of its mosaic nature, a very fine needle can easily penetrate a plasma membrane without causing it to burst, and the membrane will flow and self-seal when the needle is extracted.

The mosaic characteristics of the membrane explain some but not all of its fluidity. There are two other factors that help maintain this fluid characteristic. One factor is the nature of the phospholipids themselves. In their saturated form, the fatty acids in phospholipid tails are saturated with bound hydrogen atoms. There are no double bonds between adjacent carbon atoms. This results in tails that are relatively straight. In contrast, unsaturated fatty acids do not contain a maximal number of hydrogen atoms, but they do contain some double bonds between adjacent carbon atoms; a double bond results in a bend in the string of carbons of approximately 30 degrees ([Figure 2]).

Thus, if saturated fatty acids, with their straight tails, are compressed by decreasing temperatures, they press in on each other, making a dense and fairly rigid membrane. If unsaturated fatty acids are compressed, the "kinks" in their tails elbow adjacent phospholipid molecules away, maintaining some space between the phospholipid molecules. This "elbow room" helps to maintain fluidity in the membrane at temperatures at which membranes with saturated fatty acid tails in their phospholipids would "freeze" or solidify. The relative fluidity of the membrane is particularly important in a cold environment. A cold environment tends to compress membranes composed largely of saturated fatty acids, making them less fluid and more susceptible to rupturing. Many organisms (fish are one example) are capable of adapting to cold environments by changing the proportion of unsaturated fatty acids in their membranes in response to the lowering of the temperature.

---

Visit this site to see animations of the fluidity and mosaic quality of membranes: http://openstaxcollege.org/l/biological_memb

---

Animals have an additional membrane constituent that assists in maintaining fluidity. Cholesterol, which lies alongside the phospholipids in the membrane, tends to dampen the effects of temperature on the membrane. Thus, this lipid functions as a buffer, preventing lower temperatures from inhibiting fluidity and preventing increased temperatures from increasing fluidity too much. Thus, cholesterol extends, in both directions, the range of temperature in which the membrane is appropriately fluid and consequently functional. Cholesterol also serves other functions, such as organizing clusters of transmembrane proteins into lipid rafts.

*The Components and Functions of the Plasma Membrane*

| Component | Location |
| --- | --- |
| Phospholipid | Main fabric of the membrane |
| Cholesterol | Attached between phospholipids and between the two phospholipid layers |
| Integral proteins (for example, integrins) | Embedded within the phospholipid layer(s). May or may not penetrate through both layers |
| Peripheral proteins | On the inner or outer surface of the phospholipid bilayer; not embedded within the phospholipids |
| Carbohydrates (components of glycoproteins and glycolipids) | Generally attached to proteins on the outside membrane layer |

*Career Connection*

ImmunologistThe variations in peripheral proteins and carbohydrates that affect a cell's recognition sites are of prime interest in immunology. These changes are taken into consideration in vaccine development. Many infectious diseases, such as smallpox, polio, diphtheria, and tetanus, were conquered by the use of vaccines.

Immunologists are the physicians and scientists who research and develop vaccines, as well as treat and study allergies or other immune problems. Some immunologists study and treat autoimmune problems (diseases in which a person's immune system attacks his or her own cells or tissues, such as lupus) and immunodeficiencies, whether acquired (such as acquired immunodeficiency syndrome, or AIDS) or hereditary (such as severe combined immunodeficiency, or SCID). Immunologists are called in to help treat organ transplantation patients, who must have their immune systems suppressed so that their bodies will not reject a transplanted organ. Some immunologists work to understand natural immunity and the effects of a person's environment on it. Others work on questions about how the immune system affects diseases such as cancer. In the past, the importance of having a healthy immune system in preventing cancer was not at all understood.

To work as an immunologist, a PhD or MD is required. In addition, immunologists undertake at least 2–3 years of training in an accredited program and must pass an examination given by the American Board of Allergy and Immunology. Immunologists must possess knowledge of the functions of the human body as they relate to issues beyond immunization, and knowledge of pharmacology and medical technology, such as medications, therapies, test materials, and surgical procedures.

## SECTION SUMMARY

The modern understanding of the plasma membrane is referred to as the fluid mosaic model. The plasma membrane is composed of a bilayer of phospholipids, with their hydrophobic, fatty acid tails in contact with each other. The landscape of the membrane is studded with proteins, some of which span the membrane. Some of these proteins serve to transport materials into or out of the cell. Carbohydrates are attached to some of the proteins and lipids on the outward-facing surface of the membrane, forming complexes that function to identify the cell to other cells. The fluid nature of the membrane is due to temperature, the configuration of the fatty acid tails (some kinked by double bonds), the presence of cholesterol embedded in the membrane, and the mosaic nature of the proteins and protein-carbohydrate combinations, which are not firmly fixed in place. Plasma membranes enclose and define the borders of cells, but rather than being a static bag, they are dynamic and constantly in flux.

## Glossary

**amphiphilic:** molecule possessing a polar or charged area and a nonpolar or uncharged area capable of interacting with both hydrophilic and hydrophobic environments

**fluid mosaic model:** describes the structure of the plasma membrane as a mosaic of components including phospholipids, cholesterol, proteins, glycoproteins, and glycolipids (sugar chains attached to proteins or lipids, respectively), resulting in a fluid character (fluidity)

**glycolipid:** combination of carbohydrates and lipids

**glycoprotein:** combination of carbohydrates and proteins

**hydrophilic:** molecule with the ability to bond with water; "water-loving"

**hydrophobic:** molecule that does not have the ability to bond with water; "water-hating"

**integral protein:** protein integrated into the membrane structure that interacts extensively with the hydrocarbon chains of membrane lipids and often spans the membrane; these proteins can be removed only by the disruption of the membrane by detergents

**peripheral protein:** protein found at the surface of a plasma membrane either on its exterior or interior side; these proteins can be removed (washed off of the membrane) by a high-salt wash

# Passive Transport

> **Learning Objectives**
>
> By the end of this section, you will be able to:
> - Explain why and how passive transport occurs
> - Understand the processes of osmosis and diffusion
> - Define tonicity and describe its relevance to passive transport

Plasma membranes must allow certain substances to enter and leave a cell, and prevent some harmful materials from entering and some essential materials from leaving. In other words, plasma membranes are selectively permeable—they allow some substances to pass through, but not others. If they were to lose this selectivity, the cell would no longer be able to sustain itself, and it would be destroyed. Some cells require larger amounts of specific substances than do other cells; they must have a way of obtaining these materials from extracellular fluids. This may happen passively, as certain materials move back and forth, or the cell may have special mechanisms that facilitate transport. Some materials are so important to a cell that it spends some of its energy, hydrolyzing adenosine triphosphate (ATP), to obtain these materials. Red blood cells use some of their energy doing just that. All cells spend the majority of their energy to maintain an imbalance of sodium and potassium ions between the interior and exterior of the cell.

The most direct forms of membrane transport are passive. Passive transport is a naturally occurring phenomenon and does not require the cell to exert any of its energy to accomplish the movement. In passive transport, substances move from an area of higher concentration to an area of lower concentration. A physical space in which there is a range of concentrations of a single substance is said to have a concentration gradient.

## SELECTIVE PERMEABILITY

Plasma membranes are asymmetric: the interior of the membrane is not identical to the exterior of the membrane. In fact, there is a considerable difference between the array of phospholipids and proteins between the two leaflets that form a membrane. On the interior of the membrane, some proteins serve to anchor the membrane to fibers of the cytoskeleton. There are peripheral proteins on the exterior of the membrane that bind elements of the extracellular matrix. Carbohydrates, attached to lipids or proteins, are also found on the exterior surface of the plasma membrane. These carbohydrate complexes help the cell bind substances that the cell needs in the extracellular fluid. This adds considerably to the selective nature of plasma membranes ([Figure 1]).

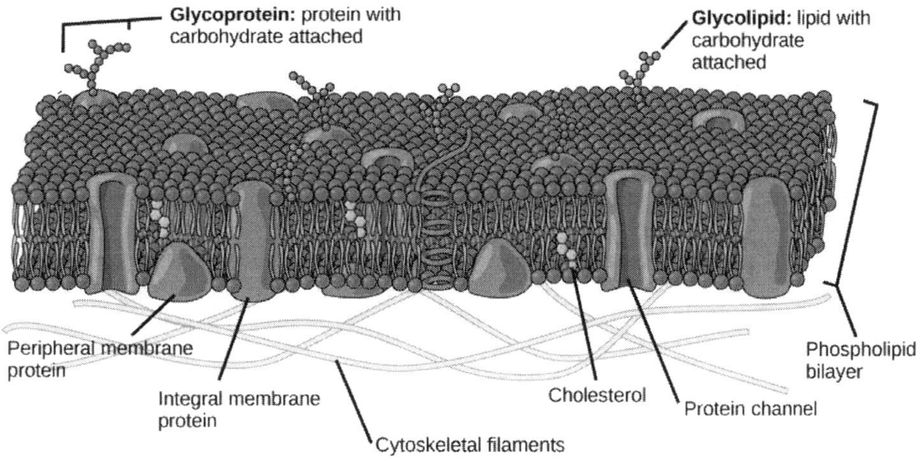

*Figure 1: The exterior surface of the plasma membrane is not identical to the interior surface of the same membrane.*

Recall that plasma membranes are amphiphilic: They have hydrophilic and hydrophobic regions. This characteristic helps the movement of some materials through the membrane and hinders the movement of others. Lipid-soluble material with a low molecular weight can easily slip through the hydrophobic lipid core of the membrane. Substances such as the fat-soluble vitamins A, D, E, and K readily pass through the plasma membranes in the digestive tract and other tissues. Fat-soluble drugs and hormones also gain easy entry into cells and are readily transported into the body's tissues and organs. Molecules of oxygen and carbon dioxide have no charge and so pass through membranes by simple diffusion.

Polar substances present problems for the membrane. While some polar molecules connect easily with the outside of a cell, they cannot readily pass through the lipid core of the plasma membrane. Additionally, while small ions could easily slip through the spaces in the mosaic of the membrane, their charge prevents them from doing so. Ions such as sodium, potassium, calcium, and chloride must have special means of penetrating plasma membranes. Simple sugars and amino acids also need help with transport across plasma membranes, achieved by various transmembrane proteins (channels).

## DIFFUSION

Diffusion is a passive process of transport. A single substance tends to move from an area of high concentration to an area of low concentration until the concentration is equal across a space. You are familiar with diffusion of substances through the air. For example, think about someone opening a bottle of ammonia in a room filled with people. The ammonia gas is at its highest concentration in the bottle; its lowest concentration is at the edges of the room. The ammonia vapor will diffuse, or spread away, from the bottle, and gradually, more and more people will smell the ammonia as it spreads. Materials move within the cell's cytosol by diffusion, and certain materials move through the plasma membrane by diffusion ([**Figure 2**]). Diffusion expends no energy. On the contrary, concentration gradients are a form of potential energy, dissipated as the gradient is eliminated.

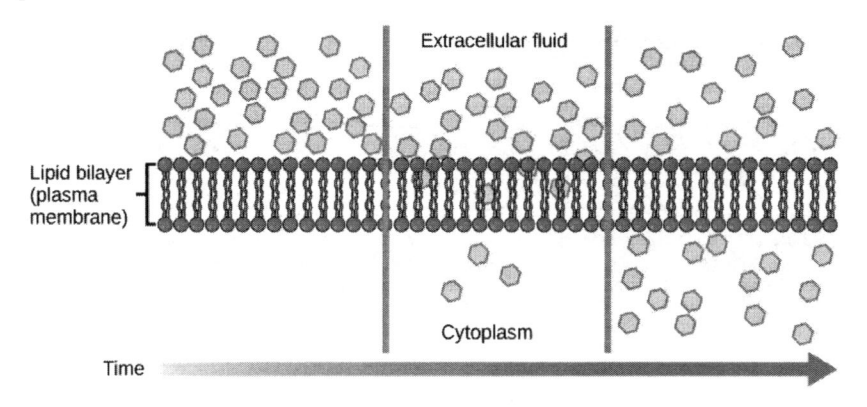

*Figure 2: Diffusion through a permeable membrane moves a substance from an area of high concentration (extracellular fluid, in this case) down its concentration gradient (into the cytoplasm). (credit: modification of work by Mariana Ruiz Villareal)*

Each separate substance in a medium, such as the extracellular fluid, has its own concentration gradient, independent of the concentration gradients of other materials. In addition, each substance will diffuse according to that gradient. Within a system, there will be different rates of diffusion of the different substances in the medium.

*Factors That Affect Diffusion*

Molecules move constantly in a random manner, at a rate that depends on their mass, their environment, and the amount of thermal energy they possess, which in turn is a function of temperature. This movement accounts for the diffusion of molecules through whatever medium in which they are localized. A substance will tend to move into any space available to it until it is evenly distributed throughout it. After a substance has diffused completely through a space, removing its concentration gradient, molecules will still move around in the space, but there will be no *net* movement of the number of molecules from one area to another. This lack of a concentration gradient in which there is no net movement of a substance is known as dynamic equilibrium. While diffusion will go forward in the presence of a concentration gradient of a substance, several factors affect the rate of diffusion.

- Extent of the concentration gradient: The greater the difference in concentration, the more rapid the diffusion. The closer the distribution of the material gets to equilibrium, the slower the rate of diffusion becomes.

- Mass of the molecules diffusing: Heavier molecules move more slowly; therefore, they diffuse more slowly. The reverse is true for lighter molecules.

- Temperature: Higher temperatures increase the energy and therefore the movement of the molecules, increasing the rate of diffusion. Lower temperatures decrease the energy of the molecules, thus decreasing the rate of diffusion.

- Solvent density: As the density of a solvent increases, the rate of diffusion decreases. The molecules slow down because they have a more difficult time getting through the denser medium. If the medium is less dense, diffusion increases. Because cells primarily use diffusion to move materials within the cytoplasm, any increase in the cytoplasm's density will inhibit the movement of the materials. An example of this is a person experiencing dehydration. As the body's cells lose water, the rate of diffusion decreases in the cytoplasm, and the cells' functions deteriorate. Neurons tend to be very sensitive to this effect. Dehydration frequently leads to unconsciousness and possibly coma because of the decrease in diffusion rate within the cells.

- Solubility: As discussed earlier, nonpolar or lipid-soluble materials pass through plasma membranes more easily than polar materials, allowing a faster rate of diffusion.

- Surface area and thickness of the plasma membrane: Increased surface area increases the rate of diffusion, whereas a thicker membrane reduces it.

- Distance travelled: The greater the distance that a substance must travel, the slower the rate of diffusion. This places an upper limitation on cell size. A large, spherical cell will die because nutrients or waste cannot reach or leave the center of the cell, respectively. Therefore, cells must either be small in size, as in the case of many prokaryotes, or be flattened, as with many single-celled eukaryotes.

A variation of diffusion is the process of filtration. In filtration, material moves according to its concentration gradient through a membrane; sometimes the rate of diffusion is enhanced by pressure, causing the substances to filter more rapidly. This occurs in the kidney, where blood pressure forces large amounts of water and accompanying dissolved substances, or solutes, out of the blood and into the renal tubules. The rate of diffusion in this instance is almost totally dependent on pressure. One of the effects of high blood pressure is the appearance of protein in the urine, which is "squeezed through" by the abnormally high pressure.

## FACILITATED TRANSPORT

In facilitated transport, also called facilitated diffusion, materials diffuse across the plasma membrane with the help of membrane proteins. A concentration gradient exists that would allow these materials to diffuse into the cell without expending cellular energy. However, these materials are ions are polar molecules that are repelled by the hydrophobic parts of the cell membrane. Facilitated transport proteins shield these materials from the repulsive force of the membrane, allowing them to diffuse into the cell.

The material being transported is first attached to protein or glycoprotein receptors on the exterior surface of the plasma membrane. This allows the material that is needed by the cell to be removed from the extracellular fluid. The substances are then passed to specific integral proteins that facilitate their passage. Some of these integral proteins are collections of beta pleated sheets that form a pore or channel through the phospholipid bilayer. Others are carrier proteins which bind with the substance and aid its diffusion through the membrane.

### Channels

The integral proteins involved in facilitated transport are collectively referred to as transport proteins, and they function as either channels for the material or carriers. In both cases, they are transmembrane proteins. Channels are specific for the substance that is being transported. Channel proteins have hydrophilic domains exposed to the intracellular and extracellular fluids; they additionally have a hydrophilic channel through their core that provides a hydrated opening through the membrane layers ([Figure 3]). Passage through the channel allows polar compounds to avoid the nonpolar central layer of the plasma membrane that would otherwise slow or prevent their entry into the cell. Aquaporins are channel proteins that allow water to pass through the membrane at a very high rate.

Channel proteins are either open at all times or they are "gated," which controls the opening of the channel. The attachment of a particular ion to the channel protein may control the opening, or other mechanisms or substances may be involved. In some tissues, sodium and chloride ions pass freely through open channels, whereas in other tissues a gate must be opened to allow passage.

324   Cell Transport

An example of this occurs in the kidney, where both forms of channels are found in different parts of the renal tubules. Cells involved in the transmission of electrical impulses, such as nerve and muscle cells, have gated channels for sodium, potassium, and calcium in their membranes. Opening and closing of these channels changes the relative concentrations on opposing sides of the membrane of these ions, resulting in the facilitation of electrical transmission along membranes (in the case of nerve cells) or in muscle contraction (in the case of muscle cells).

### Carrier Proteins

Another type of protein embedded in the plasma membrane is a carrier protein. This aptly named protein binds a substance and, in doing so, triggers a change of its own shape, moving the bound molecule from the outside of the cell to its interior ([Figure 4]); depending on the gradient, the material may move in the opposite direction. Carrier proteins are typically specific for a single substance. This selectivity adds to the overall selectivity of the plasma membrane. The exact mechanism for the change of shape is poorly understood. Proteins can change shape when their hydrogen bonds are affected, but this may not fully explain this mechanism. Each carrier protein is specific to one substance, and there are a finite number of these proteins in any membrane. This can cause problems in transporting enough of the material for the cell to function properly. When all of the proteins are bound to their ligands, they are saturated and the rate of transport is at its maximum. Increasing the concentration gradient at this point will not result in an increased rate of transport.

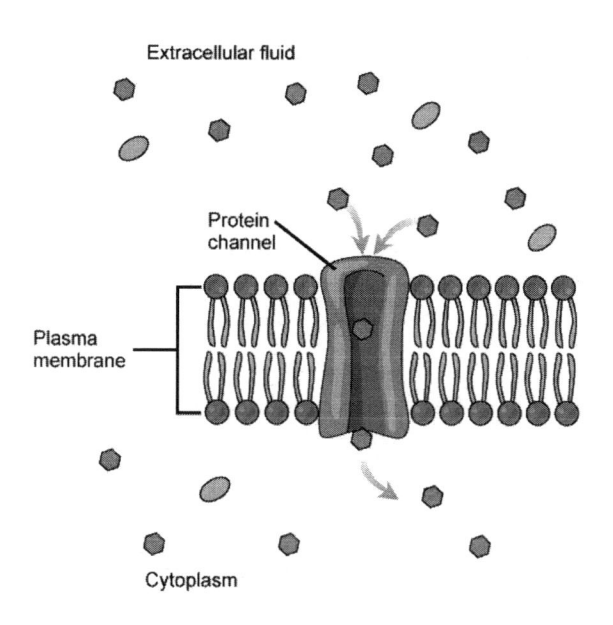

*Figure 3: Facilitated transport moves substances down their concentration gradients. They may cross the plasma membrane with the aid of channel proteins. (credit: modification of work by Mariana Ruiz Villareal)*

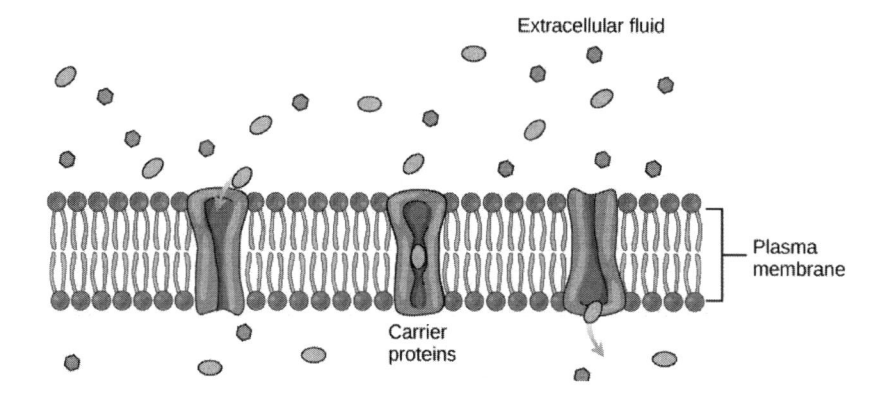

*Figure 4: Some substances are able to move down their concentration gradient across the plasma membrane with the aid of carrier proteins. Carrier proteins change shape as they move molecules across the membrane. (credit: modification of work by Mariana Ruiz Villareal)*

An example of this process occurs in the kidney. Glucose, water, salts, ions, and amino acids needed by the body are filtered in one part of the kidney. This filtrate, which includes glucose, is then reabsorbed in another part of the kidney. Because there are only a finite number of carrier proteins for glucose, if more glucose is present than the proteins can handle, the excess is not transported and it is excreted from the body in the urine. In a diabetic individual, this is described as "spilling glucose into the urine." A different group of carrier proteins called glucose transport proteins, or GLUTs, are involved in transporting glucose and other hexose sugars through plasma membranes within the body.

Channel and carrier proteins transport material at different rates. Channel proteins transport much more quickly than do carrier proteins. Channel proteins facilitate diffusion at a rate of tens of millions of molecules per second, whereas carrier proteins work at a rate of a thousand to a million molecules per second.

## OSMOSIS

Osmosis is the movement of water through a semipermeable membrane according to the concentration gradient of water across the membrane, which is inversely proportional to the concentration of solutes. While diffusion transports material across membranes and within cells, osmosis transports *only water* across a membrane and the membrane limits the diffusion of solutes in the water. Not surprisingly, the aquaporins that facilitate water movement play a large role in osmosis, most prominently in red blood cells and the membranes of kidney tubules.

### Mechanism

Osmosis is a special case of diffusion. Water, like other substances, moves from an area of high concentration to one of low concentration. An obvious question is what makes water move at all? Imagine a beaker with a semipermeable membrane separating the two sides or halves ([Figure 5]). On both sides of the membrane the water level is the same, but there are different concentrations of a dissolved substance, or solute, that cannot cross the membrane (otherwise the concentrations on each side would be balanced by the solute crossing the membrane). If the volume of the solution on both sides of the membrane is the same, but the concentrations of solute are different, then there are different amounts of water, the solvent, on either side of the membrane.

To illustrate this, imagine two full glasses of water. One has a single teaspoon of sugar in it, whereas the second one contains one-quarter cup of sugar. If the total volume of the solutions in both cups is the same, which cup contains more water? Because the large amount of sugar in the second cup takes up much more space than the teaspoon of sugar in the first cup, the first cup has more water in it.

Returning to the beaker example, recall that it has a mixture of solutes on either side of the membrane. A principle of diffusion is that the molecules move around and will spread evenly throughout the medium if they can.

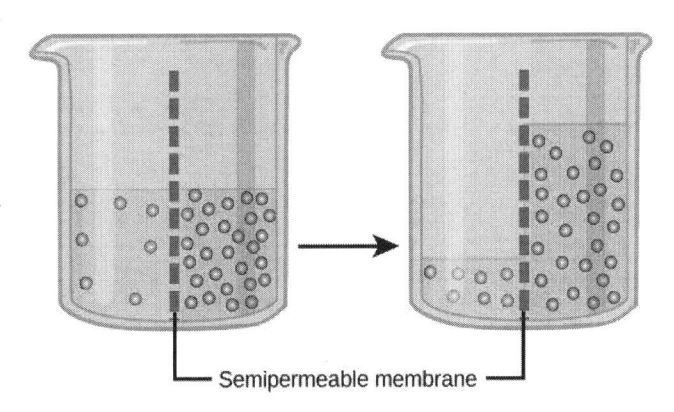

*Figure 5: In osmosis, water always moves from an area of higher water concentration to one of lower concentration. In the diagram shown, the solute cannot pass through the selectively permeable membrane, but the water can.*

326  Cell Transport

However, only the material capable of getting through the membrane will diffuse through it. In this example, the solute cannot diffuse through the membrane, but the water can. Water has a concentration gradient in this system. Thus, water will diffuse down its concentration gradient, crossing the membrane to the side where it is less concentrated. This diffusion of water through the membrane—osmosis—will continue until the concentration gradient of water goes to zero or until the hydrostatic pressure of the water balances the osmotic pressure. Osmosis proceeds constantly in living systems.

## TONICITY

Tonicity describes how an extracellular solution can change the volume of a cell by affecting osmosis. A solution's tonicity often directly correlates with the osmolarity of the solution. Osmolarity describes the total solute concentration of the solution. A solution with low osmolarity has a greater number of water molecules relative to the number of solute particles; a solution with high osmolarity has fewer water molecules with respect to solute particles. In a situation in which solutions of two different osmolarities are separated by a membrane permeable to water, though not to the solute, water will move from the side of the membrane with lower osmolarity (and more water) to the side with higher osmolarity (and less water). This effect makes sense if you remember that the solute cannot move across the membrane, and thus the only component in the system that can move—the water—moves along its own concentration gradient. An important distinction that concerns living systems is that osmolarity measures the number of particles (which may be molecules) in a solution. Therefore, a solution that is cloudy with cells may have a lower osmolarity than a solution that is clear, if the second solution contains more dissolved molecules than there are cells.

### Hypotonic Solutions

Three terms—hypotonic, isotonic, and hypertonic—are used to relate the osmolarity of a cell to the osmolarity of the extracellular fluid that contains the cells. In a hypotonic situation, the extracellular fluid has lower osmolarity than the fluid inside the cell, and water enters the cell. (In living systems, the point of reference is always the cytoplasm, so the prefix *hypo-* means that the extracellular fluid has a lower concentration of solutes, or a lower osmolarity, than the cell cytoplasm.) It also means that the extracellular fluid has a higher concentration of water in the solution than does the cell. In this situation, water will follow its concentration gradient and enter the cell.

### Hypertonic Solutions

As for a hypertonic solution, the prefix *hyper-* refers to the extracellular fluid having a higher osmolarity than the cell's cytoplasm; therefore, the fluid contains less water than the cell does. Because the cell has a relatively higher concentration of water, water will leave the cell.

### Isotonic Solutions

In an isotonic solution, the extracellular fluid has the same osmolarity as the cell. If the osmolarity of the cell matches that of the extracellular fluid, there will be no net movement of water into or out of the cell, although water will still move in and out. Blood cells and plant cells in hypertonic, isotonic, and hypotonic solutions take on characteristic appearances ([**Figure 6**]).

> ## Art Connection
>
> *Figure 6: In osmosis, water always moves from an area of higher water concentration to one of lower concentration. In the diagram shown, the solute cannot pass through the selectively permeable membrane, but the water can.*
>
> A doctor injects a patient with what the doctor thinks is an isotonic saline solution. The patient dies, and an autopsy reveals that many red blood cells have been destroyed. Do you think the solution the doctor injected was really isotonic?

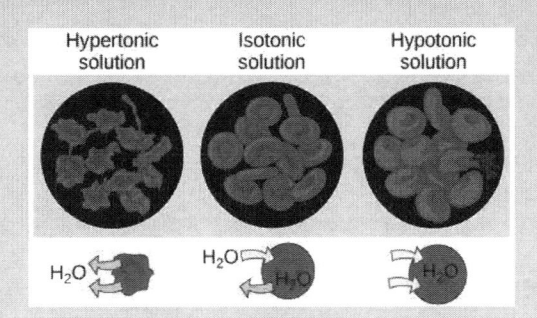

For a video illustrating the process of diffusion in solutions, visit this site: http://openstaxcollege.org/l/dispersion.

## TONICITY IN LIVING SYSTEMS

In a hypotonic environment, water enters a cell, and the cell swells. In an isotonic condition, the relative concentrations of solute and solvent are equal on both sides of the membrane. There is no net water movement; therefore, there is no change in the size of the cell. In a hypertonic solution, water leaves a cell and the cell shrinks. If either the hypo- or hyper- condition goes to excess, the cell's functions become compromised, and the cell may be destroyed.

A red blood cell will burst, or lyse, when it swells beyond the plasma membrane's capability to expand. Remember, the membrane resembles a mosaic, with discrete spaces between the molecules composing it. If the cell swells, and the spaces between the lipids and proteins become too large, the cell will break apart.

In contrast, when excessive amounts of water leave a red blood cell, the cell shrinks, or crenates. This has the effect of concentrating the solutes left in the cell, making the cytosol denser and interfering with diffusion within the cell. The cell's ability to function will be compromised and may also result in the death of the cell.

Various living things have ways of controlling the effects of osmosis—a mechanism called osmoregulation. Some organisms, such as plants, fungi, bacteria, and some protists, have cell walls that surround the plasma membrane and prevent cell lysis in a hypotonic solution. The plasma membrane can only expand to the limit of the cell wall, so the cell will not lyse. In fact, the cytoplasm in plants is always slightly hypertonic to the cellular environment, and water will always enter a cell if water is available. This inflow

# 328  Cell Transport

of water produces turgor pressure, which stiffens the cell walls of the plant (**[Figure 7]**). In nonwoody plants, turgor pressure supports the plant. Conversely, if the plant is not watered, the extracellular fluid will become hypertonic, causing water to leave the cell. In this condition, the cell does not shrink because the cell wall is not flexible. However, the cell membrane detaches from the wall and constricts the cytoplasm. This is called plasmolysis. Plants lose turgor pressure in this condition and wilt (**[Figure 8]**).

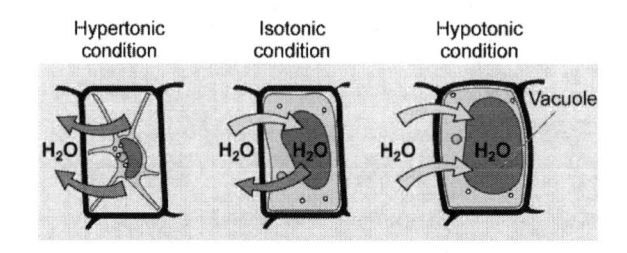

*Figure 7: The turgor pressure within a plant cell depends on the tonicity of the solution that it is bathed in. (credit: modification of work by Mariana Ruiz Villareal)*

*Figure 8: Without adequate water, the plant on the left has lost turgor pressure, visible in its wilting; the turgor pressure is restored by watering it (right). (credit: Victor M. Vicente Selvas)*

Tonicity is a concern for all living things. For example, paramecia and amoebas, which are protists that lack cell walls, have contractile vacuoles. This vesicle collects excess water from the cell and pumps it out, keeping the cell from lysing as it takes on water from its environment (**[Figure 9]**).

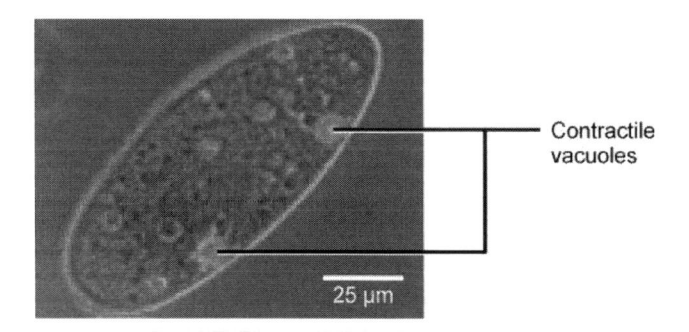

*Figure 9: A paramecium's contractile vacuole, here visualized using bright field light microscopy at 480x magnification, continuously pumps water out of the organism's body to keep it from bursting in a hypotonic medium. (credit: modification of work by NIH; scale-bar data from Matt Russell)*

Many marine invertebrates have internal salt levels matched to their environments, making them isotonic with the water in which they live. Fish, however, must spend approximately five percent of their metabolic energy maintaining osmotic homeostasis. Freshwater fish live in an environment that is hypotonic to their cells. These fish actively take in salt through their gills and excrete diluted urine to rid themselves of excess water. Saltwater fish live in the reverse environment, which is hypertonic to their cells, and they secrete salt through their gills and excrete highly concentrated urine.

In vertebrates, the kidneys regulate the amount of water in the body. Osmoreceptors are specialized cells in the brain that monitor the concentration of solutes in the blood. If the levels of solutes increase beyond a certain range, a hormone is released that retards water loss through the kidney and dilutes the blood to safer levels. Animals also have high concentrations of albumin, which is produced by the liver, in their blood. This protein is too large to pass easily through plasma membranes and is a major factor in controlling the osmotic pressures applied to tissues.

## SECTION SUMMARY

The passive forms of transport, diffusion and osmosis, move materials of small molecular weight across membranes. Substances diffuse from areas of high concentration to areas of lower concentration, and this process continues until the substance is evenly distributed in a system. In solutions containing more than one substance, each type of molecule diffuses according to its own concentration gradient, independent of the diffusion of other substances. Many factors can affect the rate of diffusion, including concentration gradient, size of the particles that are diffusing, temperature of the system, and so on.

In living systems, diffusion of substances into and out of cells is mediated by the plasma membrane. Some materials diffuse readily through the membrane, but others are hindered, and their passage is made possible by specialized proteins, such as channels and transporters. The chemistry of living things occurs in aqueous solutions, and balancing the concentrations of those solutions is an ongoing problem. In living systems, diffusion of some substances would be slow or difficult without membrane proteins that facilitate transport.

### Glossary

**aquaporin:** channel protein that allows water through the membrane at a very high rate

**carrier protein:** membrane protein that moves a substance across the plasma membrane by changing its own shape

**channel protein:** membrane protein that allows a substance to pass through its hollow core across the plasma membrane

**concentration gradient:** area of high concentration adjacent to an area of low concentration

**diffusion:** passive process of transport of low-molecular weight material according to its concentration gradient

**facilitated transport:** process by which material moves down a concentration gradient (from high to low concentration) using integral membrane proteins

**hypertonic:** situation in which extracellular fluid has a higher osmolarity than the fluid inside the cell, resulting in water moving out of the cell

**hypotonic:** situation in which extracellular fluid has a lower osmolarity than the fluid inside the cell, resulting in water moving into the cell

**isotonic:** situation in which the extracellular fluid has the same osmolarity as the fluid inside the cell, resulting in no net movement of water into or out of the cell

**osmolarity:** total amount of substances dissolved in a specific amount of solution

**osmosis:** transport of water through a semipermeable membrane according to the concentration gradient of water across the membrane that results from the presence of solute that cannot pass through the membrane

**passive transport:** method of transporting material through a membrane that does not require energy

**plasmolysis:** detaching of the cell membrane from the cell wall and constriction of the cell membrane when a plant cell is in a hypertonic solution

**selectively permeable:** characteristic of a membrane that allows some substances through but not others

**solute:** substance dissolved in a liquid to form a solution

**tonicity:** amount of solute in a solution

**transport protein:** membrane protein that facilitates passage of a substance across a membrane by binding it

# Active Transport

> **Learning Objectives**
>
> By the end of this section, you will be able to:
> - Understand how electrochemical gradients affect ions
> - Describe endocytosis, including phagocytosis, pinocytosis, and receptor-mediated endocytosis
> - Understand the process of exocytosis

Active transport mechanisms require the use of the cell's energy, usually in the form of adenosine triphosphate (ATP). If a substance must move into the cell against its concentration gradient, that is, if the concentration of the substance inside the cell must be greater than its concentration in the extracellular fluid, the cell must use energy to move the substance. Some active transport mechanisms move small-molecular weight material, such as ions, through the membrane.

In addition to moving small ions and molecules through the membrane, cells also need to remove and take in larger molecules and particles. Some cells are even capable of engulfing entire unicellular microorganisms. You might have correctly hypothesized that the uptake and release of large particles by the cell requires energy. A large particle, however, cannot pass through the membrane, even with energy supplied by the cell.

## ELECTROCHEMICAL GRADIENT

We have discussed simple concentration gradients—differential concentrations of a substance across a space or a membrane—but in living systems, gradients are more complex. Because cells contain proteins, most of which are negatively charged, and because ions move into and out of cells, there is an electrical gradient, a difference of charge, across the plasma membrane. The interior of living cells is electrically negative with respect to the extracellular fluid in which they are bathed; at the same time, cells have higher concentrations of potassium ($K^+$) and lower concentrations of sodium ($Na^+$) than does the extracellular fluid. Thus, in a living cell, the concentration gradient and electrical gradient of $Na^+$ promotes diffusion of the ion into the cell, and the electrical gradient of $Na^+$ (a positive ion) tends to drive it inward to the negatively charged interior. The situation is more complex, however, for other elements such as potassium. The electrical gradient of $K^+$ promotes diffusion of the ion *into* the cell, but the concentration gradient of $K^+$ promotes diffusion *out* of the cell (**[Figure 1]**). The combined gradient that affects an ion is called its electrochemical gradient, and it is especially important to muscle and nerve cells.

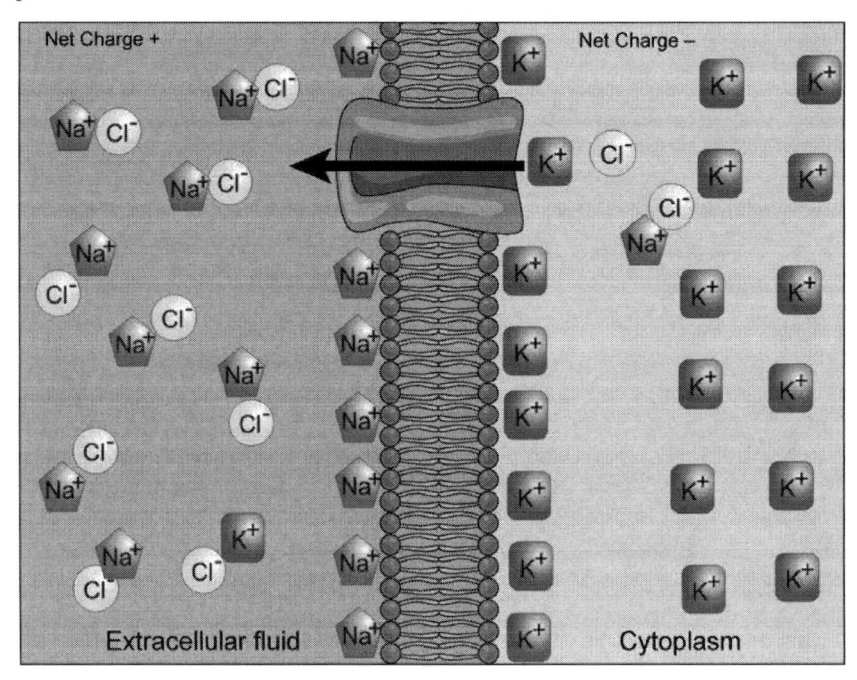

*Figure 1: Electrochemical gradients arise from the combined effects of concentration gradients and electrical gradients. (credit: modification of work by "Synaptitude"/ Wikimedia Commons)*

### Moving Against a Gradient

To move substances against a concentration or an electrochemical gradient, the cell must use energy. This energy is harvested from ATP that is generated through cellular metabolism. Active transport mechanisms, collectively called pumps or carrier proteins, work against electrochemical gradients. With the exception of ions, small substances constantly pass through plasma membranes. Active transport maintains concentrations of ions and other substances needed by living cells in the face of these passive changes. Much of a cell's supply of metabolic energy may be spent maintaining these processes. Because active transport mechanisms depend on cellular metabolism for energy, they are sensitive to many metabolic poisons that interfere with the supply of ATP.

Two mechanisms exist for the transport of small-molecular weight material and macromolecules. Primary active transport moves ions across a membrane and creates a difference in charge across that membrane. The primary active transport system uses ATP to move a substance, such as an ion, into the cell, and often at the same time, a second substance is moved out of the cell. The sodium-potassium pump, an important pump in animal cells, expends energy to move potassium ions into the cell and a different number of sodium ions out of the cell ([Figure 2]). The action of this pump results in a concentration and charge difference across the membrane.

Secondary active transport describes the movement of material using the energy of the electrochemical gradient established by primary active transport.

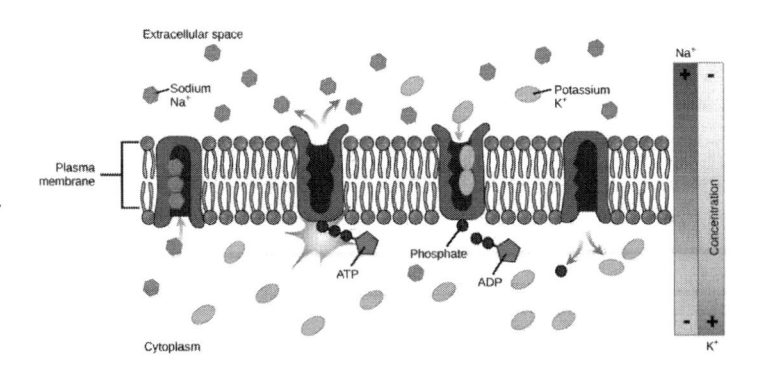

*Figure 2: The sodium-potassium pump move potassium and sodium ions across the plasma membrane. (credit: modification of work by Mariana Ruiz Villarreal)*

Using the energy of the electrochemical gradient created by the primary active transport system, other substances such as amino acids and glucose can be brought into the cell through membrane channels. ATP itself is formed through secondary active transport using a hydrogen ion gradient in the mitochondrion.

## ENDOCYTOSIS

Endocytosis is a type of active transport that moves particles, such as large molecules, parts of cells, and even whole cells, into a cell. There are different variations of endocytosis, but all share a common characteristic: The plasma membrane of the cell invaginates, forming a pocket around the target particle. The pocket pinches off, resulting in the particle being contained in a newly created vacuole that is formed from the plasma membrane.

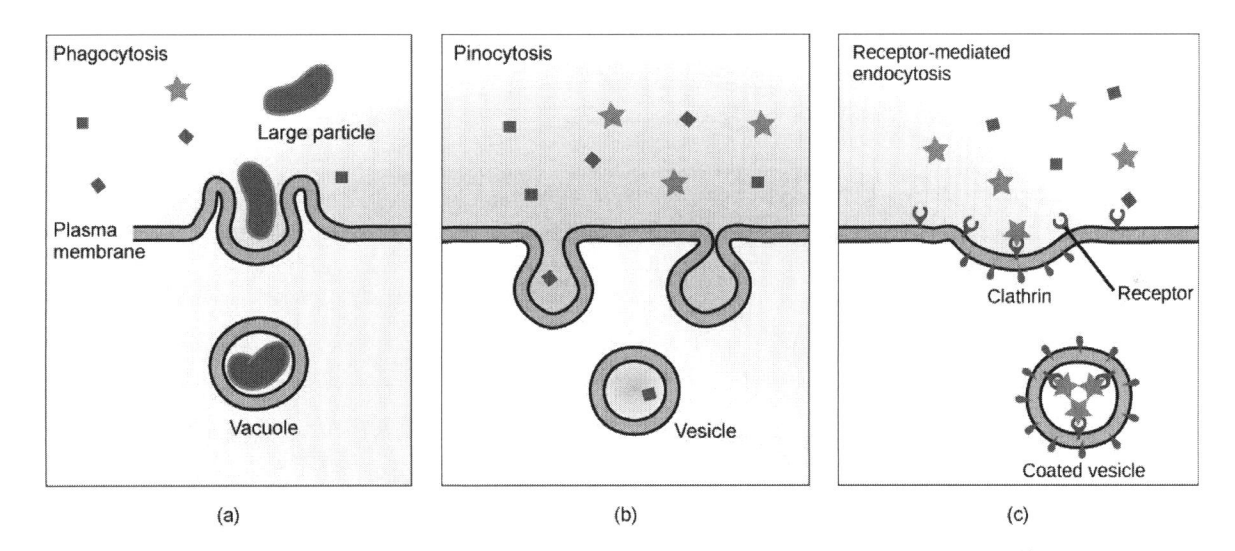

*Figure 3: Three variations of endocytosis are shown. (a) In one form of endocytosis, phagocytosis, the cell membrane surrounds the particle and pinches off to form an intracellular vacuole. (b) In another type of endocytosis, pinocytosis, the cell membrane surrounds a small volume of fluid and pinches off, forming a vesicle. (c) In receptor-mediated endocytosis, uptake of substances by the cell is targeted to a single type of substance that binds at the receptor on the external cell membrane. (credit: modification of work by Mariana Ruiz Villarreal)*

Phagocytosis is the process by which large particles, such as cells, are taken in by a cell. For example, when microorganisms invade the human body, a type of white blood cell called a neutrophil removes the invader through this process, surrounding and engulfing the microorganism, which is then destroyed by the neutrophil (**Figure 3**]).

A variation of endocytosis is called pinocytosis. This literally means "cell drinking" and was named at a time when the assumption was that the cell was purposefully taking in extracellular fluid. In reality, this process takes in solutes that the cell needs from the extracellular fluid ([**Figure 3**]).

A targeted variation of endocytosis employs binding proteins in the plasma membrane that are specific for certain substances ([**Figure 3**]). The particles bind to the proteins and the plasma membrane invaginates, bringing the substance and the proteins into the cell. If passage across the membrane of the target of receptor-mediated endocytosis is ineffective, it will not be removed from the tissue fluids or blood. Instead, it will stay in those fluids and increase in concentration. Some human diseases are caused by a failure of receptor-mediated endocytosis. For example, the form of cholesterol termed low-density lipoprotein or LDL (also referred to as "bad" cholesterol) is removed from the blood by receptor-mediated endocytosis. In the human genetic disease familial hypercholesterolemia, the LDL receptors are defective or missing entirely. People with this condition have life-threatening levels of cholesterol in their blood, because their cells cannot clear the chemical from their blood.

See receptor-mediated endocytosis in action and click on different parts for a focused animation to learn more: http://openstaxcollege.org/l/endocytosis2.

## EXOCYTOSIS

In contrast to these methods of moving material into a cell is the process of exocytosis. Exocytosis is the opposite of the processes discussed above in that its purpose is to expel material from the cell into the extracellular fluid. A particle enveloped in membrane fuses with the interior of the plasma membrane. This fusion opens the membranous envelope to the exterior of the cell, and the particle is expelled into the extracellular space ([[Figure 4]]).

### SECTION SUMMARY

The combined gradient that affects an ion includes its concentration gradient and its electrical gradient. Living cells need certain substances in concentrations greater than they exist in the extracellular space. Moving substances up their electrochemical gradients requires energy from the cell. Active transport uses energy stored in ATP to fuel the transport. Active transport of small molecular-size material uses integral proteins in the cell membrane to move the material—these proteins are analogous to pumps. Some pumps, which carry out primary active transport, couple directly with ATP to drive their action. In secondary transport, energy from primary transport can be used to move another substance into the cell and up its concentration gradient.

Endocytosis methods require the direct use of ATP to fuel the transport of large particles such as macromolecules; parts of cells or whole cells can be engulfed by other cells in a process called phagocytosis. In phagocytosis, a portion of the membrane

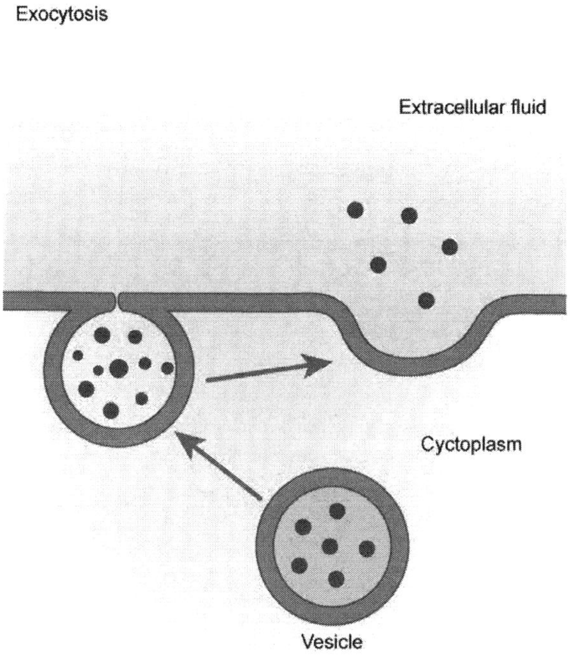

*Figure 4: In exocytosis, a vesicle migrates to the plasma membrane, binds, and releases its contents to the outside of the cell. (credit: modification of work by Mariana Ruiz Villarreal)*

invaginates and flows around the particle, eventually pinching off and leaving the particle wholly enclosed by an envelope of plasma membrane. Vacuoles are broken down by the cell, with the particles used as food or dispatched in some other way. Pinocytosis is a similar process on a smaller scale. The cell expels waste and other particles through the reverse process, exocytosis. Wastes are moved outside the cell, pushing a membranous vesicle to the plasma membrane, allowing the vesicle to fuse with the membrane and incorporating itself into the membrane structure, releasing its contents to the exterior of the cell.

## Glossary

**active transport:** the method of transporting material that requires energy

**electrochemical gradient:** a gradient produced by the combined forces of the electrical gradient and the chemical gradient

**endocytosis:** a type of active transport that moves substances, including fluids and particles, into a cell

**exocytosis:** a process of passing material out of a cell

**phagocytosis:** a process that takes macromolecules that the cell needs from the extracellular fluid; a variation of endocytosis

**pinocytosis:** a process that takes solutes that the cell needs from the extracellular fluid; a variation of endocytosis

**receptor-mediated endocytosis:** a variant of endocytosis that involves the use of specific binding proteins in the plasma membrane for specific molecules or particles

# Connections between Cells and Cellular Activities

### Learning Objectives

By the end of this section, you will be able to:

- Describe the extracellular matrix
- List examples of the ways that plant cells and animal cells communicate with adjacent cells
- Summarize the roles of tight junctions, desmosomes, gap junctions, and plasmodesmata

You already know that a group of similar cells working together is called a tissue. As you might expect, if cells are to work together, they must communicate with each other, just as you need to communicate with others if you work on a group project. Let's take a look at how cells communicate with each other.

## EXTRACELLULAR MATRIX OF ANIMAL CELLS

Most animal cells release materials into the extracellular space. The primary components of these materials are proteins, and the most abundant protein is collagen. Collagen fibers are interwoven with carbohydrate-containing protein molecules called proteoglycans. Collectively, these materials are called the extracellular matrix ([**Figure 1**]). Not only does the extracellular matrix hold the cells together to form a tissue, but it also allows the cells within the tissue to communicate with each other. How can this happen?

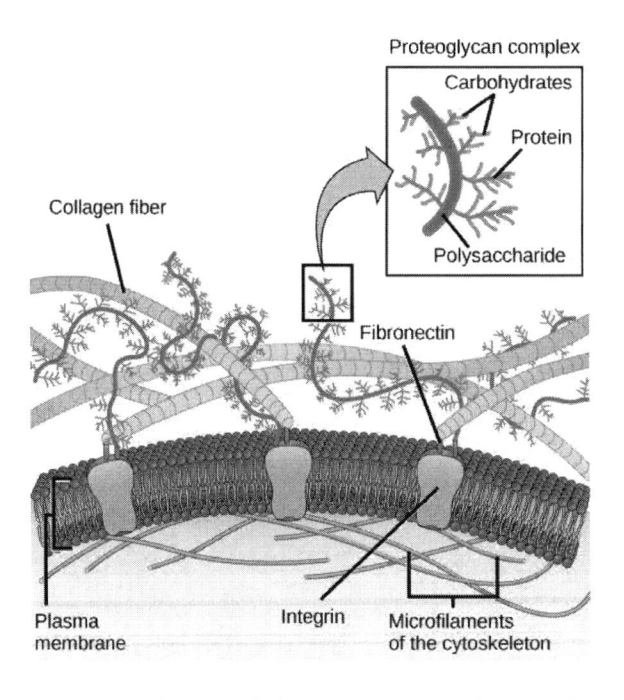

*Figure 1: The extracellular matrix consists of a network of proteins and carbohydrates.*

Cells have protein receptors on the extracellular surfaces of their plasma membranes. When a molecule within the matrix binds to the receptor, it changes the molecular structure of the receptor. The receptor, in turn, changes the conformation of the microfilaments positioned just inside the plasma membrane. These conformational changes induce chemical signals inside the cell that reach the nucleus and turn "on" or "off" the transcription of specific sections of DNA, which affects the production of associated proteins, thus changing the activities within the cell.

Blood clotting provides an example of the role of the extracellular matrix in cell communication. When the cells lining a blood vessel are damaged, they display a protein receptor called tissue factor. When tissue factor binds with another factor in the extracellular matrix, it causes platelets to adhere to the wall of the damaged blood vessel, stimulates the adjacent smooth muscle cells in the blood vessel to contract (thus constricting the blood vessel), and initiates a series of steps that stimulate the platelets to produce clotting factors.

## INTERCELLULAR JUNCTIONS

Cells can also communicate with each other via direct contact, referred to as intercellular junctions. There are some differences in the ways that plant and animal cells do this. Plasmodesmata are junctions between plant cells, whereas animal cell contacts include tight junctions, gap junctions, and desmosomes.

### Plasmodesmata

In general, long stretches of the plasma membranes of neighboring plant cells cannot touch one another because they are separated by the cell wall that surrounds each cell ([link]b). How then, can a plant transfer water and other soil nutrients from its roots, through its stems, and to its leaves? Such transport uses the vascular tissues (xylem and phloem) primarily. There also exist structural modifications called plasmodesmata (singular = plasmodesma), numerous channels that pass between cell walls of adjacent plant cells, connect their cytoplasm, and enable materials to be transported from cell to cell, and thus throughout the plant ([Figure 2]).

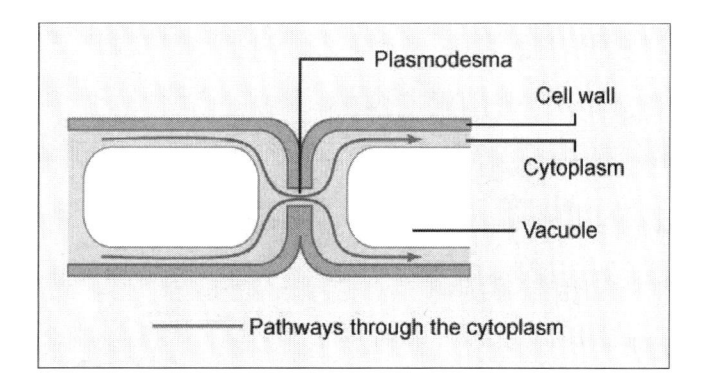

*Figure 2: A plasmodesma is a channel between the cell walls of two adjacent plant cells. Plasmodesmata allow materials to pass from the cytoplasm of one plant cell to the cytoplasm of an adjacent cell.*

### Tight Junctions

A tight junction is a watertight seal between two adjacent animal cells ([Figure 3]). The cells are held tightly against each other by proteins (predominantly two proteins called claudins and occludins).

338    Cell Transport

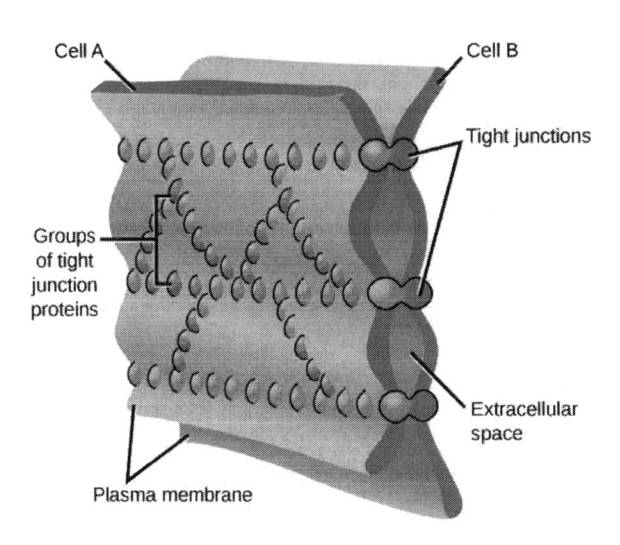

*Figure 3: Tight junctions form watertight connections between adjacent animal cells. Proteins create tight junction adherence. (credit: modification of work by Mariana Ruiz Villareal)*

This tight adherence prevents materials from leaking between the cells; tight junctions are typically found in epithelial tissues that line internal organs and cavities, and comprise most of the skin. For example, the tight junctions of the epithelial cells lining your urinary bladder prevent urine from leaking out into the extracellular space.

*Desmosomes*

Also found only in animal cells are desmosomes, which act like spot welds between adjacent epithelial cells ([Figure 4]). Short proteins called cadherins in the plasma membrane connect to intermediate filaments to create desmosomes. The cadherins join two adjacent cells together and maintain the cells in a sheet-like formation in organs and tissues that stretch, like the skin, heart, and muscles.

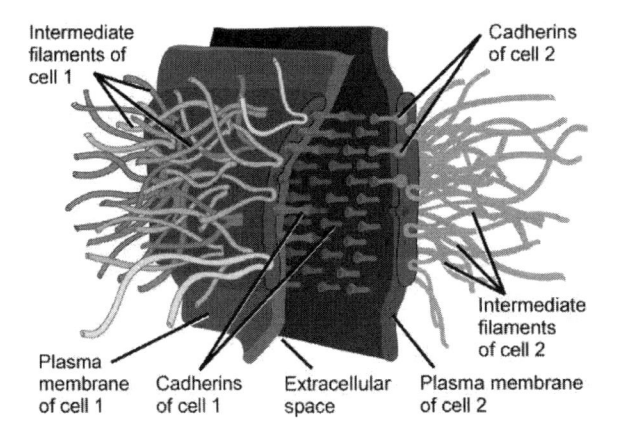

*Figure 4: A desmosome forms a very strong spot weld between cells. It is created by the linkage of cadherins and intermediate filaments. (credit: modification of work by Mariana Ruiz Villareal)*

## Gap Junctions

Gap junctions in animal cells are like plasmodesmata in plant cells in that they are channels between adjacent cells that allow for the transport of ions, nutrients, and other substances that enable cells to communicate ([Figure 5]). Structurally, however, gap junctions and plasmodesmata differ.

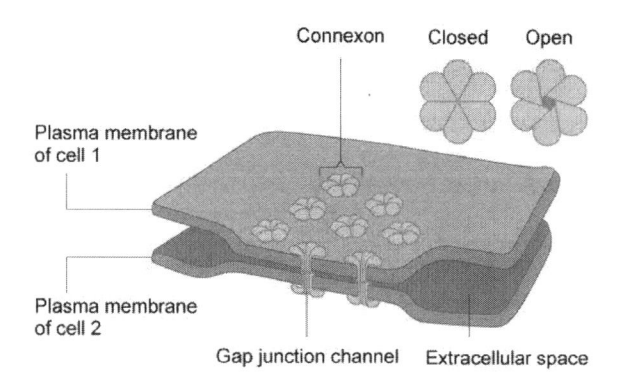

*Figure 5: A gap junction is a protein-lined pore that allows water and small molecules to pass between adjacent animal cells. (credit: modification of work by Mariana Ruiz Villareal)*

Gap junctions develop when a set of six proteins (called connexins) in the plasma membrane arrange themselves in an elongated donut-like configuration called a connexon. When the pores ("doughnut holes") of connexons in adjacent animal cells align, a channel between the two cells forms. Gap junctions are particularly important in cardiac muscle: The electrical signal for the muscle to contract is passed efficiently through gap junctions, allowing the heart muscle cells to contract in tandem.

To conduct a virtual microscopy lab and review the parts of a cell, work through the steps of this interactive assignment: http://bio.rutgers.edu/~gb101/lab1_cell_structure/index.html.

## SECTION SUMMARY

Animal cells communicate via their extracellular matrices and are connected to each other via tight junctions, desmosomes, and gap junctions. Plant cells are connected and communicate with each other via plasmodesmata.

When protein receptors on the surface of the plasma membrane of an animal cell bind to a substance in the extracellular matrix, a chain of reactions begins that changes activities taking place within the cell. Plasmodesmata are channels between adjacent plant cells, while gap junctions are channels between adjacent animal cells. However, their structures are quite different. A tight junction is a watertight seal between two adjacent cells, while a desmosome acts like a spot weld.

### Glossary

**desmosome:** linkages between adjacent epithelial cells that form when cadherins in the plasma membrane attach to intermediate filaments

**extracellular matrix:** material (primarily collagen, glycoproteins, and proteoglycans) secreted from animal cells that provides mechanical protection and anchoring for the cells in the tissue

**gap junction:** channel between two adjacent animal cells that allows ions, nutrients, and low molecular weight substances to pass between cells, enabling the cells to communicate

**plasmodesma:** (plural = plasmodesmata) channel that passes between the cell walls of adjacent plant cells, connects their cytoplasm, and allows materials to be transported from cell to cell

**tight junction:** firm seal between two adjacent animal cells created by protein adherence

# Exercises

1. Which of the following are found only in plant cells?
   1. gap junctions
   2. desmosomes
   3. plasmodesmata
   4. tight junctions
2. The principal force driving movement in diffusion is the _____.
   1. temperature
   2. particle size
   3. concentration gradient
   4. membrane surface area
3. What problem is faced by organisms that live in fresh water?
   1. Their bodies tend to take in too much water.
   2. They have no way of controlling their tonicity.
   3. Only salt water poses problems for animals that live in it.
   4. Their bodies tend to lose too much water to their environment.
4. The key components of desmosomes are cadherins and _____.
   1. actin
   2. microfilaments
   3. intermediate filaments
   4. microtubules
5. How does the structure of a plasmodesma differ from that of a gap junction?
6. Explain how the extracellular matrix functions.
7. Discuss why the following affect the rate of diffusion: molecular size, temperature, solution density, and the distance that must be traveled.
8. Why does water move through a membrane?
9. Which plasma membrane component can be either found on its surface or embedded in the membrane structure?
   1. protein
   2. cholesterol
   3. carbohydrate
   4. phospholipid

10. What is the primary function of carbohydrates attached to the exterior of cell membranes?
    1. identification of the cell
    2. flexibility of the membrane
    3. strengthening the membrane
    4. channels through membrane

# Check Your Knowledge: Self-Test

1. Why is it advantageous for the cell membrane to be fluid in nature?
2. Why do phospholipids tend to spontaneously orient themselves into something resembling a membrane?
3. Discuss why the following affect the rate of diffusion: molecular size, temperature, solution concentration, and the distance that must be traveled.
4. Why does water move through a membrane?
5. Both of the regular intravenous solutions administered in medicine, normal saline and lactated Ringer's solution, are isotonic. Why is this important?
6. Describe two ways that decreasing temperature would affect the rate of diffusion of molecules across a cell's plasma membrane.
7. Where does the cell get energy for active transport processes?
8. Why is it important that there are different types of proteins in plasma membranes for the transport of materials into and out of a cell?
9. Why do ions have a difficult time getting through plasma membranes despite their small size?
10. What happens to the membrane of a vesicle after exocytosis?
    1. It leaves the cell.
    2. It is disassembled by the cell.
    3. It fuses with and becomes part of the plasma membrane.
    4. It is used again in another exocytosis event.
11. Which transport mechanism can bring whole cells into a cell?
    1. pinocytosis
    2. phagocytosis
    3. facilitated transport
    4. primary active transport
12. In what important way does receptor-mediated endocytosis differ from phagocytosis?
    1. It transports only small amounts of fluid.
    2. It does not involve the pinching off of membrane.
    3. It brings in only a specifically targeted substance.
    4. It brings substances into the cell, while phagocytosis removes substances.

13. Many viruses enter host cells through receptor-mediated endocytosis. What is an advantage of this entry strategy?

    1. The virus directly enters the cytoplasm of the cell.

    2. The virus is protected from recognition by white blood cells.

    3. The virus only enters its target host cell type.

    4. The virus can directly inject its genome into the cell's nucleus.

14. Which of the following organelles relies on exocytosis to complete its function?

    1. Golgi apparatus

    2. vacuole

    3. mitochondria

    4. endoplasmic reticulum

15. Imagine a cell can perform exocytosis, but only minimal endocytosis. What would happen to the cell?

    1. The cell would secrete all its intracellular proteins.

    2. The plasma membrane would increase in size over time.

    3. The cell would stop expressing integral receptor proteins in its plasma membrane.

    4. The cell would lyse.

16. Which plasma membrane component can be either found on its surface or embedded in the membrane structure?

    1. protein

    2. cholesterol

    3. carbohydrate

    4. phospholipid

17. Which characteristic of a phospholipid contributes to the fluidity of the membrane?

    1. its head

    2. cholesterol

    3. a saturated fatty acid tail

    4. double bonds in the fatty acid tail

18. Water moves via osmosis _____.

    1. throughout the cytoplasm

    2. from an area with a high concentration of other solutes to a lower one

    3. from an area with a high concentration of water to one of lower concentration

    4. from an area with a low concentration of water to higher concentration

19. The principal force driving movement in diffusion is the _____.
    1. temperature
    2. particle size
    3. concentration gradient
    4. membrane surface area
20. Active transport must function continuously because _____.
    1. plasma membranes wear out
    2. not all membranes are amphiphilic
    3. facilitated transport opposes active transport
    4. diffusion is constantly moving solutes in opposite directions